World Scientific
Series in Advanced
Manufacturing

Volume 1: Recent Advances in Additive Manufacturing

Manufacturing in the Era of 4th Industrial Revolution

A World Scientific Reference

World Scientific Series in Advanced Manufacturing

Print ISSN: 2717-5901
Online ISSN: 2717-591X

Editor-in-Chief: Prof. Satyandra K. Gupta *(University of Southern California)*

The World Scientific Series in Advanced Manufacturing aims to disseminate the new knowledge being created in the area of advanced manufacturing. Books in this series are expected to serve as a reference for practicing engineers and textbooks for new courses. This series has a broad scope and will cover all potential areas related to Advanced Manufacturing technologies.

Published

Editor-in-chief: **Satyandra K Gupta**

World Scientific
Series in Advanced
Manufacturing

Volume 1: Recent Advances in Additive Manufacturing

Manufacturing in the Era of 4th Industrial Revolution

A World Scientific Reference

editors

Hugh Bruck
University of Maryland, College Park, USA

Yong Chen
University of Southern California, USA

Satyandra K Gupta
University of Southern California, USA

NEW JERSEY · LONDON · SINGAPORE · BEIJING · SHANGHAI · HONG KONG · TAIPEI · CHENNAI · TOKYO

Published by

World Scientific Publishing Co. Pte. Ltd.

5 Toh Tuck Link, Singapore 596224

USA office: 27 Warren Street, Suite 401-402, Hackensack, NJ 07601

UK office: 57 Shelton Street, Covent Garden, London WC2H 9HE

Library of Congress Cataloging-in-Publication Data
Names: Gupta, Satyandra K, editor.
Title: Manufacturing in the era of 4th industrial revolution / editor-in-chief, Satyandra K Gupta.
Other titles: Manufacturing in the era of fourth industrial revolution
Description: Hackensack, NJ : World Scientific, 2020. | Series: A World Scientific reference |
    Includes bibliographical references and index. | Contents: volume 1. Recent advances in
    additive manufacturing / editors, Hugh Bruck, University of Maryland, College Park, USA, Yong Chen,
    University of Southern California, USA, Satyandra K Gupta, University of Southern California, USA --
    volume 2. Recent advances in industrial robotics / editors, Satyandra K Gupta,
    University of Southern California, USA, Venkat N Krovi, Clemson University, USA,
    Craig Schlenoff, National Institute of Standards and Technology, USA --
    volume 3. Augmented, virtual and mixed reality applications in advanced manufacturing /
    editors, Monica Bordegoni, Politecnico di Milano, Italy, Satyandra K Gupta,
    University of Southern California, USA, James Ritchie, Heriot-Watt University, UK.
Identifiers: LCCN 2020034693 | ISBN 9789811222818 (v. 1 ; hardcover) |
    ISBN 9789811222832 (v. 2 ; hardcover) | ISBN 9789811222856 (v. 3 ; hardcover) |
    ISBN 9789811222788 (hardcover ; set) | ISBN 9789811222825 (v. 1 ; ebook for institutions) |
    ISBN 9789811222849 (v. 2 ; ebook for institutions) | ISBN 9789811222863 (v. 3 ; ebook for institutions) |
    ISBN 9789811222801 (ebook for individuals) | ISBN 9789811222795 (ebook for institutions ; set)
Subjects: LCSH: Manufacturing processes--Technological innovations.
Classification: LCC TS183 .M353 2020 | DDC 670--dc23
LC record available at https://lccn.loc.gov/2020034693

British Library Cataloguing-in-Publication Data
A catalogue record for this book is available from the British Library.

For any available supplementary material, please visit
https://www.worldscientific.com/worldscibooks/10.1142/11898#t=suppl

Printed in Singapore

Contents

About the Editors

Hugh A. Bruck

Prof. Bruck received his Bachelor of Science and Master of Science in Mechanical Engineering from the University of South Carolina, and PhD in Materials Science from Caltech. He was a Visiting Professor at Tel Aviv University, and is currently a Full Professor in the Department of Mechanical of Engineering at the University of Maryland where he also serves as Associate Dean for Faculty Affairs in the A. James Clark School of Engineering. He has authored or co-authored 250 publications involving advanced characterization and modeling techniques for understanding the thermomechanical and multifunctional behavior of materials. He is a Fellow of Society for Experimental Mechanics (SEM) and American Society of Mechanical Engineers (ASME), and has served on the executive board of SEM, as well as an associate editor for the journal *Experimental Mechanics*. He has received numerous honors and awards in recognition of his research and teaching accomplishments, including the ONR Young Investigator Program award, the A.J. Durelli Award from SEM, Best Paper Award at the 2010 ASME Mechanisms & Robotics Conference, Fulbright Scholar Award, Pi Tau Sigma Faculty Appreciation Award, Distinguished Scholar-Teacher Award at the University of Maryland, and NSF PROMISE Outstanding Faculty Mentor.

Yong Chen

Dr. Yong Chen is a Professor of Industrial and Systems Engineering and Aerospace and Mechanical Engineering and the Director of Daniel J. Epstein Institute at the University of Southern California (USC). Dr. Chen's research focuses on additive manufacturing (3D printing) in micro- and meso- scales and related modeling, control, material, and application. He has published more than 150 publications in refereed journals and conferences as well as 12 issued and pending U.S. patents. His work has been recognized by over ten Best/Outstanding Paper Awards in major design and manufacturing journals and conferences. Other major awards he received include the National Science Foundation Faculty Early Career Development (CAREER) Award, the Outstanding Young Manufacturing

Engineer Award from the Society of Manufacturing Engineers (SME), and the invitations to the National Academy of Engineering (NAE) Frontiers of Engineering Symposiums. Dr. Chen is a Fellow of the American Society of Mechanical Engineers (ASME). He has served as conference/program chairs as well as keynote speakers in several international conferences, including the conference chair of the 2017 International Manufacturing Research Conference and the program co-chair of the 2019 International Design Engineering Technical Conferences (IDETC).

Satyandra K. Gupta

Dr. Satyandra K. Gupta is Smith International Professor in the Department of Aerospace and Mechanical Engineering and Department of Computer Science in Viterbi School of Engineering at the University of Southern California. He served as a program director for the National Robotics Initiative at the National Science Foundation from September 2012 to September 2014. Dr. Gupta's interests are in the area of physics-aware decision making to facilitate and advance the state of automation. He has published more than 350 technical articles. He is a fellow of the American Society of Mechanical Engineers (ASME), Institute of Electrical and Electronics Engineers (IEEE), and Society of Manufacturing Engineers (SME). He serves as editor of the *ASME Journal of Computing and Information Science in Engineering*. Dr. Gupta has received numerous honors and awards for his scholarly contributions. Representative examples include a Young Investigator Award from the Office of Naval Research in 2000, Robert W. Galvin Outstanding Young Manufacturing Engineer Award from the Society of Manufacturing Engineers in 2001, CAREER Award from the National Science Foundation in 2001, Presidential Early Career Award for Scientists and Engineers in 2001, Invention of the Year Award at the University of Maryland in 2007, Kos Ishii-Toshiba Award from ASME in 2011, Excellence in Research Award from ASME Computers and Information in Engineering Division in 2013, and Distinguished Alumnus Award from Indian Institute of Technology, Roorkee in 2014. He has also received ten best paper awards at international conferences.

About the Contributors

1. *Introduction*

Hugh A. Bruck

Prof. Bruck received his Bachelor of Science and Master of Science in Mechanical Engineering from the University of South Carolina, and PhD in Materials Science from Caltech. He was a Visiting Professor at Tel Aviv University, and is currently a Full Professor in the Department of Mechanical of Engineering at the University of Maryland where he also serves as Associate Dean for Faculty Affairs in the A. James Clark School of Engineering. He has authored or co-authored 250 publications involving advanced characterization and modeling techniques for understanding the thermomechanical and multifunctional behavior of materials. He is a Fellow of Society for Experimental Mechanics (SEM) and American Society of Mechanical Engineers (ASME), and has served on the executive board of SEM, as well as an associate editor for the journal *Experimental Mechanics*. He has received numerous honors and awards in recognition of his research and teaching accomplishments, including the ONR Young Investigator Program award, the A.J. Durelli Award from SEM, Best Paper Award at the 2010 ASME Mechanisms & Robotics Conference, Fulbright Scholar Award, Pi Tau Sigma Faculty Appreciation Award, Distinguished Scholar-Teacher Award at the University of Maryland, and NSF PROMISE Outstanding Faculty Mentor.

Yong Chen

Dr. Yong Chen is a Professor of Industrial and Systems Engineering and Aerospace and Mechanical Engineering and the Director of Daniel J. Epstein Institute at the University of Southern California (USC). Dr. Chen's research focuses on additive manufacturing (3D printing) in micro- and meso- scales and related modeling, control, material, and application. He has published more than 150 publications in refereed journals and conferences as well as 12 issued and pending U.S. patents. His work has been recognized by over ten Best/Outstanding

Paper Awards in major design and manufacturing journals and conferences. Other major awards he received include the National Science Foundation Faculty Early Career Development (CAREER) Award, the Outstanding Young Manufacturing Engineer Award from the Society of Manufacturing Engineers (SME), and the invitations to the National Academy of Engineering (NAE) Frontiers of Engineering Symposiums. Dr. Chen is a Fellow of the American Society of Mechanical Engineers (ASME). He has served as conference/program chairs as well as keynote speakers in several international conferences, including the conference chair of the 2017 International Manufacturing Research Conference and the program co-chair of the 2019 International Design Engineering Technical Conferences (IDETC).

Satyandra K. Gupta

Dr. Satyandra K. Gupta is Smith International Professor in the Department of Aerospace and Mechanical Engineering and Department of Computer Science in Viterbi School of Engineering at the University of Southern California. He served as a program director for the National Robotics Initiative at the National Science Foundation from September 2012 to September 2014. Dr. Gupta's interests are in the area of physics-aware decision making to facilitate and advance the state of automation. He has published more than 350 technical articles. He is a fellow of the American Society of Mechanical Engineers (ASME), Institute of Electrical and Electronics Engineers (IEEE), and Society of Manufacturing Engineers (SME). He serves as editor of the *ASME Journal of Computing and Information Science in Engineering*. Dr. Gupta has received numerous honors and awards for his scholarly contributions. Representative examples include a Young Investigator Award from the Office of Naval Research in 2000, Robert W. Galvin Outstanding Young Manufacturing Engineer Award from the Society of Manufacturing Engineers in 2001, CAREER Award from the National Science Foundation in 2001, Presidential Early Career Award for Scientists and Engineers in 2001, Invention of the Year Award at the University of Maryland in 2007, Kos Ishii-Toshiba Award from ASME in 2011, Excellence in Research Award from ASME Computers and Information in Engineering Division in 2013, and Distinguished Alumnus Award from Indian Institute of Technology, Roorkee in 2014. He has also received ten best paper awards at international conferences.

2. *Support Structure Design for Selective Laser Melting Process*

Behzad Rankouhi

Behzad Rankouhi is a PhD candidate at the University of Wisconsin-Madison. His interests are in the area of metal additive manufacturing and experimental mechanics. Behzad started his research on Additive Manufacturing in 2014 when he joined SDSU graduate program. He collaborated with Made In Space, the startup that built the first 3D printer on board of the International Space Station (ISS), to characterize polymers to be printed in micro gravity. He further studied the effects of gamma radiation on 3D printed samples for aerospace and medical applications. After receiving his Master of Science degree in mechanical engineering from SDSU, he joined University of Wisconsin-Madison PhD program in 2016, where he changed his focus from polymers to metal Additive Manufacturing. Currently, Behzad is collaborating in multidisciplinary research between departments of mechanical engineering and materials science and engineering with a focus on multi-material selective laser melting.

Dan J. Thoma

Dr. Thoma is the Director of the Grainger Institute for Engineering (GIE) at the University of Wisconsin-Madison (UW) and a Professor in Materials Science and Engineering. The GIE is an accelerator within the College of Engineering to mature scientific and technical areas with significant societal impact. Prior to UW, he was the Deputy Division Leader for the Materials Science and Technology (MST) Division at Los Alamos National Laboratory (LANL). This division focused on manufacturing and novel materials research. His research and technical efforts have been devoted to new manufacturing methods and materials by design, with a particular focus on property response as a function of microstructural evolution during phase transformations. Dr. Thoma has been active within materials professional societies, where he was the president of The Minerals, Metals, Materials Society (TMS) in 2003, the American Institute of Mining, Metallurgical, and Petroleum Engineers (AIME) in 2008, and the Federation of Materials Societies (FMS) in 2009-2010. His expertise in materials and manufacturing was recognized in 2008 by being elected as a Fellow of ASM International. In 2019, he was elected Fellow of TMS.

Krishnan Suresh

Krishnan Suresh is the Philip and Jean Myers Professor of Mechanical Engineering at the University of Wisconsin-Madison. He received a Master's in Manufacturing Engineering from UCLA in 1992, and a Master's and PhD in Mechanical Engineering from Cornell in 1994 and 1998, respectively. He later served as an Engineering Manager at Kulicke and Soffa Industries, Philadelphia from 1998 to 2002. He has received numerous peer-reviewed grants, including the prestigious NSF Career award. His research interests include topology optimization, additive manufacturing, advanced finite element analysis and high-performance computing. He has co-authored over 80 peer-reviewed papers, two of which have received best-paper awards from ASME. He has also authored two textbooks on applied optimization. He is also the co-founder of SciArt, LLC, a UW- Madison spinoff that creates and supports high-performance topology optimization software solutions.

3. *Multiscale Process Modeling of Shape Memory Alloy Fabrication with Directed Energy Deposition*

Jesse M. Sestito

Jesse M. Sestito is a PhD candidate in the George W. Woodruff School of Mechanical Engineering at Georgia Institute of Technology in Atlanta, GA in the labs of Dr. Yan Wang and Dr. Tequila A. L. Harris. He received his BS in mechanical engineering at Rose-Hulman Institute of Technology in 2013. His research involves multiscale modeling using kinetic Monte Carlo, molecular dynamics, and Bayesian optimization. At Georgia Institute of Technology, he has developed a new methodology using Bayesian optimization to generate molecular dynamic force fields and has used this methodology to develop a polycaprolactone force field. He also worked with Sandia National Laboratories on examining the sintering process of nickel nanoparticles, and developed a new aluminum scandium nitride force field at Oak Ridge National Laboratory to study the effects of radiation damage on ceramics.

Dehao Liu

Dehao Liu is a PhD candidate in the George W. Woodruff School of Mechanical Engineering at Georgia Institute of Technology. He received his B.S. in mechanical engineering from Tsinghua University in 2016. His research interests include multiphysics simulation and physics-constrained machine learning. His current research goal is to construct the process-structure-property relationship in additive manufacturing for process and materials design. He has published nine high-impact journal papers and four American Society of Mechanical Engineers (ASME) conference papers.

Yanglong Lu

Yanglong Lu is a PhD candidate in the Woodruff School of Mechanical Engineering at Georgia Institute of Technology. He received his BS in mechanical engineering from Georgia Institute of Technology in 2016 and expects to receive the PhD degree in 2020. His research interests are modeling and monitoring additive manufacturing processes by introducing physical knowledge in data-driven approaches. His future research plan is to develop sensing protocols for different manufacturing systems and cyber-physical systems.

Ji-Hyeon Song

Dr. Ji-Hyeon Song received a joint PhD degree in mechanical engineering from Seoul National University and Georgia Institute of Technology in 2019. She is currently a postdoctoral researcher at Nanyang Technological University. Her research interests include micro-/nanoscale fabrication, micro-/nano electronics, and multi-scale computational simulation. She has received numerous awards, including Young researcher award from WISET (Women in Science, Engineering, and Technology) and Global PhD Fellowship from NRF (National Research Foundation) of Korea.

Anh V. Tran

Dr. Anh Tran currently is a postdoctoral appointee at Optimization and Uncertainty Quantification Department, Center of Computing Research, Sandia National Laboratories, where he conducts research on machine learning, uncertainty quantification, and optimization, with applications in multiscale computational materials science and manufacturing. He obtained his BS and PhD in mechanical engineering from Georgia Institute of Technology in 2011 and 2018 respectively, and MS in applied mathematics from Georgia Southern University in 2014.

Michael J. Kempner

Michael Kempner is a PhD student in the George W. Woodruff School of Mechanical Engineering at the Georgia Institute of Technology. He received his

BS in biomedical engineering from the University of South Carolina. Michael focuses on developing novel machine learning methodologies to accelerate the materials design process. The widely applicable work has involved collaborations with researchers in engineering, science, and medicine, including ongoing partnerships with Oak Ridge National Laboratory (ORNL) and the National Aeronautics and Space Administration (NASA). Michael's research has earned support from the National Science Foundation (NSF) From Learning, Analytics, and Materials to Entrepreneurship and Leadership (FLAMEL) Doctoral Traineeship Program and the Georgia Tech President's Fellowship.

Tequila A. L. Harris

Dr. Tequila A. L. Harris is an Associate Professor in the Woodruff School of Mechanical Engineering at Georgia Institute of Technology (Georgia Tech). She received her Bachelor of Science (BS) in Physics from Lane College, her Master of Science (MS) and Doctor of Philosophy (PhD) in Mechanical Engineering from Rensselaer Polytechnic Institute in 2003 and 2006, respectively. She joined the Georgia Tech faculty in 2006 where she manages the Polymer Thin Film Process Laboratory, which enables high throughput scalable manufacturing knowhow for a wide variety of liquids across multiple length scales. In her research, she explores the connectivity between thin film quality and its functionality, durability, and performance, based on its manufacture, with the aim of elucidating mechanisms that cause system failure. With the use of numerical simulations, experimentation, and analytical approaches, she has introduced unique models and approaches to predict and control the quality of thin films, processed on permeable and impermeable substrates. She has published more than 75 technical articles and three patents. She is the recipient of several awards and honors of note the Lockheed Martin Inspirational Young Faculty award in 2010, the National Science Foundation CAREER award in 2010, and the International Society of Coating Science and Technology (ISCST) L. E. Scriven Young Investigator award in 2018.

Sung-Hoon Ahn

Dr. Sung-Hoon Ahn is a Professor in the Department of Mechanical Engineering at Seoul National University, Korea. He received Ph.D. degree in Dept. of Aeronautics and Astronautics from Stanford University (1997). Since then, he has held professional and visiting positions at Stanford University, University of California at Berkeley, Gyeongsang National University and University of Washington. He joined Seoul National University in 2003 and has served SNU

Institute of Global Social Responsibility as a Director, Graduate School of Engineering Practice as the Associate Dean. He serves as an Outside Director of Hyundai WIA Corp., His research interests cover 3D/4D printing, smart factory, soft robotics, renewable energy and nano fabrication. He has published more than three hundred thirty journal articles. He is a fellow of the International Academy of Production Engineering (CIRP). He has served as editor-in-chief of the International Journal of Precision Engineering and Manufacturing – Green Technology since 2013 and editorial board member of Rapid Prototyping Journal, Multifunctional Materials, Advances in Manufacturing, etc. He has received numerous honors and awards for his contributions. Representative examples include: Highly Commended Award from Literati Club, UK, Sinyang Engineering Award, Gaheon Award, Seoul National University Education Award, LG Yonam International Visiting Fellowship, Certificate of Commendation from Ministry of Science, ICT and Future Planning, and the Presidential Commendation from Republic of Korea. Since 2011 Dr. Ahn and his volunteer team has constructed 9 off-grid solar/small-hydro/wind power plants in Nepal and Tanzania which provide electricity and LED light for over 3,500 people in remote villages.

Yan Wang

Dr. Yan Wang is a Professor of Mechanical Engineering at Georgia Institute of Technology. He received his BS from Tsinghua University, MS from Chinese Academy of Sciences, and PhD from the University of Pittsburgh. His research interests include computer-aided design, computer-aided manufacturing, modeling and simulation, cyber-physical systems, and uncertainty quantification. He has published over 90 archived journal papers and 80 peer-reviewed conference papers. His research is recognized with multiple Best Paper Awards at American Society of Mechanical Engineers (ASME), Institute of Industrial & Systems Engineers (IISE), Minerals, Metals and Materials Society (TMS), and CAD conferences, as well as a U.S. National Science Foundation CAREER Award. He has been regularly invited as the proposal reviewer for government agencies in North America, Europe, and Asia. He is the Chair of ASME Computers & Information in Engineering Division, and served as the Chair of ASME Advanced Modeling & Simulation Technical Committee. He also served as the editors for ASME Journal of Computing & Information Science in Engineering, Journal of Mechanical Design, Journal of Computational & Nonlinear Dynamics, Journal of Risk & Uncertainty in Engineering Systems, and Computer-Aided Design.

4. *Towards Direct Deposition of Continuous-Fibers on Curved Surfaces*

Chi Chung Li

Chi Chung Li is a PhD candidate in Mechanical Engineering at UC, Berkeley. His research interest encompasses volumetric additive manufacturing, holographic lithography, computational imaging and photonic processing of electronics materials. Prior to his graduate study, he obtained a diverse education background in mechanical, automation and aerospace engineering. He received his BEng in Mechanical and Automation Engineering from the Chinese University of Hong Kong (CUHK) and conducted a senior-level year-exchange study in Aerospace Engineering in UC, San Diego (UCSD). He has taken research internships at CUHK and Temasek Laboratories at National University of Singapore.

Chengkai Dai

Chengkai Dai is currently a PhD candidate of the Department of Sustainable Design Engineering at Delft University of Technology. His research area includes robotics, geometry computing, and computational design.

Wei-Hsin Liao

Wei-Hsin Liao received his PhD in Mechanical Engineering from The Pennsylvania State University, University Park, USA. Since August 1997, Dr. Liao has been with The Chinese University of Hong Kong (CUHK), where he is now Chairman and Professor of Mechanical and Automation Engineering. His research interests include smart structures, vibration control, energy harvesting, mechatronics, and medical devices. His research has led to publications of over 260 technical papers in international journals and conference proceedings, 18 patents in US, China, Hong Kong, Taiwan, Japan, and Korea. He was the Conference Chair for the 20th International Conference on Adaptive Structures and Technologies in 2009; the Active and Passive Smart Structures and Integrated Systems, SPIE Smart Structures/NDE in 2014 and 2015. He is a recipient of the T A Stewart-Dyer/F H Trevithick Prize 2005, the ASME 2008 Best Paper Award in Structures, the ASME 2017 Best Paper Award in Mechanics and Material Systems, and three Best Paper Awards in the IEEE conferences. At CUHK, he received the Research Excellence Award (2011), and was awarded Outstanding Fellow of the Faculty of Engineering (2014). He received the SPIE 2018 SSM Lifetime Achievement Award. Dr. Liao currently serves as an Associate Editor for

Mechatronics, Journal of Intelligent Material Systems and Structures, as well as Smart Materials and Structures. Dr. Liao is a Fellow of ASME, HKIE, and IOP.

Charlie C.L. Wang

Charlie C.L. Wang currently holds a Chair of Smart Manufacturing with the University of Manchester. Prior to this, he was a Chair of Advanced Manufacturing at Delft University of Technology and a Professor of Mechanical and Automation Engineering at the Chinese University of Hong Kong. He also worked as a visiting professor at University of Southern California during sabbatical leave. Prof. Wang received a few awards from professional societies including the ASME CIE Excellence in Research Award (2016), the ISSMO/Springer Prize (2019), the Best Paper Award (2nd Place) of Solid and Physical Modeling (2019), the NAMRI/SME Outstanding Paper Award (2013), the Best Paper Awards of ASME CIE Conferences (twice in 2008 and 2001 respectively), the Prakash Krishnaswami CAPPD Best Paper Award of ASME CIE Conference (2011), and the ASME CIE Young Engineer Award (2009). He received his B.Eng. degree (1998) in mechatronics engineering from Huazhong University of Science and Technology and his M.Phil (2000) and Ph.D. (2002) degrees in mechanical engineering from Hong Kong University of Science and Technology (HKUST). He was elected Fellow of American Society of Mechanical Engineers (ASME) in 2013.

5. *Additive Manufacturing of Magnetic Particle-Polymer Composites*

Lu Lu

Dr. Lu Lu received a PhD in Industrial Engineering and Operations Research from the University of Illinois at Chicago with a focus on multi-functional and multi-material additive manufacturing process development. She received her MS in Mechanical Engineering from Purdue University. Lu is a mechanical design engineer at General Electric Healthcare. In this role, she is leading implementation of additive manufacturing technology and applications across GE Healthcare globally. She has 13 peer-reviewed journal papers and conference papers that are published, including four first-author journal papers published in the top-tier journals of the field including Composites Part B: Engineering, 3D Printing and Additive Manufacturing, and the ASME Journal of Manufacturing Science and Engineering.

Erina Baynojir Joyee

Erina B. Joyee is a PhD candidate in the department of Mechanical and Industrial Engineering at the University of Illinois at Chicago (UIC). Under the supervision of Dr. Yayue Pan, Erina has been conducting research in the field of multi-material magnetic field assisted additive manufacturing and soft robot applications. She has published multiple journal papers in prestigious journals like Soft Robotics and Composite B and presented her work in several international conferences.

Yayue Pan

Dr. Yayue Pan holds a PhD degree from the University of Southern California. Dr. Pan is an Associate Professor in the Department of Mechanical and Industrial Engineering at the University of Illinois at Chicago (UIC). Her research focuses on multi-material and multi-functional Additive Manufacturing processes for applications in anisotropic composites, sensing and actuating devices, energy management and storage.

6. *Additive Manufacturing of Bio-inspired Structures via Nanocomposite 3D Printing*

Yang Yang

Dr. Yang Yang is an Assistant Professor in the Department of Mechanical Engineering at San Diego State University (SDSU). He completed the joint PhD at Wuhan University in the School of Physics and Technology and University of California, Los Angeles in the Department of Bioengineering in 2015. Before joining SDSU, he worked as a postdoctoral research associate in the Department of Industrial and Systems Engineering at the University of Southern California (USC). Dr. Yang's interests are in the area of bioinspired Additive Manufacturing (3D printing), 3D printing of multimaterials, bioinspired design, novel 3D printing process development, mechanism of materials and structures, high dielectric nanocomposites, wearable sensor and energy harvesting device. He has published more than thirty technical articles. Dr. Yang served as Guest Editor of Frontiers in Materials for a special topic on "Bioinspired 3D printing".

Xiangjia (Cindy) Li

Dr. Xiangjia (Cindy) Li is an assistant professor in the Department of Aerospace and Mechanical Engineering in School for Engineering of Matter, Transport and Energy at Arizona State University. Dr. Li's current research is focused on additive manufacturing process development, aiming to explore and create functional devices with bio-inspired hierarchical structures and materials. She has worked on stereolithography based multi-scale additive manufacturing with bioinspired design methodologies and programmable functional materials for potential applications in biomedical devices, interfacial devices, and flexible sensors. Dr. Li developed several novel approaches for multi-scale additive manufacturing of nanocomposite and ceramic. She published multiple journals and conference papers on the topics of nanocomposite and ceramic fabrication, bioinspired nano/microstructures design and novel additive manufacturing processes development. She is the recipient of the University of Southern California Innovation Commercialization Award in 2018 and her major research work were patented in U.S. for the novelty and innovation of additive manufacturing process.

Yong Chen

Dr. Yong Chen is a professor of Industrial and Systems Engineering and Aerospace and Mechanical Engineering and the Director of Daniel J. Epstein Institute at the University of Southern California (USC). Dr. Chen's research focuses on additive manufacturing (3D printing) in micro- and meso- scales and related modeling, control, material, and application. He has published more than 150 publications in refereed journals and conferences as well as 12 issued and pending U.S. patents. His work has been recognized by over ten Best/Outstanding Paper Awards in major design and manufacturing journals and conferences. Other major awards he received include the National Science Foundation Faculty Early Career Development (CAREER) Award, the Outstanding Young Manufacturing Engineer Award from the Society of Manufacturing Engineers (SME), and the invitations to the National Academy of Engineering (NAE) Frontiers of Engineering Symposiums. Dr. Chen is a Fellow of the American Society of Mechanical Engineers (ASME). He has served as conference/program chairs as well as keynote speakers in several international conferences, including the conference chair of the 2017 International Manufacturing Research Conference and the program co-chair of the 2019 International Design Engineering Technical Conferences (IDETC).

7. 4D Printing Based on Multi-Material Design

Devin J. Roach

Mr. Devin J. Roach is currently a candidate for a PhD degree in the Mechanics of Soft Materials and 3D Printing Laboratory under Dr. H. Jerry Qi at the Georgia Institute of Technology. He received his master's and bachelor's degrees in

Mechanical Engineering at Georgia Tech in 2018 and 2016, respectively. His research interests include soft-active "smart" polymers, multi-material 3D printing, and printed electronics. Specifically, he seeks to discover how these technologies can be integrated to fabricate novel 4D printed structures for use in soft robotics, sensor networks, and biomedical devices. In the past, he has worked on research teams at Sandia National Laboratories, Delta Air Lines, and Airbus in Germany.

Xiao Kuang

Dr. Xiao Kuang is a research scientist in the George Woodruff School of Mechanical Engineering at Georgia Institute of Technology. He received a BEng in polymer material and engineering from Beijing University of Chemical Technology in 2011. Then he was recommended to the Institute of Chemistry Chinese Academy and received a PhD in Polymer Physics and Chemistry in 2016. Afterward, he became a post-doctoral researcher at the Georgia Institute of Technology. Dr. Kuang's research is focused on the design and 3D/4D printing of adaptive programmable materials and structures.

Craig M. Hamel

Mr. Craig M. Hamel is a PhD candidate in the George W. Woodruff School of Mechanical Engineering at the Georgia Institute of Technology. He received his BS in physics from the University of Mississippi in 2012 and a MS in mechanical engineering from the New Jersey Institute of Technology in 2015. Craig's research revolves around the cross section of theoretical mechanics, material modelling, and continuum level simulations with applications to polymer based additive manufacturing and other polymer processes. Recently he has held internship positions at Sandia National Laboratories working with both the solid mechanics and computational shock physics group.

Martin L. Dunn

Dr. Martin L. Dunn is a professor and dean of the College of Engineering, Design and Computing at the University of Colorado Denver. He joined CU Denver in 2018 after serving as the founding associate provost for research at the Singapore University of Technology and Design (SUTD) where he oversaw the design and operation of the research and innovation enterprise. He was also a professor at SUTD and the founding director of the National Research Foundation-supported Digital Manufacturing and Design Center. Prior to joining SUTD, he served as a program director (mechanics of materials) in the Civil, Mechanical and

Manufacturing Innovation Division at the U.S. National Science Foundation (NSF), where he was also the founding program director for the Design of Engineering Materials Systems program. He served the NSF while on leave from the University of Colorado Boulder where he was the associate dean of research in the College of Engineering, Design and Computing, chair of the Department of Mechanical Engineering and a professor of mechanical engineering, holding the Victor Schelke Endowed Chair. Dunn's research has focused on understanding the mechanics and physics of complex heterogeneous materials through a combination of theory and experiment and using this understanding to create methods and tools to design and manufacture new materials and components. This includes constitutive modeling of the nonlinear multiphysics response (thermal, optical, mechanical) of active polymers and polymer composites, computational design automation approaches based on shape and topology optimization and additive manufacturing.

H. Jerry Qi

Dr. H. Jerry Qi has been a professor at the Georgia Institute of Technology since 2014 and was promoted to full professor in March 2016. Prior, he was an associate professor at the University of Colorado, Boulder (2004–2013) and was a postdoctoral fellow at MIT (2003–2004). He is also an ASME Fellow and a Woodruff School of Mechanical Engineering Fellow. His research primarily focuses on modeling, development, and 3D printing of soft active materials. Current research focuses on developing 3D printing technologies for high performance polymers, 4D printing of active materials, mechanics in 3D printing, and active polymer design and manufacturing.

8. Functionalized Materials for Additive Manufacturing and 3D Printing

Tarek I. Zohdi

Tarek I. Zohdi is a Professor of Mechanical Engineering and Chair of the Designated Emphasis Program in Computational and Data Science and Engineering at UC Berkeley (2012 to present). Previously, he has served as Chair of the Engineering Science Program (2008 to 2012). He is currently a Chancellors Professor and holder of the W. C. Hall Family Endowed Chair in Engineering. He also holds a Staff Scientist position at Lawrence Berkeley National Labs and an Adjunct Scientist position at the Children's Hospital Oakland Research Institute. His main research interests are in industrial simulation and advanced manufacturing processes. He has published over 175 archival refereed journal papers and seven books. He is an editor of Computational Mechanics and co-founder and editor-in-chief of Computational Particle Mechanics. In the past, he has organized or co-organized over 30 international conferences and workshops and has been appointed/invited to the Scientific Advisory Boards of over 40 international conferences. He was elected President of the United States Association for Computational Mechanics in 2012, and served from 2012 to 2014. Overall, he has given more than 200 plenary, keynote, and contributed lectures at conferences, universities, and other research institutions. In 2017, he received the UC Berkeley Distinguished Teaching Award; the highest award for teaching at UC Berkeley. In 2019 he was elected as Fellow of the American Academy of Mechanics (AAM).

9. *Machine Learning for Quality Control in Additive Manufacturing*

Qiang Huang

Dr. Qiang Huang is a Professor at the Daniel J. Epstein Department of Industrial and Systems Engineering, University of Southern California (USC), Los Angeles. Dr. Huang's research interests include Data Science and Engineering Applications, AI and Machine Learning for Advanced Manufacturing, and Nanomanufacturing and Nanoinformatics. He was the holder of Gordon S. Marshall Early Career Chair in Engineering at USC from 2012 to 2016. He received National Science Foundation CAREER award in 2011 and IEEE Transactions on Automation Science and Engineering Best Paper Award from IEEE Robotics and Automation Society in 2014. He has served as a Department Editor for IISE Transactions, Associate Editor for ASME Transactions, Journal of Manufacturing Science and Engineering, and a member of Editorial Board for Journal of Quality Technology, an Associate Editor for IEEE Transactions on Automation Science and Engineering and for IEEE Robotics.

https://doi.org/10.1142/9789811222825_0001

Chapter 1

Introduction

Hugh Bruck, Yong Chen, and Satyandra K. Gupta

Despite being around since the 1980s, Additive Manufacturing (AM) has recently become a very popular manufacturing process. The most common form of AM is referred to as "three-dimensional (3D) printing", since the AM machines are similar to conventional two-dimensional (2D) laser or inkjet paper printers, but have an extra degree of freedom to translate the print head. In practice, AM also resembles layered manufacturing (LM), in that a part is usually built layer-by-layer. In most commonly used AM processes, the planar layers are extremely thin in order to fabricate parts with high out-of-plane spatial resolution, in addition to the high resolution achieved in-plane via conventional stepper motors. There are many varieties of AM processes that have been developed, so ASTM has classified them into the following seven categories: (1) vat photopolymerization, (2) material jetting, (3) binder jetting, (4) powder bed fusion, (5) material extrusion, (6) directed energy deposition, and (7) sheet lamination. Each method has its own advantages and limitations, which guides the choice of process given the desired material and length scale of features for a part, although every process is capable of producing geometrically complex parts.

Some of the distinct advantages for AM are as follows: it does not require (a) any part specific tooling, (b) complex process planning, or (c) elaborate setup steps. Instructions for driving AM machines can be automatically generated from 3D computer-aided design (CAD) models in matters of seconds, similar to computer numerical control (CNC) machines. Using most AM machines does not require any specialized skills, which makes it extremely attractive to individuals with no prior manufacturing knowledge or experience. There is also less time between design generation and prototyping, since fabrication can begin within a few minutes after generating a 3D CAD model. These attributes make AM a much more attractive option compared to traditional processes, such as CNC machining and injection molding.

In the future, AM is envisioned to be used both for in-home manufacturing and factory-level production. In fact, AM has been used to make tooling (e.g., mold and patterns) for traditional manufacturing processes for many years. It has already transformed global e-commerce in the manufacturing sector, by enabling "digital manufacturing", which is centered around a 3D CAD model that can be easily submitted to any AM process throughout the world. Designers are now able to buy customized parts fabricated using AM over the internet in a matter of days. This is enabling democratization of innovation, where the barriers to accessing manufacturing processes have been removed for the masses. Therefore, AM is expected to continue to fuel a revolution in manufacturing that will focus on taking ideas from any individual, and turning them into reality within an extremely short period of time. To illustrate the revolution that is taking place, this introductory chapter will present an overview of the recent advances that have been made in AM. Subsequent chapters will present details of some specific advances, and how these advances are contributing to new AM capabilities.

As previously mentioned, material plays an important role in selecting an appropriate AM process. Polymers have been the most popular material to use in AM processes, because of the low cost and relatively safe operating temperatures. In particular, a wide variety of thermoplastic materials have become available for the extrusion-based processes. Thermoset resin-based polymers, such as photopolymers, are widely used in vat photopolymerization. There has been considerable interest in metal AM as well, but the machines have been substantially more expensive than those used for polymers, and there are more significant safety issues, since high-powered lasers are used. Over the last few years, several AM machines for metals have emerged on the market. Powder bed fusion has been the most popular process for fabricating metal parts, since it is fairly easy to set-up the bed and control the build of parts layer-by-layer. Metal AM machines have also been developed based on material extrusion and directed energy deposition, which mimic the processes used for polymers. Metal AM process has been gaining popularity primarily in the aerospace industry, where customized parts are more frequently made in lower quantities and at higher costs than other industries. The aerospace industry, as well as the automotive, has already been a primary consumer of composites for years, using them in layup and continuous filament winding processes to create a wide range of products with geometrically complex features that have high strength, light weight, and corrosion resistance. Therefore, there has been an affinity for these industries to expand their use of composites in new AM processes that have been developed for composites made from both short and continuous fibers.

The basic approach to the AM process has been to treat the build of a part layer-by-layer, even though the process may be depositing material at a point. Therefore, layer thickness has determined the part accuracy, as well as dictating the quality of the surface finish. It is possible to achieve reasonably high accuracy in AM processes by using a very small layer thickness, but this leads to higher processing times and costs. Recent work in conformal printing allows the use of non-planar layers, which mitigates some of the surface finish issues associated with fabrication using planar layers. Conformal printing also enables orienting fibers in optimal orientations to improve mechanical strength and stiffness of parts. Work in multi-resolution printing is also circumventing the bottleneck for throughput using conventional printers, by adjusting the printing resolution in situ to realize either fine or coarse features in different regions of the part.

Traditional AM processes have imposed some restrictions on the size of parts that can be built, due to the length of the screws or rails used to control the traversal of the print heads. However, recent developments in material extrusion-based systems are enabling printing of much larger parts. By using a large gantry type of system instead of screws or rails for moving the extrusion tool, these systems are enabling fabrication of parts whose size is limited only by the mobility of the gantry. Researchers are also exploring the use of other mobile platforms, such as mobile robot arms with extrusion heads, to further increase the size of parts that can be printed using AM. On the other end of the spectrum, vat photopolymerization based process is enabling printing of really small parts. For example, two-photon stereolithography has been used to print parts with nanoscale features.

One of the most significant drawbacks to traditional AM processes has been their slow speed. This has meant that fabricating a large part will require a time that scales linearly with the part volume. Traditional manufacturing processes that involve forming and consolidation, such as stamping, molding, and casting, are much faster in terms of processing time for making large parts. However, as the size of a part is reduced, AM becomes quite competitive, because the overall processing times will be inherently quicker. The processing time for AM is also quite attractive when making large parts with a larger number of intricate features that would require significant amounts of subtractive processing using much slower traditional manufacturing processes, such as milling. For such parts, subtractive processes are inherently slow because they are only capable of removing small volumes of material at a time without causing damage to the part, such as cracks or plastic deformation. Also, some of the time constraints for AM processes are being alleviated by recent work in vat photopolymerization based materials, where large volumes of material can be quickly cured using holographic images of the part.

Currently, many AM machines use proprietary materials in order to prevent issues during processing, such as stiction and clogging of print heads. Therefore, the material costs tend to be higher than those associated with traditional manufacturing processes, where the raw form of the products does not require a great deal of refinement. However, this is not an inherent limitation for the economics of AM processes, since prices will be driven down through the economy of scale as more companies start competing in this space. We have already seen this trend in the extrusion-based AM, where material costs have been reduced considerably. In fact, to preserve some of the price point for raw materials, manufacturers of AM machines have developed customized features for use of raw materials in their machines, such as embedded RFIDs containing the optimal machine parameters for the raw material. However, it is anticipated that these measures will eventually be displaced by new automation standards for machine, such as machine intelligence for enabling machines to adjust their operating parameters in situ, so the trend towards lower raw material costs is already being observed in other processes.

Open source designs of AM machines have also led to a drastic reduction in acquisition prices for certain types of AM machines. Currently, there are several AM machines on the market that cost less than $1,000, which are cost competitive with many conventional manufacturing systems. These machines tend to work with thermoplastic materials with lower melting points, making them safer and cheaper to build. Several low cost machines are also available for vat photopolymerization, but their cost is still higher than those for thermoplastic AM machines. This development has made AM technology accessible to a wide variety of users. High-end AM machines are still very expensive, typically because of the higher resolution and build quality that they can achieve through more expensive precision components. There have also been many efforts to develop low cost AM machines that work with metals and higher strength, higher temperature polymers, but they are still not as reliable and user-friendly as higher cost machines.

To support certain 3D features, such as overhangs, AM processes require use of sacrificial support materials or structures, which need to be removed before the part can be used. In some processes, removal of the support structures is easy, because they are water-soluble or very brittle. However, in several processes, such as powder bed fusion process used to create metal parts, the removal of the support is a very time-consuming process requiring conventional subtractive manufacturing processes, such as Electrical Discharge Machining (EDM). Support structures should be very carefully designed to ensure they can be easily removed. There is also considerable interest in automating the post-processing operations. This includes polishing the part to improve the surface finish.

Recent advancements in AM processes have also allowed prefabricated structures to be embedded during the fabrication process. A wide variety of components, such as sensors and actuators, have been embedded in structures created using AM processes. Usually, a robot is needed to perform precise insertion of the prefabricated components during the printing process without damaging the already built regions of the part. Therefore, the presence of prefabricated component requires changes in the fabrication process to make sure that the components of the AM machine do not interfere with the prefabricated component.

For traditional manufacturing, the price of parts will be a function of human labor cost, equipment cost, hourly operation cost, and processing time. However, AM processes are inherently more automated, requiring less human labor. The open source movement has also brought down the equipment cost, so the energy cost remains as the main component of the hourly operational cost for AM processes. Since power consumption for AM processes, such as 3D printing, are comparable to other manufacturing processes, the main driver for the total energy cost has been the processing time. As previously mentioned, AM tends to be a slow process due to the point-by-point deposition of material. Therefore, processing costs tend to be much higher when making larger parts. Most AM processes also require post-processing operations to finish parts, which leads to additional costs.

A drawback for AM processes has been a lack of in-situ monitoring capabilities. This has meant that parts need to be closely inspected after fabrication to determine if there are any defects. On the other hand, AM fabricated parts have very complex geometries with complex interior shapes, which makes it very difficult to inspect these parts using traditional inspection methods. Therefore, AM fabricated parts require destructive testing, which is difficult given that parts are made individually and not in batch, or expensive using high-resolution Computer-aided Tomography (CT) imaging techniques. This has led to an increased interest in developing in-situ monitoring techniques that can be integrated with AM processes to enable online inspection, and alleviate possible defects, as the part is being fabricated. Increasingly machine learning is being used to detect defects from a variety of different online sensors, and to modify process parameters in real-time to mitigate defects.

AM processes are also more amenable than traditional manufacturing processes to a variety of new material concepts and structural designs. For example, multi-functional materials are enabling new design possibilities by integrating multiple functional capabilities into a single structure to reduce the parasitic mass and enhance performance. The combination of AM with multi-functional materials and embedded prefabricated components is creating new product possibilities using advanced design concepts, such as "Design for No Assembly (DNA)", in many different areas. For example, entire quadrotor robots have been built with batteries, motors, wires, sensors, and microcontrollers — no assembly required. Several different AM processes have already been developed to print multi-functional materials, usually by hybridizing the AM processes with a conventional manufacturing process, such as fiber inlay. Furthermore, AM processes are capable of working with a variety of material states. For example, in the vat photopolymerization processes, resins can be used that are deactivated, so structures can be fabricated that have liquid regions that are encapsulated by solids. This enables the realization of functional structures, such as hydraulics, that are hermetically sealed to minimize oxidation and loss of the liquid under pressure due to leaking or vaporization.

The unique capability of AM to control the material distribution in structures presents the potential of using a class of multifunctional materials known as "adaptive" materials, whose properties can be varied using internal or external stimuli, in order to create structures with the capability of programming a desired set of properties, or even dimensions. Based on this capability, the term *four-dimensional* (4D) printing has been used to describe a class of AM technologies applied to building structures that can evolve their shapes over time (the fourth dimension). By arranging adaptive materials in an anisotropic architecture, the shape-changing structures can exhibit different shape-transformation behaviors, such as shrinkage/expansion, bending, folding, twisting, and even more complex shape-changing mechanisms when exposed to external stimuli. In recent years, various 4D printing methods have been developed using stimuli including moisture, heat, light, magnet, etc. 4D printing can overcome the traditional fabrication limitations by designing heterogeneous materials to enable programmed shape variation over time. These types of programmable, shape-changing structures have numerous applications, including reconfigurable electronics, actuators, sensors, implantable devices, smart packaging, and deployable structures. While the technology has great potential, several limitations also exist. For example, it is difficult to control the shape change accurately or reproducibly. A more challenging question is the inverse design problem of creating a material distribution that can be transformed into a desired 3D shape base. The inverse design is governed by the constitutive behavior of the adaptive material, which is often coupled with the resulting stress distributions when dimensional changes in the material are activated.

At present, many research groups are exploiting the capabilities of AM in the directions of multi-scale, multi-material, multi-functional, and multi-dimensional (4D printing) fabrication. The efforts will lead to a powerful fabrication tool to build engineering products that have similar complexity to biological structures. This is evident through several recent developments on bioinspired AM technology, which have demonstrated the potential to fabricate more sustainable materials and structures inspired by nature. Overall, the revolution of additive manufacturing has led to many opportunities in fabricating complex, customized, and novel products. As the number of printable materials increases and AM processes evolve, manufacturing capabilities for future engineering systems will expand rapidly, resulting in a completely new paradigm for solving a myriad of global problems.

© 2020 World Scientific Publishing Company
https://doi.org/10.1142/9789811222825_0002

Chapter 2

Support Structure Design for Selective Laser Melting Process

Behzad Rankouhi, Dan J. Thoma, and Krishnan Suresh

2.1. Introduction

It has been almost twenty years since the first metal-based powder-bed system was introduced to the world by EOS® in early 2000's. Before the advent of these systems, polymer-based sintering machines, also known as selective laser sintering (SLS) machines, were the only powder-bed systems in existence. In a way, SLS is the backbone of today's selective laser melting (SLM) systems. These SLM machines were part of the new generation of additive manufacturing systems that were the result of more than ten years of experience in producing polymer-based sintering machines.[1] Figure 2.1 illustrates the SLM process. Although the details of the process can vary slightly, the governing principles of the process remain the same, i.e. layer-wise melting of the powder bed by a laser beam.

SLM opened a window to new possibilities in design and manufacturing of metallic parts that we are still exploring. Ability to produce parts with high complexity and precision, competitive mechanical properties, and a wide selection of materials are the main advantages of this additive manufacturing process. In SLS, polymer powder is sintered to create a solid object. Sintering is the process of heating the powder to temperatures slightly below a material's melting point to allow solid state diffusion and bonding of particles. On the other hand, in SLM process, metal powder is melted to create the object. Melting is the process of heating the powder to or above the material's melting point to allow formation of melt pool and fusion of powder particles across grain or particle boundaries. It is the formation of melt pool in SLM process that makes all the difference. Complications in the design of support structures – the main topic of this chapter – for SLM is mostly due to formation of this melt pool.

Unlike SLS where the loose powder provides sufficient support for the part, SLM requires additional support structure to secure the part on the substrate, support the weight, and reduce the temperature gradient of the part by transferring the heat of the melt pool to the substrate. Any surface with a normal angled at or greater than a specific range with respect to the build axis (Z axis in Cartesian system is commonly assigned to the build direction) requires a support structure. It should be noted that this *overhang angle* is material dependent. For SS316L, a common material used in SLM, the overhang angle is around 45°. Lighter metals with lower melting point in powder form will allow for slightly larger overhang angles. For consistency, we choose SS316L as the material of choice throughout this chapter. The need for support structure is arguably one of the major drawbacks of SLM process. Considering the overhang angle, one can see how addition of support structures brings about more complications to the design process. Orientation of the part, and location of the support with respect to the part become crucial factors when setting up a build.[2] More support structure means more material and energy consumption. Moreover, unsuitable support structure design can cause part failure, which in turn can lead to higher cost per part. Therefore, an optimum support structure design is critical for the SLM process.

An essential purpose of support structure in SLM process is transferring heat from the part to the substrate, therefore reducing the temperature gradient of the part during the manufacturing process. Smaller temperature gradient corresponds to lower residual stresses in the part which are the primary cause of deformation and warping. The secondary purpose of support structures is maintaining the surface quality of the part by preventing undesirable physical phenomena such as dross. Figure 2.2 shows an example of a typical support structure generated for SLM process that meets these requirements. Based on the role that support structures play in design process, it is evident that solely geometrical design solutions will not adequately address the problem; a comprehensive understanding of the physics of SLM process is required. Nevertheless, most of the solutions provided for support structure design for SLM are based on geometry and simplified physics.[3–5] The reason we tend to overlook the influence of physics of melt-pool perhaps stems from our relatively good knowledge of Fused Deposition Modeling (FDM) or desktop 3D printing process. To design support structures for FDM, we do not need to consider the heat transfer in the part, as a result the problem simply becomes a geometrical-structural design problem, and there are numerous solutions provided for design of support structure for FDM process.[6–8] Consequently, we are biased toward these designs that neglect the effects of heat transfer and melt pool formation. Therefore, while these solutions look elegant on paper, they cannot be regarded as designs that meet SLM requirements.

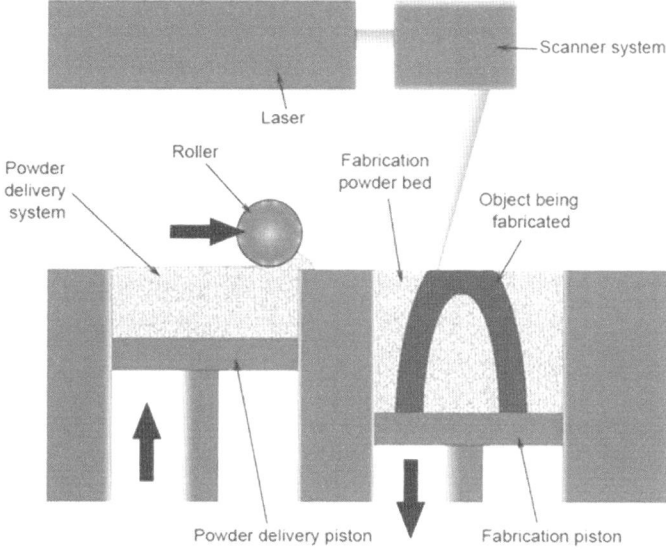

Fig. 2.1. Graphical representation of SLM process. Some systems allow for more than one laser or reflective mirrors and some use blades instead of rollers to lay down a layer of powder.[9]

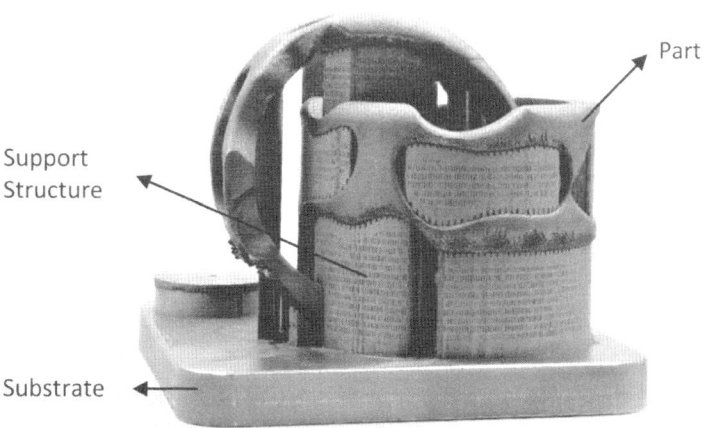

Fig. 2.2. Stainless steel 316 object manufactured using SLM process. All overhanging surfaces are supported by a common support structure called block support (adopted from Lefky et al.[10]).

Conceptually, support structure design for SLM lies at the intersection of three independent requirements, see Fig. 2.3. First, we must understand the limits of manufacturing with SLM process. For example, what is the smallest achievable thickness or diameter? In other words, we must verify if our support structure design is *manufacturable*. Second, we must choose a support geometry that possesses enough *structural integrity* to be able to support the overhanging surfaces while being efficient in terms of build time and material consumption. Finally, and most importantly, the support structure should maximize the rate of *heat transfer* from melt pool to the substrate. Any support structure design that does not adequately address these three requirements, is not a suitable design for the SLM process.

In this chapter we aim to provide an overall view of support structure design for SLM. We start by defining the vocabulary that will be used to describe different aspects of a support structure. Then, we explore the metrics that should be considered when designing support structures. These metrics will provide a base line for comparing the effectiveness of different types of designs. Furthermore, we investigate conventional and novel designs. We explore the advantages and disadvantages of each design and provide examples. Finally, we discuss the possible directions support structure design can take in the future.

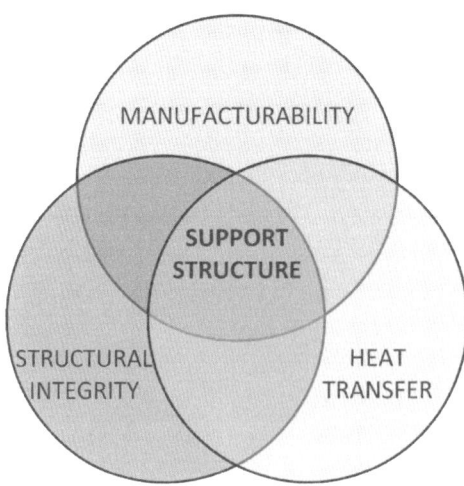

Fig. 2.3. Support structure design for SLM cannot be addressed without considering three main requirements that govern the design process.

2.1.1. *Support Structure Nomenclature*

Before we delve into support structure design, we need to establish terminology that defines different features of a support structure. These are features that are common among most types of support designs. Features that are specific to a particular type of support structure will be defined in the designated section for that support type.

As mentioned in the introduction section, surfaces with normals that are oriented at or greater than a specific angle with respect to the build axis are called *overhang surfaces* and the specific angle is called the *overhang angle*. The overhang angle is material dependent and is determined experimentally. It is worth mentioning that in SLM process, parts are typically placed a few millimeters above the substrate which means that parts will be printed on a support structure and not directly on the substrate. This makes removing the part from the substrate more convenient, because a support structure is not nearly as dense as a solid part. However, there are exceptions where placing the part directly on the substrate proves to be advantageous. These scenarios are mostly dictated by the geometry of the part.

The subset of the support close to the part is of special interest. It determines the main functionalities of the support structure and it typically has a more complicated design compared to rest of the support structure. Therefore, we propose the body and comb concepts to better understand and distinguish between these two regions. Figure 2.4 shows a part together with a typical support structure. The support structure consists of two segments which serve different purposes. *Support body* refers to the larger segment of support structure which usually starts from the substrate and ends close to the overhang surface of the part. In addition to supporting the weight of the part, support body is responsible for transferring the heat from the part to the substrate, ensuring reduced temperature gradient within the part. *Support comb* refers to the upper segment of support structure where support and part meet. Support comb design plays a crucial role in the surface roughness of the part, ease of support removal, heat transfer, and overall success of the design. In scenarios where the part contains surfaces overhanging itself, support structure can contain two support comb segments and a connecting support body. This special case is depicted in the upper section of Fig. 2.4.

From a CAD perspective, support structures can be represented as solids or surfaces. *Solid supports* are designs that contain a volume, while *surface supports* are designs with zero volume. Figure 2.5 shows two examples of solid and surface designs. In the cone design (left), each support occupies a volume (solid design). On the other hand, in the web design (right), the support is represented as a surface (the thickness will be implicitly assigned through process parameters). We will take advantage of this distinction later in the chapter where we introduce support structure metrics.

Finally, some designs are *open structures*, i.e. there is no border wall surrounding the structure. The cone design in Fig. 2.5 is an example of an open structure. On the other hand, *closed structures* are structures with a surrounding wall, such as the web design in Fig. 2.5. Open structures allow for the un-melted powder to escape while closed structures trap the un-melted powder inside. The trapped powder can only be removed after the part is separated from the substrate. We will revisit the concept of open and closed structures when we introduce the metrics in support structure design.

2.2. Support Structure Metrics in SLM

Before we can discuss metrics to gauge support structure performance, we briefly review the governing physics of the problem that is dictated by the melt pool. There are various characteristics associated with melt pool in SLM, among them, temperature and size are the most influential in support structure design. The two most important parameters affecting the temperature and size of the melt pool are laser speed and power. Often, these are combined into a single parameter:

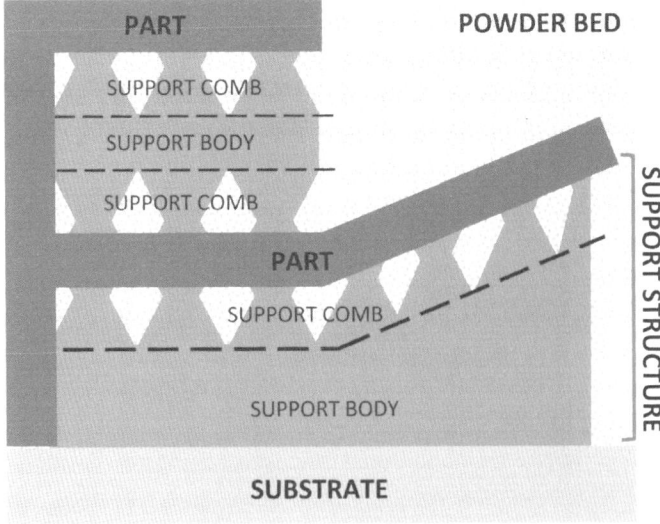

Fig. 2.4. Support structure segments. The design strategy and requirements for each segment is unique to the type of support structure.

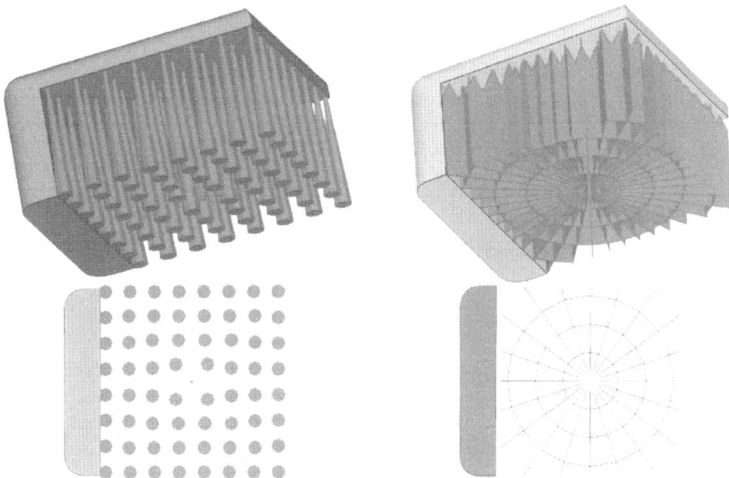

Fig. 2.5. Cone design (left) is an example of solid support design, and web design (right) is an example of surface support design.

Linear energy density (LED) defined as:[11]

$$E_L \approx \frac{P}{v} \tag{2.1}$$

where E_L is linear energy density in *J/mm*, P is laser power in *W* and v is laser speed in *mm/sec*. In practice, surface oxide, reflectivity, plasma, vaporization, etc., all affect the effectiveness of the linear energy density.

Temperature of the melt pool increases significantly by increasing the laser power or decreasing the laser speed. Melt pool temperature is in the order of thousands of degrees Celsius and it depends on the material's melting point in powder form and powder particle shape and size distribution. Dimensions of the melt pool are also a function of laser power and speed and they are in the order of hundreds of micrometers. Figure 2.6 shows a simplified drawing of melt pool dimensions in SLM process. Length, width, and depth of the melt pool increase with an increase in LED or laser power. However, an increase in laser speed alone will reduce the width while elongating the melt pool.[12,13]

Next, we consider the relationship between these two melt pool characteristics and design of support structure features. We discuss the defects that a proper support structure can help prevent and establish the metrics that are common between all support structure types.

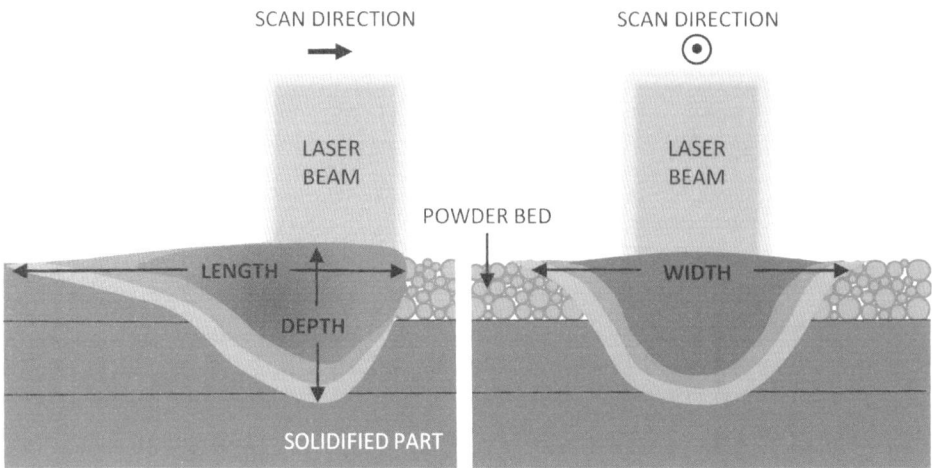

Fig. 2.6. Simplified schematic of melt pool dimensions and shape in SLM.

2.2.1. *Part Quality*

The first metric in support structure evaluation is part quality. This metric measures the quality of the manufactured part after the removal of support structure. There are certain defects that can affect the quality of the part that an effective support structure can prevent.

Residual stress induced defects

Residual stresses are stresses that remain within the part after the manufacturing process is completed and the part is in equilibrium with its environment. It is an inherent consequence of SLM process. These stresses can cause warping, delamination, distortion, and cracking.[14]

(1) *Warping* is defined as plastic deformation due to thermal stresses caused by rapid solidification of the part.[15] This plastic deformation occurs when the thermal stresses exceed the yield strength of the material. Therefore, it is more likely to happen in regions with thin features or overhangs, where each layer is slightly shifted forward, resulting in a small thin surface (Fig. 2.7). Moreover, the staircase effect of SLM process exacerbates the warping issue. As shown in Fig. 2.7(a), deflection of each layer accumulates as consecutive layers solidify, causing the last layer to protrude the powder bed. If this protrusion exceeds the layer thickness, it collides with the recoater blade when a new layer is being deposited, causing the entire build to fail. The reason the thermal stresses exceed the yield strength of the

material in overhang surfaces is that heat from the melt pool does not transfer to the substrate which acts as a heat sink. It is expected that thermal conductivity of a metal in powder form is approximately 100 times lower than its solid form,[16] as a result, the absence of support structure means that the generated heat stays in the part to cause high thermal stresses.

(2) *Delamination and cracking* is defined as separation of two consecutive layers due to residual stresses. Similar to warping, delamination occurs due to high stresses at the layer interface.[17] Figure 2.8 shows examples of defects caused by residual stresses.

Physical phenomena responsible for the origin of residual stresses are still being investigated. High temperature gradient within the part due to traveling melt pool, thermal expansion/contraction and non-uniform plastic deformation are the main factors in accumulation of residual stress in the part.[12,14] In addition, an inadequate support structure can aggravate this intrinsic consequence of SLM process.

It should be noted that utilizing support structure is not the sole remedy for the above-mentioned defects, nor will it fully prevent the occurrence of these defects. There are other contributing factors, such as change in process parameters that can help with residual stress induced defects. Some examples of other remedial approaches include: (1) reducing layer height to minimize the staircase effect which can lead to reduction in warping, (2) increasing the substrate preheat temperature to reduce the temperature gradient in the part which can lead to lower thermal stresses, and (3) controlling the process

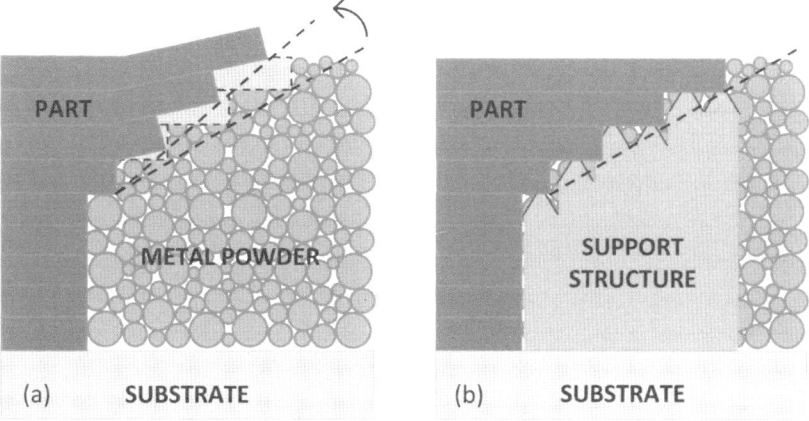

Fig. 2.7. (a) Warping in an unsupported overhang surface. (b) An effective support structure can prevent warp from happening.

Fig. 2.8. (a) Delamination at layer interface during cooling.[18] (b) crack formation in heat affected zone.[19]

parameters of SLM; for example, shorter laser scanning vectors or island scanning patterns, increase in scanning speed, and increase in powder bed temperature have shown to reduce distortion and residual stresses in the part.[20,21]

Dross

This defect is defined as unintentional partial or complete melting of powder particles below the current layer. As explained earlier, during melting of an overhang surface with no support structure, the heat conduction rate is very slow. Therefore, the absorbed energy of powder bed will be much higher resulting in much larger and heavier melt pool. Rayleigh-Taylor instability in the gravity field along with capillary forces will cause the melt pool to sink into the powder bed. Dross will form when the heat from the sunk melt pool causes unintentional melting of surrounding powder particles.[22–24] As shown in Fig. 2.9, dross can affect the top surface of the overhanging layer as well. Severe dross will cause high surface roughness on top of the overhanging layer which in turn can lead to build failure if the recoater collides with the uneven surface.

Presence of support structure underneath the overhanging layer prevents melt pool enlargement and sink into the powder bed. Further, it is important to note that the gap between each tooth of support comb is directly related to the melt pool size. If this gap is too large, the melt pool will have enough space to grow and form dross. This is one of the main reasons why an effective support comb design is crucial in preventing dross.

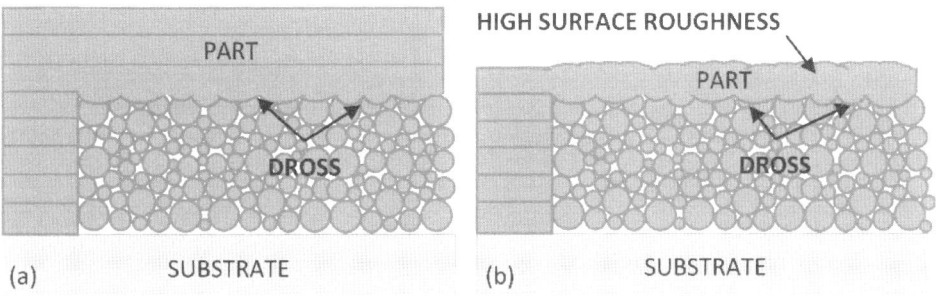

Fig. 2.9. (a) Dross formation in the absence of support structure. (b) Severe dross can cause high surface roughness on top of the overhanging layer causing failure of the entire build.

2.2.2. *Support Material Volume*

Another important metric in assessing support structures is its volume. Between two different support structures with identical performance, the one with lower volume is preferable. There are however two important observations to make.

First, recall that a support structure can be represented as a solid, or a surface with process-dependent thickness. Figure 2.10 shows the two different support structure representations. In the former, the volume is easily computable from the CAD geometry. In the latter, the thickness is determined by laser power and speed that can be combined into a single parameter: *planar energy density* (PED) defined as:

$$E_P \approx \frac{P}{v.t} \tag{2.2}$$

where E_P is planar energy density in J/mm^2, P is laser power in W, v is laser speed in mm/sec, and t is layer thickness in mm. Higher PED's will result in slightly thicker supports. For example, for SS316L, PED of 0.15 J/mm will result in wall thickness of ~0.12 mm while PED of 0.19 J/mm will result in wall thickness of ~0.15 mm.

The second observation is that for a given volume of support, the material consumed (i.e., the weight of the support) is once again process dependent. If the laser power per unit volume is low, then there will be incomplete melting of powder particles, leaving pores inside the material. This leads to lower density of the support structure and reduced material consumption.[25] A typical process parameter used to evaluate full/incomplete melting is the *volumetric energy density* (VED), defined as the average applied energy per volume of material during the scanning of one layer:[25,26]

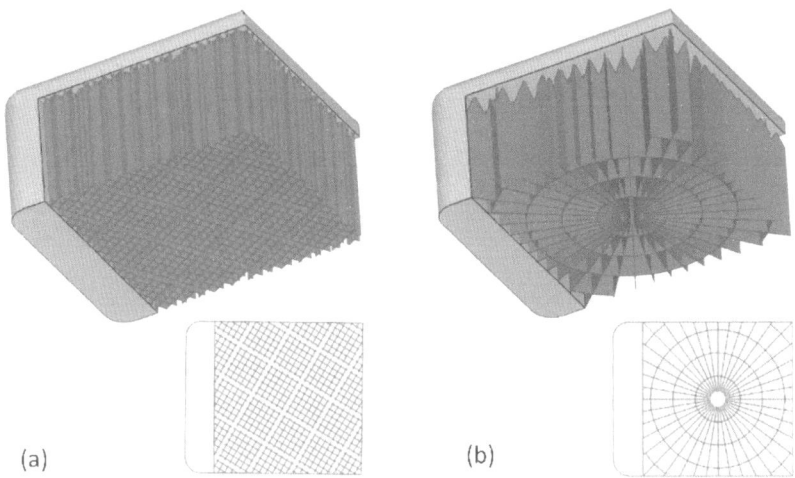

Fig. 2.10. (a) Block support structure. (b) Web support structure. Given the same process parameters, the web design has 42% less volume.

$$E_V \approx \frac{P}{v.h.t} \tag{2.3}$$

where E_V is volumetric energy density in J/mm^3, P is laser power in W, v is laser speed in mm/sec, h is hatch distance in mm, and t is the layer thickness in mm. Choosing different laser power, laser speed, layer thickness, or hatch distance can result in different density or fill-ratio.

2.2.3. *Support Manufacturability*

This brings us to our next metric in support structure design, manufacturability of the support structure. The first rule ironically states that the supporting structure should be self-supporting, both in terms of overhanging surfaces and its susceptibility to defects. We know that defects such as distortion due to residual stresses can be caused by excessive absorption of energy by the powder bed. Therefore, support structure immunity to defects is tightly tied to process parameters, especially VED and PED. To understand how these two parameters can affect the manufacturability, we need to revisit the representation of support structures, namely, solids and surfaces. Most common designs fall in the surface category, such as the designs shown in Fig. 2.10. In these designs, every feature of the model is a surface, therefore there is no infill scan strategy, there is only single scan vectors. For this group of support structures, PED is the important process parameter. On the other hand, designs such as the cone (Fig. 2.11), contain solid features that require infill scan strategies. In other words, we should consider hatch distance, hence the VED becomes the important process parameter for this group of supports. Now, we can apply the same concepts that we introduced before regarding melt pool, residual stress and defects, to support structure manufacturing.

Consider the scan strategies for solids and surfaces. Figure 2.12 shows a cross sectional view of tree and block support structure designs. Movement of a laser beam is depicted by an arrow which we call a *scan vector*. These vectors inform us of *direction* and *orientation* of laser beam's movement. In the surface representation, there is a single laser beam pass, i.e., a single scan vector, for each wall segment, whereas for a cone design, multiple passes of laser beam are needed to create a solid geometry.

Figure 2.13 shows a failed lattice support structure as an example that highlights the importance of manufacturability metric. In this design, lattice strut diameter was set to 0.1 mm. Coincidentally, the machine that manufactured the part had a minimum feature size of 0.1 mm. Therefore, although the support structure in this example is represented as a solid, there is no room for any infill scan strategies. In scenarios such as this, using VED will apply excessive energy to the powder bed which can cause lattice strut deformation. In Fig. 2.13 deformed struts protruded the powder bed and collided with the recoater. Accumulated deformation finally destroyed the entire support structure leaving the overhang surface of the part unsupported. Finally, severe dross and instability led to failure of the entire part. To avoid failures such as this, one can treat a solid representation as a surface representation, and use appropriate process parameters such as PED.

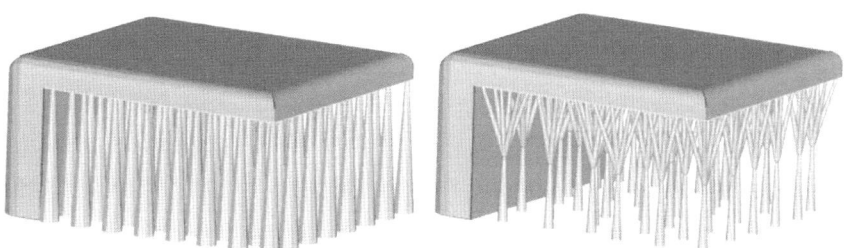

Fig. 2.11. Cone (left) and tree (right) support structure design. Both designs are examples of 3D solid designs.

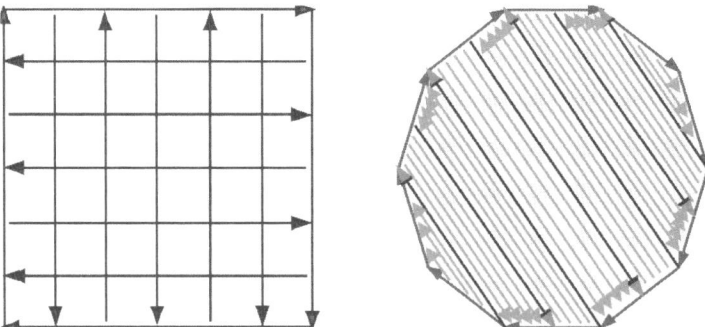

Fig. 2.12. Cross sectional view of a block design and its scanning vectors (left), and a single cone of a cone design and its scanning vectors (right).

Fig. 2.13. Using improper process parameters caused the lattice support structure to fail which led to the failure of the entire build. This example highlights the importance of manufacturability metric in support structure design.

2.2.4. *Post Processing and Removal*

The final metric in support structure design is the ease of removal, and required post processing steps. Removing support structure is a tedious task, and it is usually done manually. A support structure design that expedites removal is greatly beneficial. But support removal is not the only post-processing concern. During the manufacturing process, unused powder can get trapped inside the support structure. Further, since the structure is usually enclosed by the substrate at the bottom, and the part at the top, this trapped powder cannot be released unless the part and its support structure is removed from the substrate. Since removal is typically done by means of electric discharge machining (EDM), or use of power saws, the released trapped powder becomes contaminated and not reusable. For large parts and high production rates, loss of trapped powder can harm the economics of SLM process.

To reduce support removal time, effective support comb designs are required. Figure 2.14. (a) shows an example of a lattice support structure with an effective comb design that can be easily removed using simple hand tools while Fig. 2.14. (b) shows an example of a support structure that is impossible to remove without the use of power tools due to lack of support comb structure.

Few attempts have been made by researchers to automate or expedite the removal process. In one study, Lefky et al.[10] attempted to automate the support removal process by using self-terminating electrochemical etching. They incorporated a sensitizing agent during the heat treatment process to chemically destabilize 100-200 μm of the part's surface. The part is then etched with a high selectivity toward the sensitized surface over the substrate. The etching process self-terminates when the sensitized layers are removed. This approached proved to be effective for 17-4 PH stainless steel and SS316L. Figure 2.15 shows the proposed removal process applied to a SS316L part with a complex geometry. It should be noted that during this process 120 μm of the part were removed along with the support structure.

In another study, Wei et al.[27] demonstrated an easy-to-remove support structure for SS316L made with SiC-SS316L composite. The composite material with 40% volume fraction and 320 grit SiC produced enough mechanical defects during the SLM process that the transition zone between part and support was easily broken by applying a low external force. Addressing cross-contamination and high surface roughness proved to be challenging in the proposed method. Figure 2.16 shows a simple overhanging surface that is supported by SiC-SS316L composite support structure.

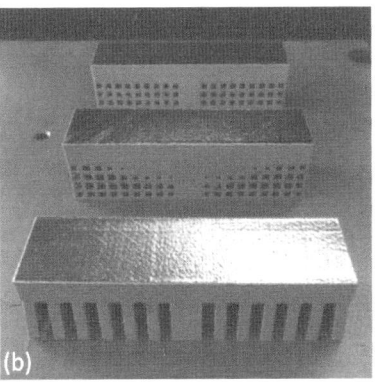

(a)   (b)

Fig. 2.14. (a) An example of a lattice support structure with effective comb design. Removal of this support structure is relatively simple.[28] (b) An example of support structure with no comb design. It is impossible to remove this support structure without extra machining.[29]

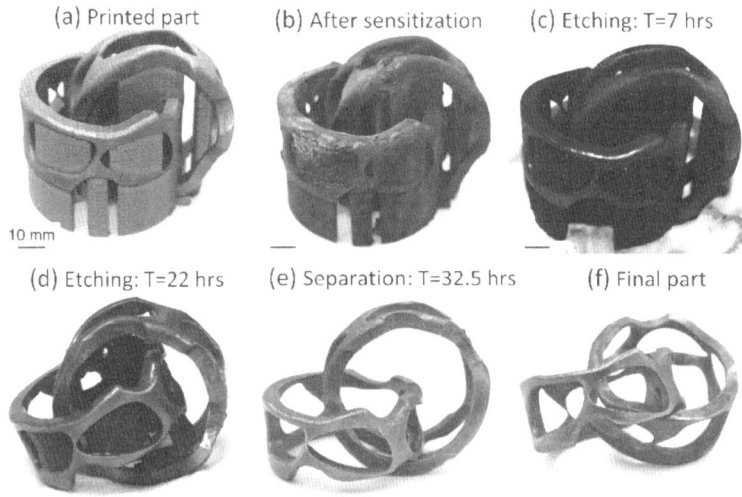

Fig. 2.15. Support structure removal using electrochemical etching applied to SS316L. The geometry of the part is mostly preserved after 33 hours of etching and support material removal.[10]

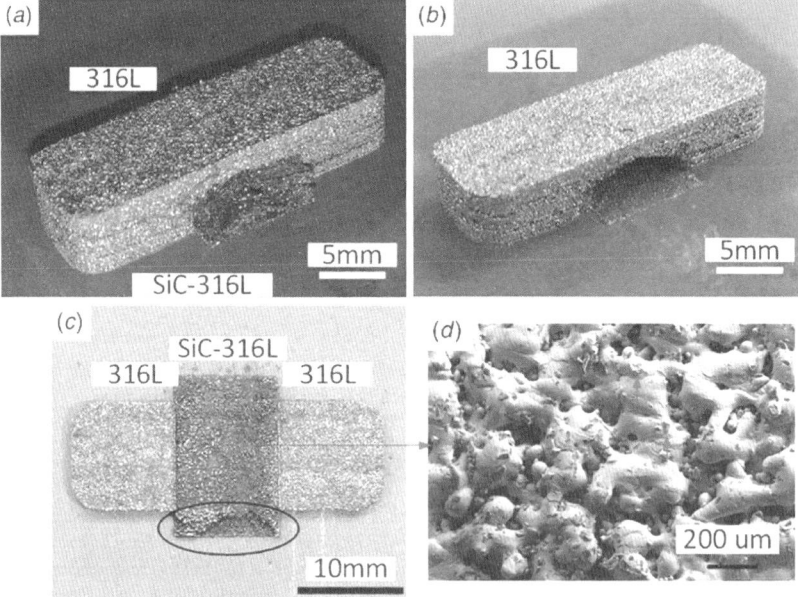

Fig. 2.16. (a) A bridge structure using SiC-316L as the support material at the aperture position, (b) the support structure removed, (c) cross section of the bridge structure, and (d) the SEM image of the top surface of the laser sintered SiC-316L support structure.[27]

Although these efforts show great promise, they remain highly material dependent and cannot be widely applied. Today, most support removal is still done manually, therefore there is no proper way of quantifying this metric.

Support removal will remain a qualitative metric for assessing different types of support structures.

On the other hand, trapped powder is quantifiable. Consider the closed support structures in Fig. 2.10. The actual support volume is the volume of the support structure plus the volume of the trapped powder, whereas in an open support structure (see Fig. 2.11), the actual support volume is the same as the volume of the support structure. To mitigate the trapped powder problem, use of open structures is recommended. Further, there are some approaches that will allow closed structures to release the entrapped powder. We will discuss this feature in the next section.

2.3. Support Structure Design and Manufacturability

In the previous section, we introduced certain metrics for evaluating support structure designs. In our view, any design that is part of this landscape is an acceptable design. Some might be more effective, robust, or efficient than others. This section is dedicated to introducing existing support structure designs. We will start by discussing conventional designs that are industry standards. Later we introduce novel designs, and discuss their advantages and disadvantages.

2.3.1. *Conventional Support Designs*

Conventional support structure designs are the most robust and widely utilized designs in the industry today. They are not optimal structures in terms of material consumption or build time, but they are the most reliable. Unfortunately, there is no universal terminology to compare these designs and define their features. Therefore, we adopt the terminology from Materialise® Magics®, an industry leading data and build preparation suite for AM.[30] Based on this terminology, the conventional support designs include: block, web, contour, line, gusset, and point. We introduce the block design and its features below. The remaining, are variations of the block design, created to accommodate certain geometrical scenarios. We will briefly discuss these scenarios, but readers are referred to commercial support generation packages for further details.

Block design

Arguably the most widely used, and reliable support structure design, is the block design. It consists of zero-thickness walls that are arranged to form a grid, and it is suitable for large overhang surface areas; see Fig. 2.10(a). Main advantages of block design are good manufacturability and high heat transfer rate. Its shortcomings are high volume, and difficulty in removal. Figure 2.17 shows a schematic of the block design grid. Critical dimensions of the grid are defined in Table 2.1. These values are material dependent and are determined experimentally.

Hatch distances are critical in preventing dross, and they depend on the melt pool characteristics. Moreover, they correlate to the relative density of the structure which in turn determines the heat transfer rate of the support structure. If the hatch distances are too large, dross will form within each block of the grid, and residual stresses will increase due to low heat transfer rates. On the other hand, small hatch distances will unnecessarily increase the support volume and hinder support removal. The *separation width* is another critical dimension that helps ease support removal by fragmenting the grid.

The next feature within the support body is *perforation*, see Fig. 2.18 and Table 2.2. Perforation allows trapped powder retrieval. Perforations are usually

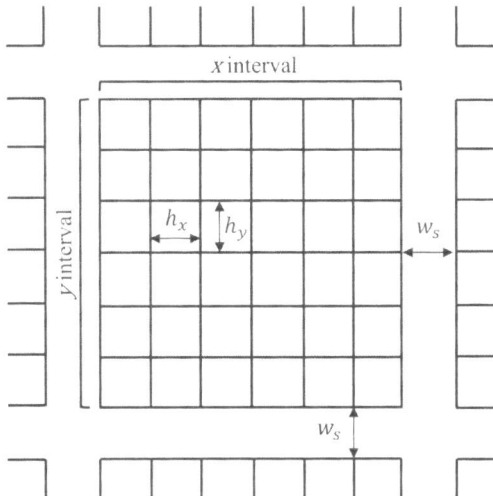

Fig. 2.17. Grid schematic of the block support structure design and its critical dimensions.

Table 2.1. Critical dimensions and their description for a grid in block design.

Dimensions	Description
h_x	Hatch distance in x direction.
h_y	Hatch distance in y direction.
x interval	Number of squares in x direction for each block.
y interval	Number of squares in y direction for each block.
w_s	Separation width which determines the fragmentation distance.

diamond or rectangle shaped. Although they turn the block design into an open structure, some trapped powder will still remain inside the support structure. Making the perforations larger can cause build failure.[23]

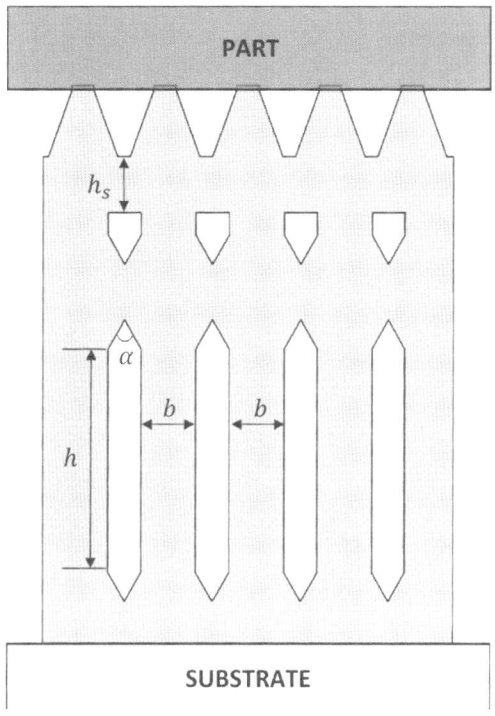

Fig. 2.18. Schematic of perforations and critical dimensions in block support structure design.

Table 2.2. Critical dimensions and their description for perforations in block design.

Dimensions	Description
h_s	Solid height, distance between the last perforation and support comb or first perforation and substrate.
h	Height of the diamond perforation.
α	Angle of the diamond perforation.
b	Distance between each perforation, known as beam.

Figure 2.19 shows the support comb for the block design. Critical dimensions are defined in Table 2.3. This comb design facilitates the removal of support structure. Moreover, it ensures that the surface quality of the part is within acceptable range. Perhaps the most critical dimension in support comb is Z_{offset}, the amount of support penetration into the part. A common rule states that

Z_{offset} cannot be smaller than the thickness of two layers. This ensures adequate support-part fusion.

Figure 2.20 shows different views of block support structure manufactured using SS316. Figure 2.20(a) reveals perforations on the border wall of the support structure. Figure 2.20(b) gives a clear view of hatch intervals and fragmentation of the blocks, and Fig. 2.20(c) illustrates the support comb. Observe that the manufactured comb bears little resemblance to the design in Figure 2.16. The reason is that critical dimensions of support comb are in the order of hundreds of microns which is in the same order as powder particle size. At this scale, model's geometry cannot be fully preserved during manufacturing, but the resulting structure can still alleviate support removal.

Recommended values for critical dimensions of the block design are provided in Table 2.4. These are conservative values to ensure a successful build. Different values for most cases are permissible but it requires experimental validation. It is important to note that these numbers are validated for a particular SLM machine and a particular metal powder vendor. Observe the subtle differences in critical dimensions for each material. These differences are driven by machine's laser and optics and melt pool characteristics.

Other conventional designs

Other conventional designs are similar in concept to the block design. Support body and support comb serve the same purpose as in block design. Perforations and fragmentation can be implemented in support body while the same comb design can be used for every design. Instead of grids, surfaces are arranged to form different patterns to better accommodate different geometries.

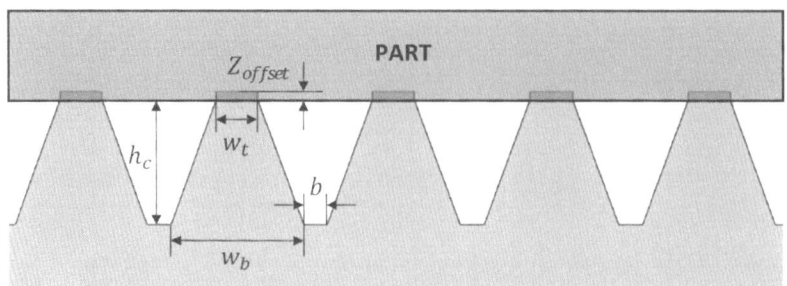

Fig. 2.19. Support comb and its critical dimensions for block design.

Table 2.3. Critical dimensions and their description for support comb in block design.

Dimensions	Description
h_c	Comb height.
w_t	Comb width at top of the comb structure.
w_b	Comb width at base of the comb structure.
b	Base interval.
Z_{offset}	Penetration distance of support comb into the part.

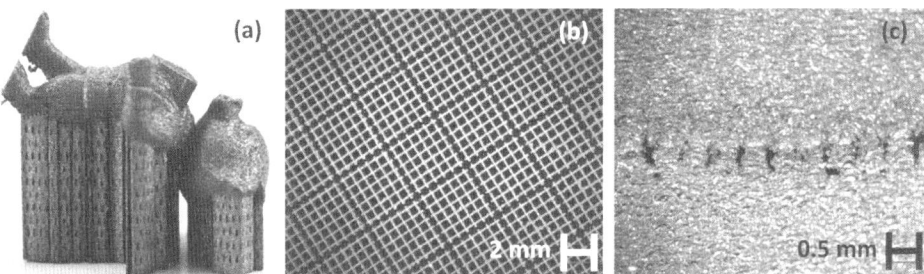

Fig. 2.20. Three different views of block support structure, emphasizing different features of the design.

For example, the *web design* is recommended for circular surfaces due to its circular pattern. Users can change the number of ribs and circles based on the material (Fig. 2.21(a)).

Contour design is created by offsetting a pattern, repeated to cover the entire overhang surface (Fig. 2.21(b)). This pattern depends on the geometry of the overhang surface. Since there are no orthogonal surfaces intersecting the pattern to buttress the structure, this design is susceptible to build failures. On the other hand, it is easy to remove.

Line design is created for thin surfaces or edges, as shown in Fig. 2.21. (c). A single wall follows the path of the overhang surface and provide adequate support. Ribs can be added to the wall to provide reinforcement and decrease failure probability. One of the main disadvantages of this design is the likelihood of residual stress induced defects. This design is recommended for delicate parts such as small surgical instruments, where support removal can irreversibly damage the surface of the part.

If there is a small overhang surface far from the substrate in the build direction, gusset design (Fig. 2.21(d)) can be used to save time and material. The advantage of this design is its minimal volume and ease of removal. Gusset design takes advantage of the fact that the body of the part itself can be used as a

Table 2.4. Recommended values of block support design for popular materials. Adjusted for 400W Yb (Ytterbium) fiber laser.[30]

Dimension	Recommended Value [mm]			
	SS316L	AlSi10Mg	Ti64	Inconel718
		Grid		
h_x	0.8	0.7	0.7	0.8
h_y	0.8	0.7	0.7	0.8
w_s	0.8	0.7	0.7	0.8
		Perforation		
h_s	0.5	0.254	0.254	0
h	2	1	1	0.6
α	60	60	60	25
b	0.6	1.2	1.2	0.4
		Comb		
Z_{offset}	0.04	0.03	0.03	0.12
w_t	0.2	0.2	0.2	0.2
h_c	0.6	1	1	0.8
w_b	0.7	0.6	0.6	0.6
b	0.1	0.1	0.1	0.1

heat sink. A compromise must be made between saving time/material and altering the thermal history of the part.

Finally, *point design* is meant for singular, small, and down-facing overhang surfaces as illustrated in Fig. 2.21(e). Situations where a point design is required seldom occurs but when they do, using a point design helps save time and material.

Currently, there are no tools available to automate the generation of different types of support designs based on the overhang geometry. Hence, choosing the appropriate support design and its critical dimensions rests on user experience.

2.3.2. *Novel Support Designs*

Besides conventional designs, researchers have also experimented with novel designs to improve on some of the metrics introduced earlier. In this section we will review some of the proposed designs.

Tree structures

Gan and Wang[24] proposed three different designs, shown in Fig. 2.22. They concluded that an effective support structure should promote uniform heat

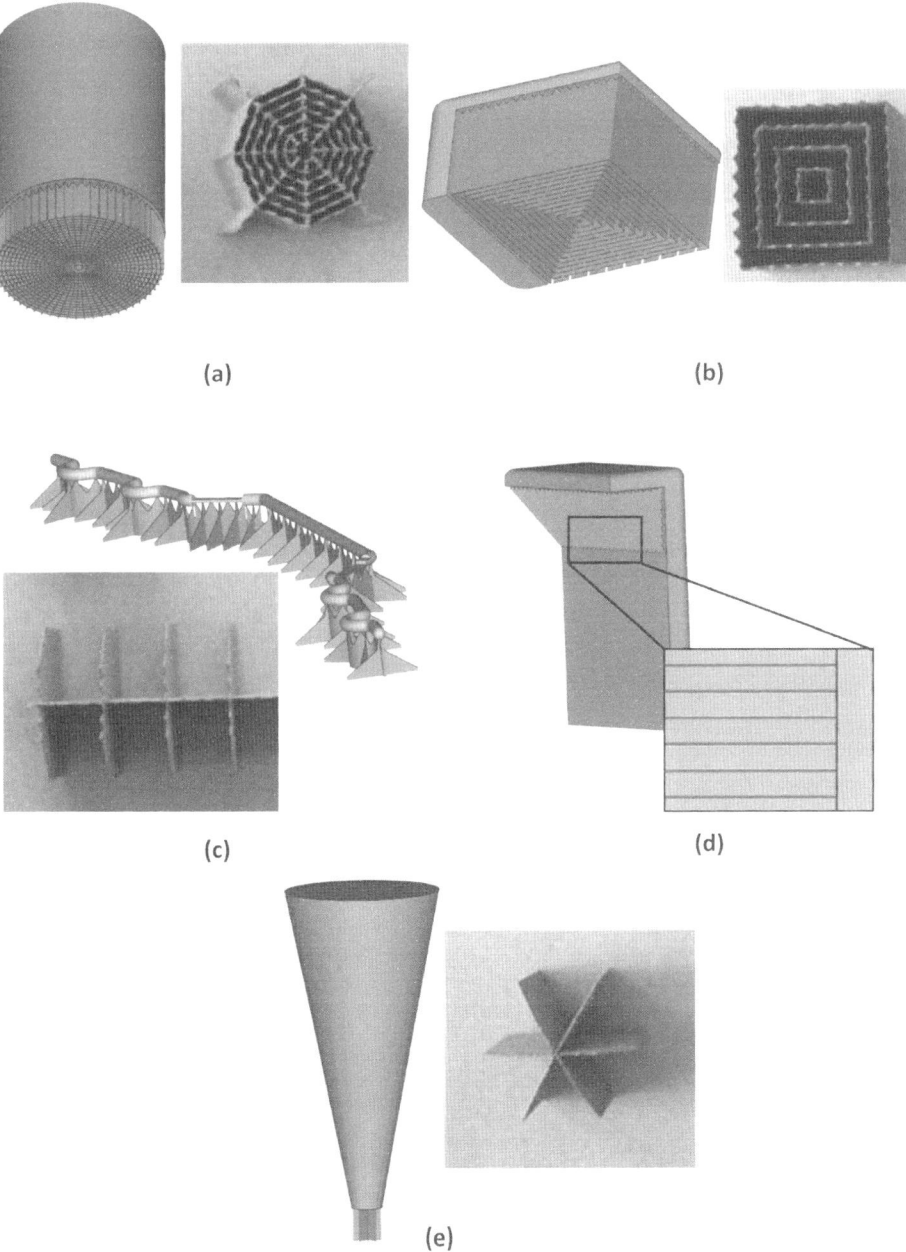

(a)

(b)

(c)

(d)

(e)

Fig. 2.21. Conventional support structure designs and their corresponding cross sectional view as fabricated by SLM.[31] (a) web, (b) contour, (c) line, (d) gusset, and (e) point.

dissipation and have a maximum for support comb spacing. Design (b) has more connection points compared to design (a) but the part experienced warping nonetheless. They hypothesized that the inclined connection points are the cause of warpage. Design (c) consists of an array of pin supports. Among the three proposed designs, design (c) yielded the poorest surface finish. A closer look at the manufactured examples in Fig. 2.22 reveals the dross formation in all three designs. It is fair to say that although these designs performed well on support removal, manufacturability, and volume metrics, they performed poorly on part quality metric.

In another study, Zhang et al.[32] proposed a tree structure design (Fig. 2.23) with the ability to change the branch diameter, angle, and number. They showed that their design can support large overhanging planes. Moreover, compared to conventional support structures, the proposed tree structure can save 23% on material and about 30% on scanning time. Although this design showed good performance on support volume, and manufacturability metrics, it did not perform well on post processing metric. Lack of support comb in this design compels the user to use excessive machining for support removal.

Lattice structures

From an engineering point of view, a lattice is defined as a pattern, known as a unit cell, which is repeated regularly in all directions. These light-weight structures have large surface area to volume ratio which provide them with good

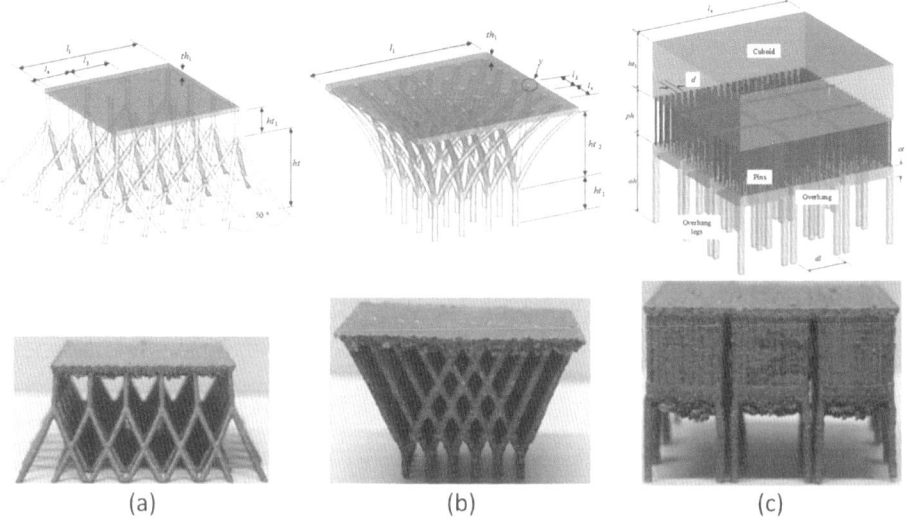

(a)   (b)   (c)

Fig. 2.22. (a) Inverse Y design, (b) Y design, and (c) pin design. Proposed by Gan and Wang[24] as practical support structure designs.

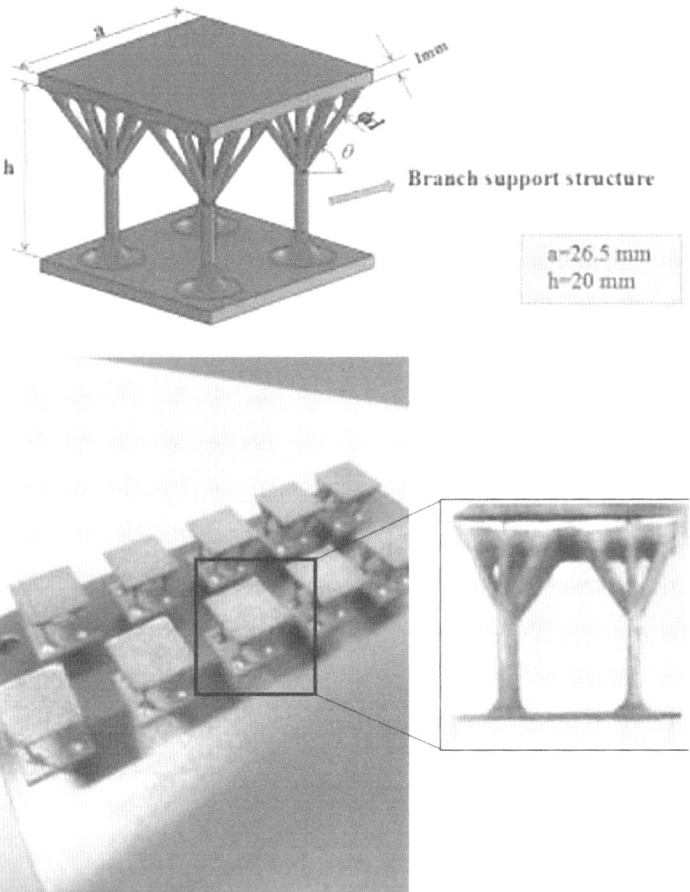

Fig. 2.23. Tree (branch) support structure design proposed by Zhang et al.[32]. Manufactured by SLM from SS316.

thermal dissipation property. These properties make lattice structures a potent candidate for support structures.

Cloots et al.[33] studied a variation of BCC lattice structure with an emphasis on SLM process parameters. They were able to successfully implement their design and manufacture an overhang surface (Fig. 2.24). High heat transfer rate of the structure helped prevent any residual stress induced defects. However, there was dross formation on the overhang surface. The advantages of this design are acceptable part quality, low volume, and ease of removal. Its drawback is in manufacturability. Authors reported that to successfully build the overhang on top of their support structure, they had to change the PED during the process resulting in part densities between 89.4% and 98.4%.

In another study, Hussein et al.[34] investigated two types of lattice support structures, Schwartz diamond and Schoen gyroid shown in Fig. 2.25. They focused on openness of lattice structures and trapped powder retrieval along with build time for each design. They concluded that smaller cell sizes will result in longer print times and more material consumption. They achieved excellent part quality and support volume, but support removal was not considered in their study. Compared to other unit cells, diamond and gyroid created larger contact areas with the part. That means breaking the support structure from the part would be difficult.

Fig. 2.24. Lattice support structure made with SLM. (a) PED = 0.85 J/mm^2 and density of 89.4%, and (b) PED = 2.35 J/mm^2 and density of 98.4%.[33]

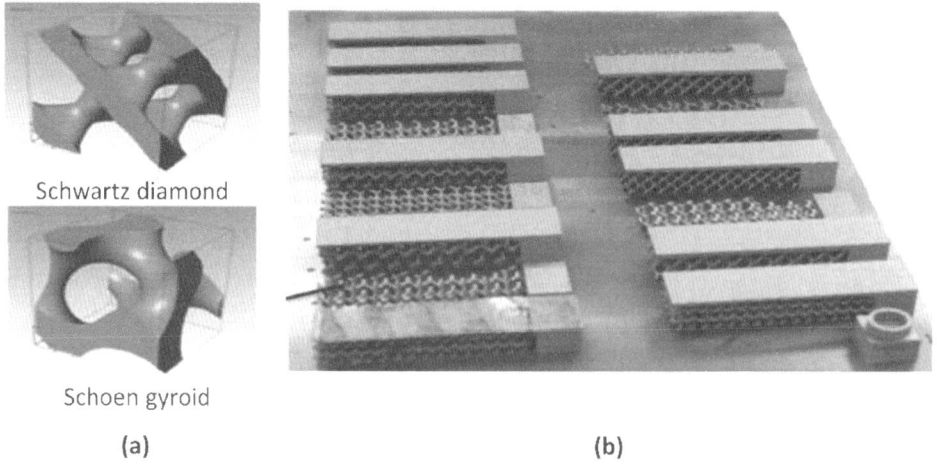

Fig. 2.25. (a) CAD models of diamond and gyroid structures. (b) Manufactured cantilever parts and their lattice support structures.[34]

2.3.3. *Customized, Simulation Based Support Design*

Both the conventional and novel support structure designs discussed above require the user to make critical decisions. Poor decisions can lead to build failure, or, at best, an inefficient design. With increased computational power and fundamental understanding of SLM physics, *we foresee an increase in use of simulation tools to create customized and optimized support structures.*

Numerous efforts have been made to simulate the SLM process at a micro scale, where laser interaction with powder, powder melting, and evolution of melt is considered.[13,35–37] Other approaches are in the macro scale where laser heating and melting is treated as a thermal source, part shape, and laser scan strategies are taken into account, and residual stresses and local effective material properties can be calculated.[38–41] The above-mentioned approaches are computationally expensive and they are not applied to the support structure. Only recently have researchers tried to implement simulation based support structure design. For example, Cheng et al.[29] investigated the feasibility of using topology optimization in support structure design to mitigate residual stress induced defects. They used the inherent strain method for residual stress calculations during the SLM process to significantly reduce the computational cost of their approach. Variable density lattice structures were used due to their open design with the objective to minimize the mass of the support structure under stress constraints. Figure 2.26 shows the resulting design. They were able to reduce the weight of the support structure by 60% while mitigating cracking and warping induced by SLM process. Design shown in Fig. 2.26 proved to satisfy all the metrics introduced in this chapter, except for ease of removal. The lattice structure lacks support comb to alleviate the removal process.

In another similar study,[42] a voxel-based fictitious domain method is used to calculate residual stresses in the design domain which includes the support structure. This approach reduced the computation time and allowed for residual stress minimization through part orientation optimization based on process modeling. Moreover, the yield strength is computed using a multi-scaled model. Finally, a multi-objective optimization is carried out to minimize both residual stresses and support volume. Figure 2.27 shows the proposed method and support structure design. Similar to previous work, the proposed design performed well on all metrics except the ease of removal.

Based on the metrics introduced in this chapter, a simulation tool suitable for support structure design requires a multiscale modelling approach, wherein melt pool dimensions and temperature, and residual stress development in the part are both implemented,[43] and coupled with support structure build simulation. Results of such simulations should drive design of support structure and the choice of critical dimensions.

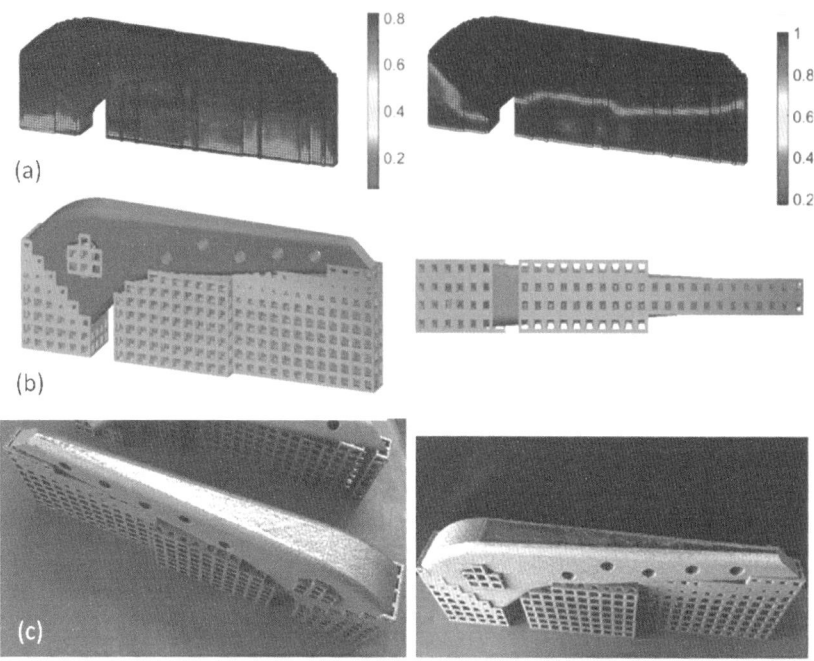

Fig. 2.26. (a) Optimization results of normalized residual stress distribution in the model.[29] (b) Reconstruction of the optimal support structure design using variable-density lattice structure.[29] (c) Manufactured designs using SLM process and Ti6Al4V.[29]

2.4. Summary and Conclusion

Support structure for SLM process is an enabler and a challenge at the same time. They allow for manufacturing of complex designs, but if used inappropriately, they may cause defects in the part, or inefficiencies in the process. By introducing support structure metrics, we proposed a systematic way to evaluate the performance of a support structure. The metrics show that the performance of a support structure is tightly tied to the physics of the SLM process, and designs that ignore this link are often ineffective. Conventional designs provide solutions for most common scenarios, but there is no automated tool for choosing/ generating these designs. Therefore, most support structure implementations rely on user experience. Some designs consume less material but are cumbersome to remove, while other designs are the exact opposite.

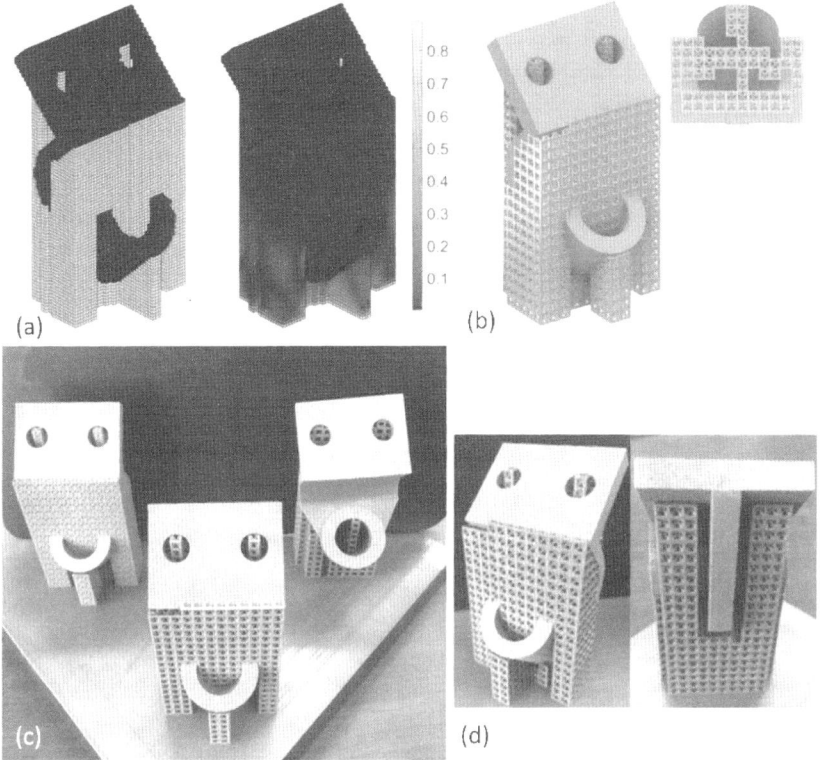

Fig. 2.27. (a) Orientation optimization of maximum residual stress minimization for diagonal lattice structure support.[42] (b) Design reconstruction. (c) Manufactured part and support with different lattice support structures.[42] (d) Different views of the final design.[42]

Compromise should be made in utilizing any support structure design based on the metric most relevant to the user.

Future efforts on support structure design for SLM should focus on performance optimization and simulation based, automatic generation. These efforts will eliminate guesswork and inefficiencies. In summary, critical advances in support structure design are needed today to drive the growth and adoption of SLM technology.

References

1. T. Wohlers and T. Caffrey, *Wohlers Report 2014: 3D Printing and Additive Manufacturing State of the Industry.* Annual Worldwide Progress Report. Wohlers Associates (2014).
2. I. Gibson, D. Rosen and B. Stucker, *Additive Manufacturing Technologies.* New York: Springer (2015).

3. Y-H. Kuo, C-C. Cheng, Y-S. Lin and C-H. San, Support structure design in additive manufacturing based on topology optimization, *Struct. Multidisc. Optim.* **58**, 183–195 (2018). https://doi.org/10.1007/s00158-017-1743-z

4. G. Strano, L. Hao, R.M. Everson and K.E. Evans, A new approach to the design and optimisation of support structures in additive manufacturing, *Int. J. Adv. Manuf. Technol.* **66**, 1247–1254 (2013). https://doi.org/10.1007/s00170-012-4403-x

5. A.M. Mirzendehdel and K. Suresh, Support structure constrained topology optimization for additive manufacturing, *Comput. Des.* **81**, 1–13 (2016).

6. J. Vanek, J.A.G. Galicia and B. Benes, Clever support: Efficient support structure generation for digital fabrication, *Comput. Graph. Forum.* **33**, 117–125 (2014).

7. M. Langelaar, Topology optimization of 3D self-supporting structures for additive manufacturing, *Addit. Manuf.* **12**, 60–70 (2016).

8. F. Mezzadri, V. Bouriakov and X. Qian, Topology optimization of self-supporting support structures for additive manufacturing, *Addit. Manuf.* **21**, 666–682 (2018).

9. Selective Laser Melting (Online). Available at: https://en.wikipedia.org/wiki/Selective_laser_melting (Accessed March 15, 2020).

10. C.S. Lefky, B. Zucker, D. Wright, A.R. Nassar, T.W. Simpson, and O.J. Hildreth, Dissolvable supports in powder bed fusion-printed stainless steel, *3D Print. Addit. Manuf.* **4**, 3–11 (2017).

11. I. Yadroitsev, P. Bertrand and I. Smurov, Parametric analysis of the selective laser melting process, *Appl. Surf. Sci.* **253**, 8064–8069 (2007).

12. S. Sun, M. Brandt and M. Easton, Powder bed fusion processes: An overview. In *Laser Additive Manufacturing,* Milan Brandt (ed). Elsevier (2017).

13. I. Yadroitsev, A. Gusarov, I. Yadroitsava and I. Smurov, Single track formation in selective laser melting of metal powders, *J. Mater. Process. Technol.* **210**, 1624–1631 (2010).

14. T. DebRoy, H.L. Wei, J.S. Zuback, T. Mukherjee, J.W. Elmer, J.O. Milewski, A.M. Beese, A. Wilson-Heid, A. De and W. Zhang, Additive manufacturing of metallic components — Process, structure and properties, *Prog. Mater. Sci.* **92**, 112–224 (2018).

15. D. Wang, Y. Yang, Z. Yi and X. Su, Research on the fabricating quality optimization of the overhanding surface in SLM process, *Int. J. Adv. Manuf. Technol.* **65**, 1471–1484 (2013).

16. M. Rombouts, L. Froyen, A.V. Gusarov, E.H. Bentefour and C. Glorieux, Photopyroelectric measurement of thermal conductivity of metallic powders, *J. Appl. Phys.* **97**, 024905 (2005).

17. T. Mukherjee, W. Zhang and T. DebRoy, An improved prediction of residual stresses and distortion in additive manufacturing, *Comput. Mater. Sci.* **126**, 360–372 (2017).

18. K. Kempen, L. Thijs, B. Vrancken, S. Buls, J. Van Humbeeck and J.-P. Kruth, Producing crack-free, high density M2 HSS parts by selective laser melting: Pre-heating the baseplate. In *Solid Freeform Fabrication Symposium* (2013). Available at: http://sffsymposium.engr.utexas.edu/Manuscripts/2013/2013-10- Kempen.pdf (Accessed June 27, 2018).

19. X. Zhao, X. Lin, J. Chen, L. Xue and W. Huang, The effect of hot isostatic pressing on crack healing, microstructure, mechanical properties of Rene88DT superalloy prepared by laser solid forming, *Mater. Sci. Eng. A.* **504**, 129–134 (2009).

20. M. Shiomi, K. Osakada, K. Nakamura, T. Yamashita and F. Abe, Residual stress within metallic model made by selective laser melting process, *CIRP Ann.* **53**, 195–198 (2004).

21. D. Buchbinder, W. Meiners, N. Pirch, K. Wissenbach and J. Schrage, Investigation on reducing distortion by preheating during manufacture of aluminum components using selective laser melting, *J. Laser. Appl.* **26**, 012004 (2014).

22. Y. Chivel and I. Smurov, Temperature monitoring and overhang layers problem, *Phys. Procedia.* **12**, 691–696 (2014).

23. F. Calignano, Design optimization of supports for overhanding structures in aluminum and titanium alloys by selective laser melting, *Mater. Des.* **64**, 203–213 (2014).

24. M.X. Gan and C.H. Wong, Practical support structures for selective laser melting, *J. Mater. Process. Technol.* **238**, 474–484 (2016).

25. S. Siddique, M. Imran, E. Wycisk, C. Emmelmann and F. Walther, Influence of process-induced microstructure and imperfections on mechanical properties of AlSi12 processed by selective laser melting, *J. Mater. Process. Technol.* **221**, 205–213 (2015).

26. L. Thijs, F. Verhaeghe, T. Craeghs, J. van Humbeeck and J-P. Kruth, A study of the microstructural evolution during selective laser melting of Ti-6Al-4V, *Acta Mater.* **58**, 3303–3312 (2010).

27. C. Wei, Y-H. Chueh, X. Zhang, Y. Huang, Q. Chen and L. Li, Easy-to-remove composite support material and procedure in additive manufacturing of metallic components using multiple material laser-based powder bed fusion, *J. Manuf. Sci. Eng.* **141**, 071002 (2019).

28. e-Stage Metal | 3D Printing Software. Available at: https://www.materialise.com/en/software/e-stage/product-information-metal (Accessed June 27, 2019).

29. L. Cheng, X. Liang, J. Bai, Q. Chen, J. Lemon and A. To, On utilizing topology optimization to design support structure to prevent residual stress induced build failure in laser powder bed metal additive manufacturing, *Addit. Manuf.* **27**, 290–304 (2019).

30. Materialise Magics. Available at: https://www.materialise.com/en/software/magics (Accessed July 9, 2018).

31. J-P. Järvinen, V. Matilainen, X. Li, H. Piili, A. Salminen, I. Mäkelä and O. Nyrhilä, Characterization of effect of support structures in laser additive manufacturing of stainless steel, *Phys. Procedia* **56**, 72–81 (2014).

32. Z. Zhang, C. Wu, T. Li, K. Liang and Y. Cao, Design of internal branch support structures for selective laser melting, *J. Rapid Prototyp.* **24**(4), 764–773 (2018).

33. M. Cloots, A.B. Spierings KW. Assessing new support minimizing strategies for the additive manufacturing technology SLM. In Solid Freeform Fabrication Symposium (2013).

34. A. Hussein, L. Hao, C. Yan, R. Everson and P. Young, Advanced lattice support structures for metal additive manufacturing, *J. Mater. Process. Technol.* **213**, 1019–1026 (2013).

35. A.V. Gusarov and E.P. Kovalev, Model of thermal conductiity in powder beds, *Phys. Rev. B* **80**, 024202 (2009).

36. Y-C. Wu, C-H. San, C-H. Chang, H-J. Lin, R. Marwan, S. Baba and W. Hwang, Numerical modeling of melt-pool behavior in selective laser melting with random powder distribution and experimental validation, *J. Mater. Process. Technol.* **254**, 72–78 (2018).

37. Y. Yang, D. Gu, D. Dai and C. Ma, Laser energy absorption behavior of powder particles using ray tracing method during selective laser melting additive manufacturing of aluminum alloy, *Mater. Des.* **143**, 12–19 (2018).

38. N.E. Hodge, R.M. Ferencz and J.M. Solberg, Implementation of a thermomechanical model for the simulation of selective laser melting, *Comput. Mech.* **54**, 33–51 (2014).

39. J. Schilp, C. Seidel, H. Krauss and J. Weirather, Investigations on temperature fields during laser beam melting by means of process monitoring and multiscale process modelling, *Adv. Mech. Eng.* **6**, 217584 (2015).

40. H. Ali, H. Ghadbeigi and K. Mumtaz, Residual stress development in selective laser-melted Ti6Al4V: A parametric thermal modelling approach, *Int. J. Adv. Manuf. Technol.* **97**, 2621–2633 (2018).

41. Y. Li, K. Zhou, P. Tan, S.B. Tor, C.K. Chua and K.F. Leong, Modeling temperature and residual stress fields in selective laser melting, *Int. J. Mech. Sci.* **136**, 24-35 (2018).

42. L. Cheng and A. To, Part-scale build orientation optimization for minimizing residual stress and support volume for metal additive manufacturing: Theory and experimental validation, *Comput. Des.* **113**, 1–23 (2019).

43. W. King, A.T. Anderson, R.M. Ferenez, N.E. Hodge, C. Kamath and S.A. Khairallah, Overview of modelling and simulation of metal powder bed fusion process at Lawrence Livermore National Laboratory, *Mater. Sci. Technol.* **31**, 957–968 (2015).

Chapter 3

Multiscale Process Modeling of Shape Memory Alloy Fabrication with Directed Energy Deposition

Jesse M. Sestito, Dehao Liu, Yanglong Lu, Ji-Hyeon Song, Anh V. Tran,
Michael J. Kempner, Tequila A. L. Harris, Sung-Hoon Ahn, and Yan Wang

3.1 Introduction

A shape memory alloy (SMA) is a functional material which can return to its pre-deformed shape when experiencing external thermal or mechanical loads. This so-called shape memory effect is caused by the solid-state phase transition between the martensitic and austenitic phases. SMAs also display pseudoelasticity or superelasticity, which is characterized by a reversible stress-strain behavior with higher strain values than classic alloys. These unique properties make SMAs attractive for applications such as medical implants, stents, actuators, sensors, foldable devices, among others.[1,2] Copper zinc aluminum (CuZnAl), copper aluminum nickel (CuAlNi), and Nitinol (NiTi) are the three most studied SMAs and are commercially available. Particularly, Nitinol, which is a blend of nickel and titanium, is the most commonly used one because of its stability and biocompatibility.[3]

In spite of its great potential for new product concepts, several challenges exist in the fabrication of SMAs in traditional manufacturing methods such as casting and rolling, including difficulty of controlling material impurities, localized concentrations of materials for specific applications, fine tuning mechanical properties, as well as eliminating or introducing porosity in the structures.[4,5] In recent years, researchers have examined the properties of SMAs made by additive manufacturing (AM) techniques. AM overcomes some of these challenges and potentially allows engineers to customize the structures and control the localized compositions and processing temperatures for SMAs. Metallic AM techniques such as selective laser melting (SLM) and directed energy deposition (DED) have been applied to fabricate SMAs. Researchers investigated the impact of AM process parameters on the properties of Nitinol by tweaking the process parameters,[6,7] varying material concentration,[6] and exploring various porosities for biomedical applications.[8] There is also research being performed to examine the structural integrity of Nitinol[1] and the impacts of porosity on the material properties.[9] Despite the knowledge elicited by these studies, there is still a lack of fundamental understanding of the AM processes and their influence on the physical properties of SMAs. Systematically exploring the process-structure-property (P-S-P) relationships and understanding the microstructure formation in complex AM processes will be necessary for us to design, optimize, and control these processes. Tools developed for integrated computational materials engineering (ICME) can be helpful in materials and process design.

The P-S-P relationships for SMAs made by AM are very complex with many different factors involved. For instance, different biomedical implants of Nitinol require different levels of porosity, depending on the specific requirements of local transport and mechanical properties. The abundant combination of process parameters such as laser power, scanning strategy, and layer thickness can produce a variety of Nitinol alloys with SLM or DED. Different compositions of Ni, Ti, and dopants can alter the growth of grain structures during solidification, as well as the dynamic behaviors of shape change and phase transition with different characteristics of hysteresis. Simply relying on physical experiments to test different combinations of all process and material factors to explore the P-S-P relationships is prohibitively expensive and not feasible for process and materials design. Simulation is an efficient alternative to facilitate the design and optimization process.

Researchers have used simulations to study SMAs. For instance, molecular dynamic simulations have been used to study the melting curve of Nitinol under pressure[10] and examine the transformation mechanical properties under the presence of Ni_4Ti_3 precipitates.[11] Phase field simulations have been used to examine the addition of free surfaces and boundary condition changes to the material[12] and the impact of thermal cycling on the plasticity.[13] Finite element simulations have been used to examine properties of Nitinol composite beam[14] and multiaxial loading on the SMA.[15] Kinetic Monte Carlo simulations have been used to study the magnetic behavior and its dependence on static temperature[16] and variable temperature.[17] These simulation models were developed to understand the structure-property relationship of SMAs. To establish the complete P-S-P relationships and facilitate the design of AM processes for SMAs, a multiscale modeling and simulation approach is needed. To understand the effects of material compositions on detailed microstructures and physical properties, atomistic simulations are necessary. However, atomistic simulations are only capable of simulating processes with very short time scales (e.g. nanosecond or less) and cannot simulate manufacturing processes in order to construct process-structure relationships.

In this chapter, a multiscale multi-physics simulation framework is described to help establish the P-S-P relationships for SMAs processed with AM. The simulation framework includes finite element analysis (FEA) to predict thermal and multiphase flow phenomena in AM processes, controlled kinetic Monte Carlo (cKMC) as a reduced-order AM process simulation model for porosity and morphology predictions, phase field method (PFM) coupled with thermal lattice Boltzmann method (TLBM) to simulate the solidification of SMAs and understand the effects of process parameters and material compositions on grain formation, as well as molecular dynamics (MD) to predict thermal, transport, and mechanical properties that are used as the inputs of FEA, cKMC, PFM, and TLBM simulations. The proposed multiscale process modeling and simulation framework is shown in Fig. 3.1. Here, the generic framework is illustrated with the DED process to fabricate Nitinol.

The remainder of this chapter is organized as follows. In Section 2, MD simulations to predict material properties of Nitinol are demonstrated. Structure-property relationships such as the effects of material compositions on melting temperature, thermal conductivity, viscosity, and other physical properties are predicted. This allows for materials design to choose appropriate compositions. The predicted properties are also necessary input parameters for simulations at larger scales. In Section 3, multi-physics PFM and TLBM are demonstrated with the elucidation of the detailed solidification process at mesoscale. The effect of grain growth pattern and final crystallographic texture on the compositions of Ni and Ti, melt pool flow, cooling rate, and other process parameters are shown. In Section 4, the continuum level multi-physics FEA simulation predicts the temperature and velocity fields in the melt pool so that the details of the melting process can be revealed. The evolutions of temperature and velocity fields are also important input parameters for PFM and TLBM simulations at mesoscale. In Section 5, a cKMC model to simulate the DED process at the system level is demonstrated, where porosity and microscopic morphology can be predicted at a much lower cost than FEA. This reduced-order model is an efficient alternative for system-level simulation for path planning and large-scale process optimization.

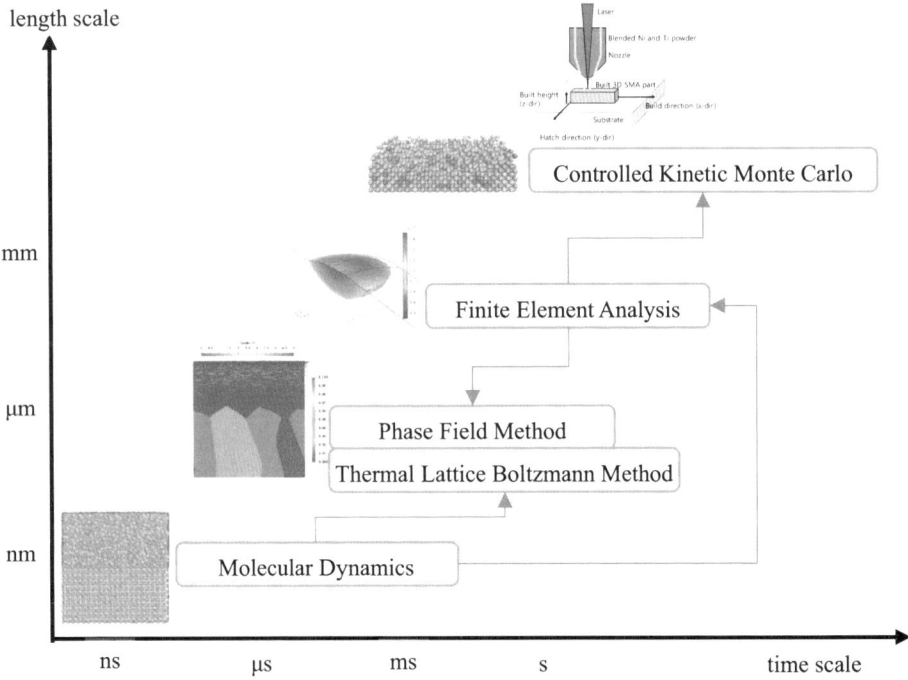

Fig. 3.1. The multiscale multiphysics simulation framework to explore process-structure-property relationship for additively manufactured shape memory alloys.

3.2   Property Calculation with MD Simulations

MD is a simulation method focused on the dynamics of individual particles in a material system, where each particle is an atom or a group of atoms. During simulation, each particle interacts with nearby particles, which is characterized by inter-atomic potential functions depending on particle positions. Inter-atomic potentials determine the inter-atomic forces. The dynamics of particles is simulated with continuous update of the positions and velocities of all particles at each time step. These time steps are normally on the order of femtoseconds. In addition, a few million atoms can only fill out a cube of one hundred nanometers. Therefore, most MD simulations predict material systems at the nanometer scale for only a period of nanoseconds. MD is not optimal for simulating processes such as melting and solidification of the crystalline structures at the micrometer and second scales. Instead, MD simulations are employed to gain an understanding of how atomic movement dictates material properties.

In the following subsections, it will be shown that using MD one can calculate properties of Nitinol. Not only will these properties be useful for gaining insight of the material's structure-property relationship, but they are necessary inputs of other simulation methods such as PFM, FEA, and cKMC to obtain more accurate results if there is a lack of experimental data to perform model calibration empirically.

Here, MD will be used to estimate the melting temperature of Nitinol. The compositions of Ni, Ti, and other elements in Nitinol can be tailored for specific device design with targeted phase transition behavior, because their melting temperature could be different. Melting temperature is an important parameter to consider in designing AM processes. Therefore, MD calculation of melting temperature is useful for materials design and process design. Similarly, thermal conductivity is an important material property to design manufacturing processes. The Green-Kubo method was used in MD simulations to calculate the lattice thermal conductivity of Nitinol 55 and 60. MD simulations can be applied to calculate many other material properties, such as dynamic viscosity, anisotropy, interfacial energy, and interface mobility.

3.1.1 *Simulation Setup*

MD simulations are performed via Large-scale Atomic/Molecular Massively Parallel Simulator (LAMMPS) using a body-centered cubic (BCC) crystal with a lattice constant of 3.512 Angstroms, subject to triaxially periodic boundary conditions. An established second nearest-neighbor (2NN) modified embedded-atom method (MEAM) potential for Nitinol is chosen for its consideration of the directional bonds characteristic of martensitic Nitinol and its ability to relax the material to its empirically observed lattice spacing. The 2NN MEAM potential is a more locally inclusive variation of the classical first nearest-neighbor MEAM potential, which in turn is a version of the traditional embedded-atom model (EAM) potential that has been updated to account for angular forces. Similar to the EAM potential, the 2NN MEAM potential relies on an embedding-energy and a short-range pair potential to describe molecular cohesion within metals. As an MEAM potential, the 2NN MEAM potential is further able to replicate shearing behavior. This allows for a single potential to be used for multiple crystalline configurations, including BCC and face-centered cubit (FCC), enabling simulations of phase changes such as the martensitic-austenitic shift observed in Nitinol. The extension of the potential to consider 2NN interactions corrects for the inadequately reproduced BCC surface energies that have been noted in simulations that neglect the influence of all but the first nearest-neighbors. Two initial setups are used. One setup consists of 16,000 atoms in a one-to-one atomic ratio between Ni and Ti, such that it has approximately 55.08% Ni by mass. The other setup is formed by random deletion of Ti atoms such that the material has approximately 60.00% Ni by mass and consists of 14,540 atoms.

3.1.2 *Melting Temperature Prediction*

The melting temperature is calculated using the interface method.[18-24] The melting point is estimated to be the temperature at which a discontinuity of a quantity of interest (QoI) is observed. QoIs, including kinetic energy, potential energy, volume, density, temperature, and pressure are monitored. Constant pressure MD simulations are carried out, and the QoI is averaged out over a period.

The simulation cell is split in two regions at the beginning, according to the z coordinate, as shown in Fig. 3.2. Both regions are equilibrated using isothermal isobaric (NPT) statistical ensemble. The temperature corresponding to two regions are different however. In the lower region with smaller z coordinate values, the temperature varies from 1000 K to 1500 K, with 20 K increments. In the region with larger z coordinate values, the temperature is fixed at 5000K, far above the melting temperature. The equilibration phase is carried for 200 ps, where the pressure is set at zero. Figure 3.2 shows the two-phase solid-liquid of the simulation cell during the first phase of equilibration.

Fig. 3.2. Creation of solid-liquid interface in the molecular dynamics simulation cell.

In the second phase, the simulation cell is reunited, where the NPT ensemble is again applied on the simulated system. The pressure is set to zero, and the temperature varies. The second equilibrium phase is carried out for another 200 ps. The calculation of QoIs is only based on the last 100 ps of the simulation, where the QoIs are averaged with respect to the time.

Presented in Fig. 3.3 (a) and (b), respectively, are the ensemble average of the density and mean square displacement for the system with 55.08% of Ni, by the end of the second equilibration phase, as a function of temperature. At about 1400 ± 10 K, a clear discontinuity is observed in both of the monitored QoIs. The density decreases dramatically, and the mean square displacement starts increasing with the simulated temperature. It is concluded that the melting temperature of Nitinol 55 is 1400 ± 10 K.

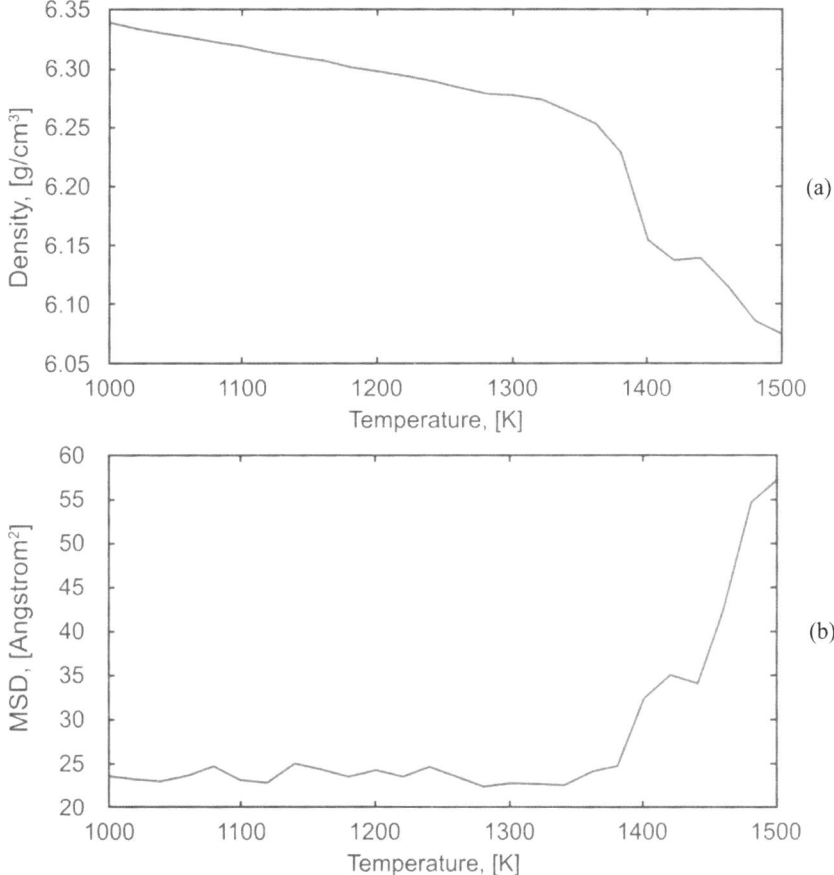

Fig. 3.3. Ensemble averaged density and averaged mean square displacement as functions of the simulated temperature for Nitinol 55.

3.1.3 *Thermal Conductivity and Viscosity Prediction*

The thermal conductivity, k_m, of a conductor includes the electronic component, k_e, and lattice component, k_g, such that

$$k_m = k_e + k_g \tag{3.1}$$

is the total thermal conductivity of the material.[25] In MD simulations, the electronic effects can only be indirectly included in the inter-atomic potentials. So for materials with a high electronic component, the calculation of total thermal conductivity based on MD only could be inaccurate.[26] However, MD does allow the lattice component to be calculated and compared for these materials, such as Nitinol. The electronic thermal conductivity instead can be analytically calculated based on the Wiedemann-Franz law[27,28] given by,

$$\frac{k_e}{\sigma} = L\,T \tag{3.2}$$

where σ is the electric conductivity, T is temperature, and L is the Lorenz number.

With the atomistic setups mentioned in Section 2.1, the Green-Kubo method can be applied to predict lattice thermal conductivity.[29] The system is equilibrated at the simulated temperature using a constant pressure and temperature (NPT) ensemble and then constant volume and temperature (NVT) ensemble for 10 ps and 1 ps, respectively. The velocities of atoms are scaled to the simulated temperature prior to the second simulation. This is because the Green-Kubo method requires an equilibrated simulation. The simulation with the NVT ensemble is then run where the average of the auto-correlation of the heat flux is related to thermal conductivity. A 100 ps time average integral value of heat flux auto-correlation gives an estimation of the lattice thermal conductivity. The three calculated lattice thermal conductivities along the x-, y-, and z- axes are shown in Fig. 3.4 (a), where the average of the three is the overall lattice thermal conductivity. These thermal conductivities are calculated from approximately 100 to 1800 K and shown in Fig. 3.4 (b) for 55% and 60% Nitinol using both the MEAM and EAM potentials.

The lattice thermal conductivity results are similar to those of copper-nickel alloys.[28] The lattice thermal conductivity is expected to dissipate at $T^{-0.5}$.[28] This is shown to be the case in Fig. 3.4 (b). The differences between Nitinol 55 and 60 are also seen. Nitinol 55 has a lower lattice constant than Nitinol 60 in both instances of the EAM and MEAM potentials. With regards to the differences between the MEAM and EAM potentials, at low temperatures the difference is significant, whereas at 400 K and above they produce similar results. Since the focus of the materials in AM processes is for high temperatures, the results of thermal conductivity for temperatures above 1200 K are most important, where both potentials give relatively accurate results.

3.1.4 *Solidification Process Simulation with Coupled PFM and TLBM*

During the rapid solidification process in DED, solute diffusion, heat transfer, fluid dynamics, as well as their interactions in the melt pool have significant effects on the formation of final solid microstructures. A fundamental understanding of the process allows us to predict the solidified microstructures and the physical properties of the solids for process design and optimization. To understand the solidification of Nitinol alloy, multi-physics simulations at the mesoscale are cost-effective alternatives to expensive experiments for in-situ observation.

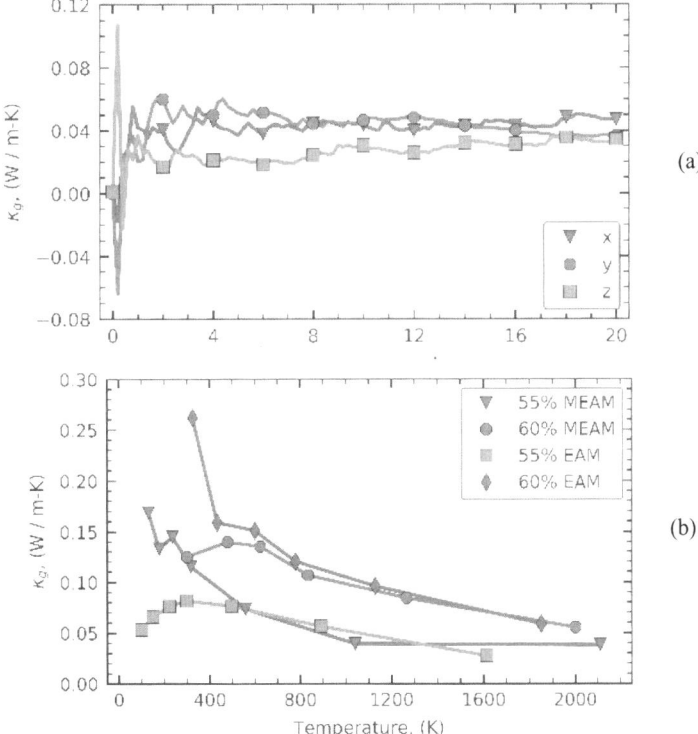

Fig. 3.4. (a) Integration of the auto-correlation heat flux to give the lattice thermal conductivities along the x, y, and z axes over the entire simulation at 1040 K for 55% Nitinol using the MEAM potential. (b) The lattice thermal conductivity with respect to temperature using the MEAM and EAM potentials.

Compared to atomistic scale simulations, mesoscale modeling such as PFM[30] is more efficient to simulate the solidification. PFM simulates a much longer time scale than what MD is able to do and provides more fine-grained details than macroscopic simulations. To distinguish between liquid and solid phases, a continuous variable, namely *phase field* or *order parameter* ϕ, is used in PFM. The evolution of microstructures in solidification is modeled by partial differential equations of phase field, ϕ. In this study, a mesoscale multi-physics model named phase field and thermal lattice Boltzmann method (PF-TLBM)[31,32] is used to simulate the microstructure evolution of Nitinol during the rapid solidification. In this integrated simulation model, the phase field method for the dendritic growth of a binary alloy is coupled with the thermal lattice Boltzmann method (TLBM) for the heat transfer and melt flow. The detailed formulation of PF-TLBM can be found in Refs. 31 and 32.

3.1.5 *Simulation Settings*

Here, PF-TLBM is used to simulate the directional solidification of Nitinol alloy during the DED process. The physical properties of Nitinol alloy[32-34] are listed in Table 3.1.

In all simulations, the grid spacing is $\Delta x = 1 \times 10^{-7}$ m, the time step is $\Delta t = 2 \times 10^{-7}$ s, and the simulation period is 3 ms. The length and width of the simulated domain are $L_x = 100$ μm and $L_y = 100$ μm in x- and y- directions, respectively. The initial diameter of the seed is $D = 2$ μm, and the width of interface is $\eta = 0.5$ μm. The setup of boundary conditions for all simulations is schematically illustrated in Fig. 3.5. Zero Neumann conditions are set at the bottom boundary $y = 0$ and top one $y = L_y$ for phase field, ϕ, and composition, C. A fixed heat flux $q_H = \rho c_p L_y \dot{T}$ [36] is set at the bottom boundary given the constant cooling rate $\dot{T} = 5 \times 10^4$ K/s, while adiabatic boundary condition is set at the top boundary. When the dendrite grows in a forced flow, a constant flow velocity $|\mathbf{u}_w| = 0.1$ m/s is imposed at the top boundary of the domain. Periodic boundary conditions are set at the left boundary $x = 0$ and right one $x = L_x$ for the phase field (ϕ), composition (C), temperature (*T*), and flow ($\mathbf{u}_l$). The nuclei are located at the bottom cold wall with constant heat flux to simulate the directional dendrite growth during DED process. The locations of five nuclei with different orientations are $x = 10$ μm, 30 μm, 50 μm, 70 μm, and 90 μm respectively.

Table 3.1. Physical properties of Nitinol alloy in the PF-TLBM models.

Physical properties	Nitinol 55	Nitinol 60
Liquidus temperature, T_l [K]	1583	1487
Solidus temperature, T_s [K]	1583	1443
Liquidus slope, m_l [K/wt%]	0.0	-19.2
Partition coefficient, k	1.0	0.70
Prefactor of interfacial energy stiffness, σ_0^* [J/m^2]	0.24	0.24
Interfacial energy stiffness anisotropy, ε^*	0.35	0.35
Interface mobility, M_ϕ [m^4/J·s]	1.0×10^{-8}	1.0×10^{-8}
Kinematic viscosity, v [m^2/s]	8.9×10^{-7}	8.9×10^{-7}
Thermal diffusivity, α [m^2/s]	3.33×10^{-6}	3.33×10^{-6}
Latent heat of fusion, L_H [J/kg·K]	1.0×10^5	1.0×10^5
Specific heat capacity, c_p [J/(kg·K)]	836.8	836.8
Density, ρ [kg/m^3]	6450	6700

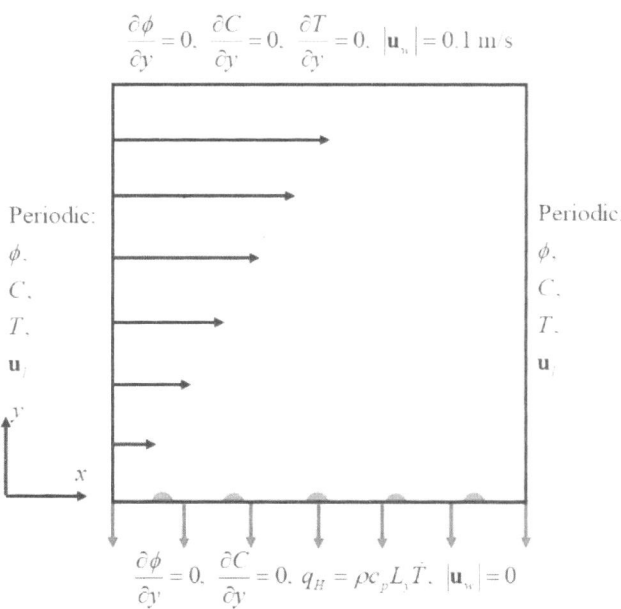

Fig. 3.5. Setup of boundary conditions.

3.1.6 *Dendrite Growth of Nitinol 55*

The initial concentration of Ni for Nitinol 55 is $C_0 = 55$ $wt\%$ in the liquid phase. It is important to note that the liquidus and solidus temperatures of Nitinol 55 are the same, which means that the solidification of Nitinol 55 is congruent and there is nearly no solute redistribution at the solid-liquid interface during solidification. The initial temperature in the simulation domain is $T = 1580\ K$, which means that the undercooling is 3 K given the initial composition. Fig. 3.6 shows the simulation results. The grain identification (ID) 0 represents the liquid phase, while other grain IDs represent the solid phase with different orientations.

During the congruent solidification of Nitinol 55, there is nearly no solute redistribution at the solid-liquid interface. A cellular growth pattern is observed during the rapid solidification of Nitinol 55, as shown in Fig. 3.6 (a-c). When the grains grow competitively, curved grain boundaries are formed because of different orientations of grains. As shown in Fig. 3.6 (d), the composition of Ni in Nitinol 55 is kept to 55 wt%, and no solute segregation is observed at the grain boundaries. Because of the release of latent heat, the vertical temperature distribution is not monotonic, and the temperature of dendrite tip is higher than liquid melt, as shown in Fig. 3.6 (e).

3.1.7 Dendrite Growth of Nitinol 60

The initial concentration of Ni for Nitinol 60 is $C_0 = 60\ wt\%$ in the liquid phase. The initial temperature in the simulation domain is $T = 1485\ K$, which means that the undercooling is 2 K given the initial composition. The results are shown in Fig. 3.7.

During the congruent solidification of Nitinol 60, solute redistribution occurs at the solid-liquid interface. As shown in Fig. 3.7 (a-b), the cellular growth pattern is observed from 0 to 2 ms. When the solute is ejected from the solid phase to liquid phase, solute segregation is observed at the grain boundaries, where some small portions of liquid are trapped, as shown in Fig. 3.7 (d). Under the influence of solute segregation and released latent heat, the growth velocity of dendrite tip changes significantly. Therefore, the solid-liquid interface is unstable, and consequent transition from a cellular to dendritic microstructure can be observed in Fig. 3.7 (c). Similarly, the vertical temperature distribution is not monotonic, and the temperature of dendrite tip is higher than liquid melt as shown in Fig. 3.7 (e). Because the temperature gradient is along the vertical direction, the vertical secondary arms are dominant in dendrite growth, as shown in Fig. 3.7 (c).

The solute distribution of Ni in Nitinol 55 and Nitinol 60 at the location $y = 10\ \mu m$ at 3 ms are shown in Fig. 3.8. It is observed that the composition of Nitinol 55 keeps 55 wt% along the line, whereas an obvious solute segregation is observed at the grain boundaries of Nitinol 60. It is worthwhile to note that the

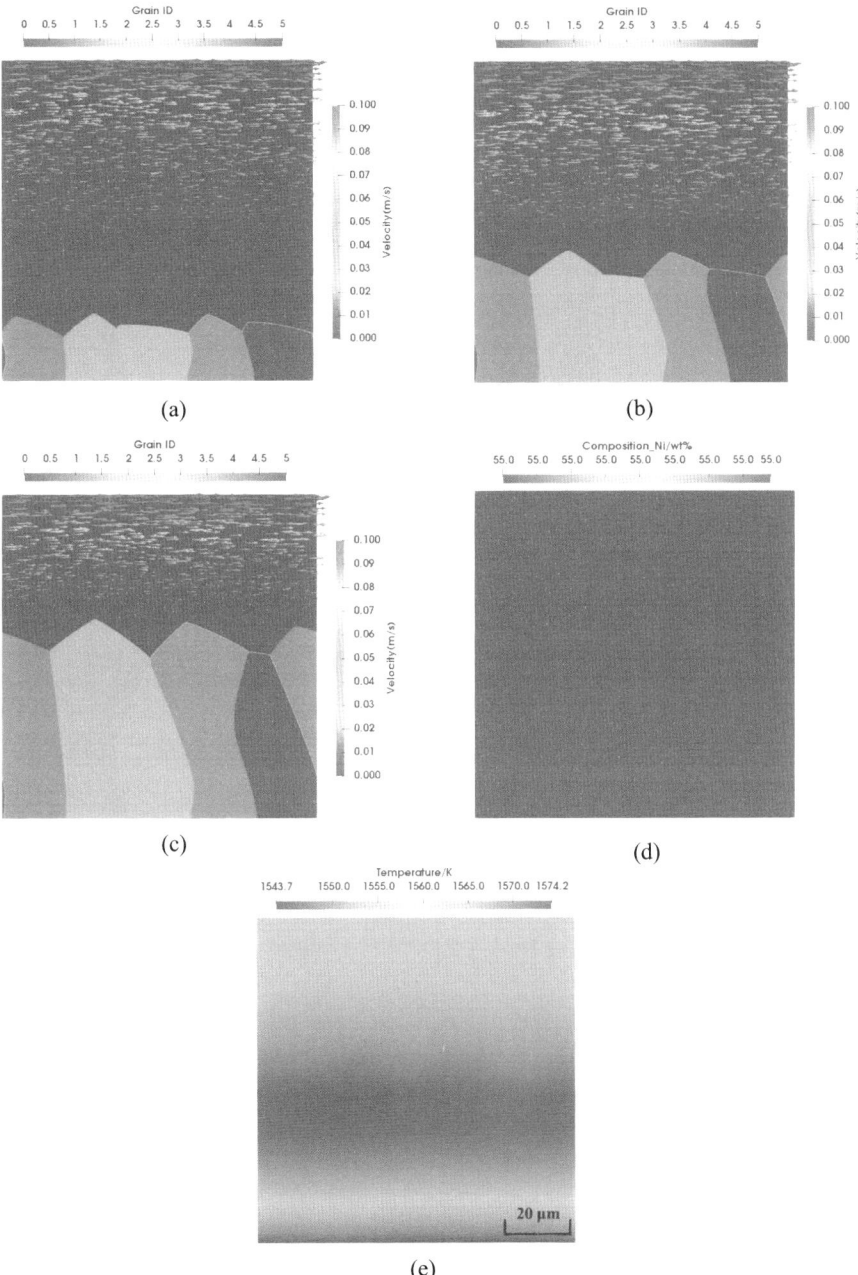

Fig. 3.6. Dendrite growth of Nitinol 55 with latent heat in a forced flow. Phase field and flow field at (a) 1 ms, (b) 2 ms, (c) 3 ms, (d) composition field at 3 ms, and (e) temperature field at 3 ms.

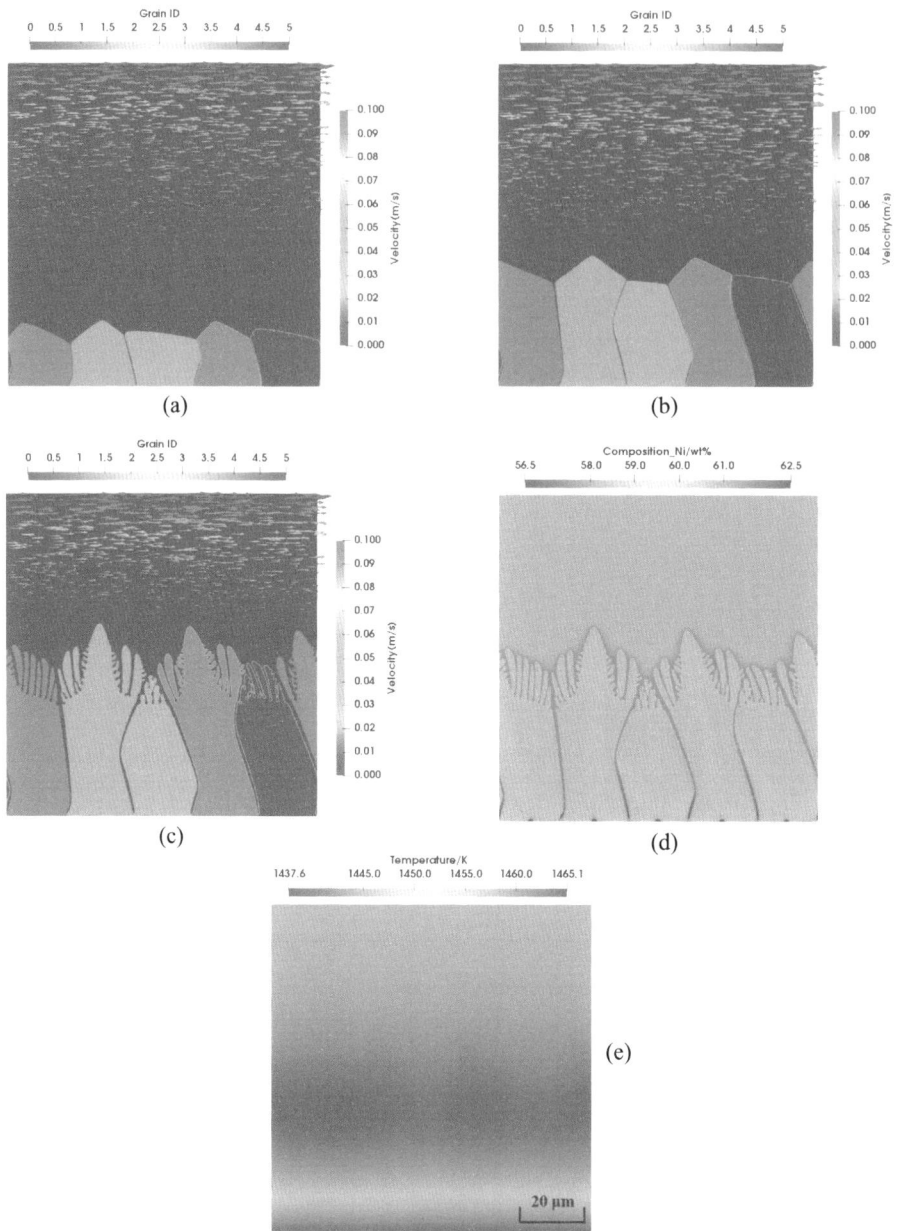

Fig. 3.7. Dendrite growth of Nitinol 60 with latent heat in a forced flow. Phase field and flow field at (a) 1 ms, (b) 2 ms, (c) 3 ms, (d) composition field at 3 ms, and (e) temperature field at 3 ms.

composition at the grain boundaries of Nitinol 60 is about 62 wt%, which corresponds to the composition of secondary phase Ni_4Ti_3. As shown in Ref. 35, about 11% of Ni_4Ti_3 shows up at the grain boundaries of Nitinol 60, which verifies the simulation results to some extent.

3.1.8 *FEA Simulation of Thermal Distribution and Fluid Flow in Melt Pool*

The microstructures and properties of build parts are highly correlated with the geometry and dynamics of the melt pool. The simulation of the melt pool during the DED process is an important tool to develop a better understanding of the characteristics and mechanisms of the process. Furthermore, it is helpful for the optimization and control of process parameters.[37] Many researchers have been involved with simulation and analysis of the DED process including models of the powder stream process, models of the melt pool and models of microstructure, stress and final geometry.[38] For powder stream modeling, Tabernero et al.[39] simulated the powder flux distribution on a coaxial nozzle.

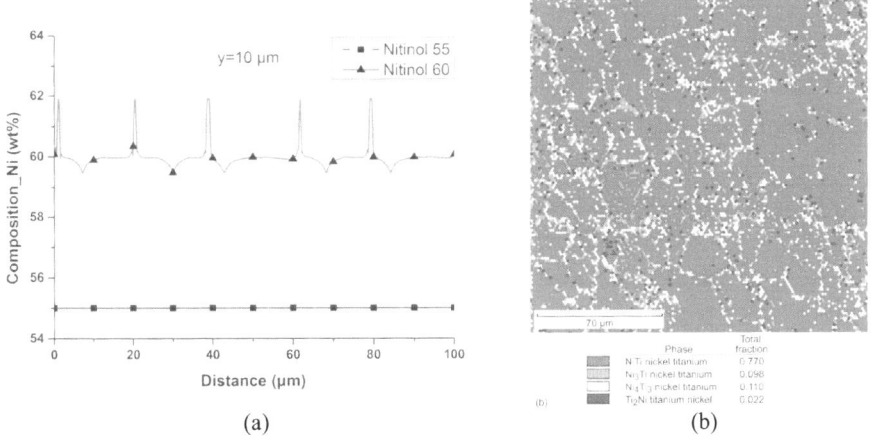

(a)   (b)

Fig. 3.8. Solute distribution of Ni (a) Comparison of Nitinol 55 and Nitinol 60 at the location $y = 10\ \mu m$ at 3 ms in simulation, (b) experimental observation of phase distribution in Nitinol 60 produced by high temperature proprietary powder metallurgy process. Courtesy of Corte et al.[35]

A powder flow model was developed to predict the powder distribution shape, particle velocities and trajectories. Zhu et al.[40] used a 2D discrete phase model of the gas-powder flow to compute the powder flow field distribution. For melt pool modeling, simplified heat transfer models were built to simulate the thermal history in earlier work.[41,42] Hoadley and Rappaz[43] used a thermal model to demonstrate the linear relationship among the laser power, processing velocity, and the thickness of the deposited layer. Zhu et al.[44] analyzed the effect of curvature change on temperature distribution when building thin-walled rings. More complex models were also built by accounting for latent heat in the phase transformation,[45] convection at the evolving interface,[46] and fluid flow.[47] To predict the free surface evolution between liquid and gas phases, the level set method was also introduced.[48-50] Wen and Shin[51] adopted the level set method to predict the free surface and extended the thermal and fluid flow models to include the mass source term in the continuity equation and additional source terms of enthalpy and momentum due to the moving interface. Lee and Farson[52] analyzed the correlations between material parameters and fluid flow patterns and showed that the hemispherical melt pool free surface has different shaping mechanisms from the laser weld melt pool formed on the flat surface. Temperature profile is also helpful to investigate the microstructure affected by the processing parameters,[53-55] and residual stresses.[56] Costa et al.[57] developed a thermo-kinetic finite element model to show the microstructural transformations and hardness variations in the deposition process. Toyserkani et al.[58] developed a thermal model to predict the dependency of the final geometry on the laser pulse shaping. The final geometry can also be affected by processing parameters[59] and the thermal stress and strain field.[60,61]

In addition to the DED process, SLM is a popular way to process alloys, which uses a high power-density laser to melt and fuse metallic powders on a pre-placed powder bed. Khairallah et al.[62,63] explained the importance of including the stochastic nature of the powder bed. Furthermore they showed that the powder has lower thermal conductivity than bulk stainless steel, because the particles are at point contact and heat diffusion in gaps strongly depends on the thermal conductivity of the gas. Shen and Chou[64] explored the effect of powder porosity on the melt pool size and found that temperatures in the melt pool become higher with increasing porosity. The simulation with the thermal model can be used to analyze the size of the melt pool[65] and the effect of laser intensity, preheating temperature, and laser beam spot size on the temperature distribution.[66-69] The thermal model has been further coupled with the mechanical model to predict the stress field[70-72] and the distortion of the part.[73] Similar to simulations of the DED process, models of latent heat of fusion and fluid flow are important for simulating the liquid-solid interface in the SLM process.[74] Shiomi et al.[75] considered the latent heat and shrinkage due to solidification in the process and found that the amount of solidified parts and the maximum temperature of powders are significantly affected by the peak laser power. Gürtler et al.[76] used the multiphysics model to describe melting, wetting and solidification phenomena. The effects of the powder-layer thickness, moving heat source intensity, scan spacing and scanning velocity on the process dynamics are shown. Dai and Shaw[77,78] considered the effect of the powder-to-solid transition and investigated the transient temperature, transient stress and residual stress fields. It is also demonstrated that the volume shrinkage because of the transformation from a powder compact to dense liquid has a negligible effect on the temperature pattern.[79]

In this study, FEA is used to discretize equations of thermal distributions and fluid flows, and COMSOL Multiphysics is used to model the melt pool in the DED process. Heat transfer, phase change, and fluid flow in the melt pool are included in the physics model. Some material properties used in the model are obtained from MD simulations. The geometry, temperature distribution and velocity field of the melt pool are analyzed. Temperatures and the velocity field are used in the phase field simulation.

3.1.9 *Model Construction*

In the DED process, the substrate is stationary and powder particles melted by the laser beam are deposited to a narrow and focused region on the substrate. The laser beam is assumed to move with a constant velocity and power level. The melt pool is formed within a small area under the laser beam, and the melt particles are solidified quickly given the high cooling rate. The movement of the laser beam creates a track of solidified material. The computational domain is set to be 9 mm ×12 mm × 6 mm as shown in Fig. 3.9 (a). The laser beam is initially at location (0, 0, 6) and then moves along the positive y-direction. Only a half of the domain is modeled to reduce the computational cost with the symmetry assumption.

3.1.9.1 *Continuum models*

The evolution of the melt pool during a short time period is analyzed. The effect of the free surface between liquid and gas phases is negligible. The liquid-solid domain is characterized as pure solid, pure liquid and a mixture of solid and liquid (mushy zone). The density, thermal conductivity, and viscosity of the mushy zone can be defined respectively as

$$\rho_m = f_s \rho_s + f_l \rho_l \tag{3.3}$$

$$k_m = f_s k_s + f_l k_l \tag{3.4}$$

$$\mu_m = \mu_l \frac{\rho_m}{\rho_l} \tag{3.5}$$

where ρ_s and k_s are density and thermal conductivity in solid phase, and ρ_l and k_l are those in liquid phase, respectively. μ_l is the dynamic viscosity in the liquid phase. f_s and f_l are the solid and liquid mass fractions, which can be expressed as

$$f_l = \begin{cases} 1 & T > T_l \\ \frac{T-T_s}{T_l-T_s} & T_s \leq T \leq T_l \\ 0 & T < T_s \end{cases} \qquad (3.6)$$

$$f_s = 1 - f_l \qquad (3.7)$$

where T_s and T_l are solidus and liquidus temperatures, respectively. Some material properties of molten Nitinol 60 alloy are difficult to measure directly. The rule of mixtures

$$M_{Nitinol} = \omega_{Ni} M_{Ni} + \omega_{Ti} M_{Ti} \qquad (3.8)$$

is applied, where M_{Ni} and M_{Ti} are material properties of Nickel and Titanium respectively, ω_{Ni} and ω_{Ti} are weight fractions of Nickel and Titanium respectively, and $M_{Nitinol}$ is the corresponding material property of Nitinol alloy. The material properties of solid Nitinol 60 can be found in Table 3.1. The properties of molten Nickel, Titanium, and Nitinol 60 alloy are listed in Table 3.2.

Table 3.2. Material properties of molten Nickel, Titanium and Nitinol alloy.[80-86]

Material properties	Pure Nickel	Pure Titanium	Nitinol alloy
Surface tension gradient (mJ m^{-2} K^{-1})	-0.33	-0.27	-0.306
Surface tension (N m^{-1})	1.588	1.6264	1.6034
Viscosity (mPa s)	4.859	1.37	3.4634
Density (g cm^{-3})	7.85	4.106	6.3524
Thermal conductivity (W m^{-1} K^{-1})	69	28	52.6
Thermal expansion (K^{-1})	5.49×10^{-5}	5.2024×10^{-5}	5.3750×10^{-5}
Specific heat capacity (J kg^{-1} K^{-1})	735	790	757

3.1.9.2 *Heat transfer*

Conservation of energy in the solid-liquid domain can be written as

$$\frac{\partial(\rho_m C_{pm} T)}{\partial t} + \vec{u} \cdot \nabla(\rho_m C_{pm} T) - k_m \nabla^2 T - \dot{S} = 0 \qquad (3.9)$$

where T is the temperature, $\vec{u}$ is the velocity field, and $\dot{S}$ is the source term. Since there is a phase change in the modeling process, the apparent heat capacity method is used by including the latent heat as an additional term in the heat capacity. The effective heat capacity can be expressed as

$$C_{pm} = \frac{1}{\rho_m}\left(f_s\rho_s C_{ps} + f_l\rho_l C_{pl}\right) + L\frac{\partial\alpha_m}{\partial T} \tag{3.10}$$

where L is the latent heat and

$$\alpha_m = \frac{1}{2}\frac{f_s\rho_s - f_l\rho_l}{f_l\rho_l + f_s\rho_s}. \tag{3.11}$$

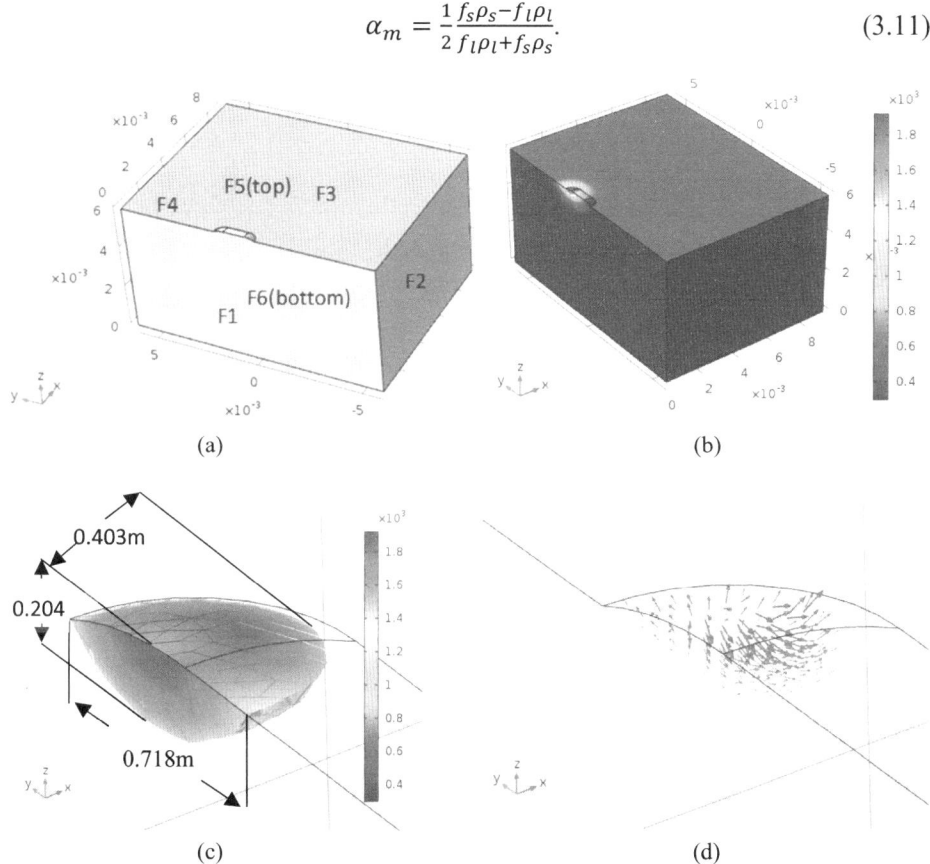

Fig. 3.9. FEA model of temperature distribution, phase change, and fluid flow of the melt pool. (a) The computational domain, (b) Temperature distribution of the computational domain, (c) Temperature distribution and geomtry of the melt pool at 41 ms, (d) The velocity field in the melt pool at 41 ms.

A set of boundary conditions should be satisfied for the heat transfer model. Boundary conditions on the top face F5 in Fig. 3.9 (a) include heat flux from the laser source, convection, and radiation, which can be represented as

$$q''_{in\_top} = q''_{LS} - h(T - T_0) - \sigma\varepsilon(T^4 - T_0^4) \tag{3.12}$$

where h is the heat transfer coefficient, T_0 is the ambient temperature, and σ and ε represent the Stefan-Boltzmann constant and the emissivity respectively. The laser source is assumed to be Gaussian distributed, as

$$q''_{LS} = \frac{2P\eta}{\pi R_L} \exp\left(-\frac{2r^2}{R_L^2}\right) \tag{3.13}$$

where P is the laser power, η is the laser absorbability, r is the radial distance to the center of the laser spot, and R_L is the effective laser beam radius.

Since only half of the domain is used, the symmetric boundary condition is applied on the front face F1. Other faces F2-F4 and F6 are assumed to be adiabatic.

3.1.9.3 *Fluid flow*

The fluid flow in this model is assumed to be incompressible and laminar flow. The conservation of mass is therefore formulated as

$$\nabla \cdot \vec{u} = 0 \tag{3.14}$$

and the conservation of momentum is

$$\rho_m \frac{\partial \vec{u}}{\partial t} + \rho_m(\vec{u} \cdot \nabla)\vec{u} = \nabla \cdot [-p + \vec{u}(\nabla\vec{u} + (\nabla\vec{u})^T)] + \vec{F}_b + \vec{F}_D \tag{3.15}$$

where p is the pressure, and $\vec{F}_b$ is the buoyancy force due to the difference of the density in the melt pool and expressed as

$$\vec{F}_b = \rho_r \vec{g}[1 - \beta(T - T_r)] \tag{3.16}$$

where ρ_r and T_r are reference density and temperature, respectively, $\vec{g}$ is the gravity field, and β is the thermal expansion coefficient. $\vec{F}_D$ is a Darcy term representing the damping force when fluid passes through a porous medium and is formulated as

$$\vec{F}_D = -\frac{\mu_m}{K}\vec{u} \tag{3.17}$$

where K is the isotropic permeability and can be expressed as

$$K^{-1} = \frac{K_0^{-1}(1-f_l)^2}{f_l^3 + \tau} \tag{3.18}$$

where K_0 is a constant determined by the morphology of the mushy zone, and τ is a small number to avoid the singularity. For the pure solid, K approaches zero. For the pure liquid, K approaches to infinity.

A set of boundary conditions should be satisfied for the fluid flow model. On the top face F5, the capillary force is given by

$$\sigma_n = -\gamma\hat{n}\kappa \tag{3.19}$$

and Marangoni forces is given by

$$\sigma_t = \frac{d\gamma}{dT}\nabla_s T \tag{3.20}$$

where γ is the surface tension, $\hat{n}$ is the normal direction, κ is the face curvature, and ∇_s is the gradient in the tangent plane. The capillary force acts in the normal direction, whereas the Marangoni force acts in the direction tangent to the surface. Symmetric boundary conditions are applied on the front face F1. Boundary conditions on other faces are assumed to be zero velocity.

3.1.10 *Results and Discussion*

The melt particles solidify quickly due to the high cooling rate, and the solid track is formed after the laser moves away. The heat absorbed by the melt pool dissipates into the environment and the substrate. Here, the geometry of a single track is predefined. The height and width of the single track on the top face F5 are assumed to be 0.1 mm and 0.48 mm. The laser spot moves along the positive y-direction with a velocity of 13 mm/s. The power of the laser is 800 W, and the radius of the laser beam is 1.25 mm. The laser absorbability is 0.4.[87] The initial temperature is assumed to be 300 K, and the initial velocity is 0 m/s.

The finite element method is used to discretize equations in heat transfer and fluid flow models. The mesh gird of the region near the laser beam is locally refined to ensure accurate computational results. The minimum element size of the mesh grid is set to be 1.48 μm. The segregated approach is used to solve the coupled models iteratively. That is, the heat transfer model and fluid flow model are solved sequentially until convergence. This method requires less memory than the fully coupled approach, where two physics models are solved simultaneously. The adaptive time stepping is used and the maximum time step is 0.1 ms to improve the convergence of results. It is found that the velocity result converges much slower than the temperature because the fluid flow model is highly nonlinear. The melt pool is simulated for a short time period. The temperature and the size of the melt pool increase in this period. The melt pool is formed at 25 ms when the temperature of the substrate reaches the melting temperature. Fig. 3.9 (b) shows the temperature field at 41 ms when the peak temperature is 1926 K. The geometry of the melt pool can be determined by setting the temperature range above the solidus temperatures. The length in the y-direction, height in the z-direction and width in the x-direction of the melt pool are 0.718 mm, 0.204 mm, and 0.403 mm respectively, as seen in Fig. 3.9 (c). The velocity field is shown in Fig. 3.9 (d), where the maximum velocity is 1.6 m/s. The fluid within the melt pool flows outward from the center due to the Marangoni force, which indicates the large gradient of the temperature on the surface. It is consistent with other findings that Marangoni force caused by the surface tension is dominant, whereas the buoyancy force is insignificant in the velocity field. Simulated results from the proposed model can be verified with results provided by Qi et al.[49] in terms of the peak temperature and velocity magnitudes, and the geometry of the melt pool. The detailed values may vary for different materials. The direct measurement of the temperature and velocity fields in melt pools is difficult, which makes the experimental validation a challenge.

The results from the FEA simulation can provide information to analyze the formation and evolution of the melt pool. The temperature and velocity fields can be further integrated with the phase field simulation to investigate the melting or solidification process in future work. To further improve the model, the level-set approach will be included to model the evolution of the liquid-gas interface.

3.2 Controlled Kinetic Monte Carlo as the Reduced-Order Model of DED Process

Simulation-based process optimization is the ultimate goal for process modeling, where the effective P-S-P relationship is established and the process parameters are optimized. This may require thousands of sampling points, each of which is a simulation run. Given that the simulations of high-fidelity FEA or PFM are computationally expensive, reduced-order models will be a more viable approach for process optimization. Here, cKMC modeling is proposed to simulate DED process at mesoscale. cKMC[88,89] is a generalization of kinetic Monte Carlo (KMC). In KMC simulation, system state changes are triggered by random events. Therefore, KMC is not capable of simulating complete manufacturing processes that include external force or energy that deterministically affect system state changes. cKMC can solve these issues by including controlled events and controlled species. Controlled events and controlled species allow us to introduce deterministic events by specifying process direction or starting time.

KMC has been widely used to study self-assembly processes such as chemical vapor deposition and film growth. For instance, Zhu et al.[90] studied the growth of NiTi alloy thin films using KMC. The roles of diffusion, substrate temperature, and deposition rate on forming the microstructures were studied. However, KMC is not able to simulate laser effects in the DED process. There has been some attempts on simulating AM processes with models similar to KMC. Rodgers et al.[91] studied grain growth in metal additive manufacturing using a Potts model. Grain microstructure evolution in the heat affected zone was simulated. Note that Potts model is also known as Ising model and does not capture the time of physical systems as KMC. Song et al.[92] used cKMC to simulate the laser-enhanced nanoparticle deposition process, where laser effect and film porosity were studied.

In this study, cKMC is used as a reduced-order model. Instead of the details in melting and solidification, cKMC models the fusion of Ni and Ti powders, and the geometry and porosity of the build. Each particle in cKMC is at the scale of powder. Particles are deposited on a lattice, fusion and diffusion occur during heating-cooling cycles and the geometry of part is formed. With the computational efficiency, process-structure relationships such as the one between laser power level and porosity can be established easily. Therefore, process parameters can be optimized to achieve the desirable deposition morphology and conversion rate of NiTi prior to experimentation.

3.2.1 *Methodology*

3.2.1.1 *cKMC model setup*

Figure 3.10 (a) shows the schematic diagram of the DED process, where Ni and Ti powders are injected, melted, fused into NiTi, and solidified. Here, x is the build direction and y is the hatch direction. The cKMC model size is $20 \times 20 \times 20$ lattice units and consists of 30,400 particles. The bottom of the simulation domain is the substrate, and the top contains Ni and Ti sources which are blended with Ni and Ti powders. Ni and Ti sources are powders that are being injected above the substrate. They are deposited sequentially corresponding to the scanning path. Therefore, they are defined as different species and named as *source*11, *source*12, *source*13, *source*21, *source*22, *source*23, etc. as shown in Fig. 3.10 (b). The first notation means the layer number, and the second notation means the order of deposition within the layer. These sources are deposited one by one in a group. In this model, 6 groups of sources are used and the deposition sequence is 11, 12, 13, 21, 22, and 23. In the model, 12 different types of species are defined in total, as *Ni_source*, *Ni_active*, *Ni_liquid*, *Ni_solid*, *Ti_source*, *Ti_active*, Ti_liquid, *Ti_solid*, *NiTi_liquid*, *NiTi_solid*, *substrate*, *vacuum/void*. After the initial setup, the system evolves with events. Events for deposition, fusion, diffusion, activation, reheat, and solidification are defined in the following subsections.

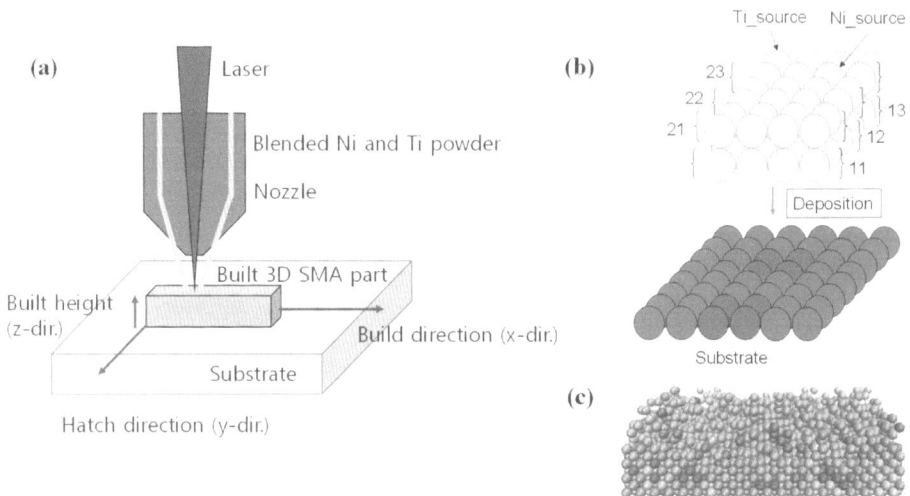

Fig. 3.10. cKMC model of DED process. (a) Schematic diagram of DED process, (b) Initial setup where particles represent powder scale geometry, (c) A snapshot of cKMC model where different species are denoted by different colors.

3.2.1.2 *Events*

Events in cKMC are divided into two types. One is stochastic, and the other is deterministic. Stochastic events are random events, and deterministic events are events that are affected by the external force or energy. The external energy in this process is the laser irradiation, and the external force is due to the air pressure. Stochastic events are defined in the same way as the normal events in KMC. Deterministic events are defined with controlled events and controlled species. Controlled species are the species that start transitions or reactions only at deterministic time. Controlled events are those events that only occur in certain neighborhoods with a specific direction or location. In controlled species command:

control_species reactant rate product x y z init_time

(x, y, z) specifies the direction or sequence that the reactant is converted to product, and init_time specifies when the conversion starts. In controlled events command:

control_event reactant1 reactant2 rate product1 product2 dx dy dz theta (dx, dy, dz) specifies the direction in which the first and second reactants should be aligned. Theta is an angular allowance such that the direction formed by the first and second reactants can be within the range of +/− theta, where the reaction still occurs. The details of cKMC formalism can be found in Ref. 89.

In the DED process model, the regular events, controlled species, and controlled events are defined and listed in Table 3.3. Events (1) to (7) are defined for Ni. Similar events are also defined for Ti. The events of activation, deposition, fusion, diffusion, and solidification are illustrated in Fig. 3.11. When multiple layers are deposited, the previously deposited layer will be reheated and some of the solid state material can become liquid state again. In this case, reheating events (9)-(11) will occur. The reheated Ni, Ti, and NiTi will be involved in the fusion and diffusion processes again. The events in multiple layer deposition are illustrated in Fig. 3.12.

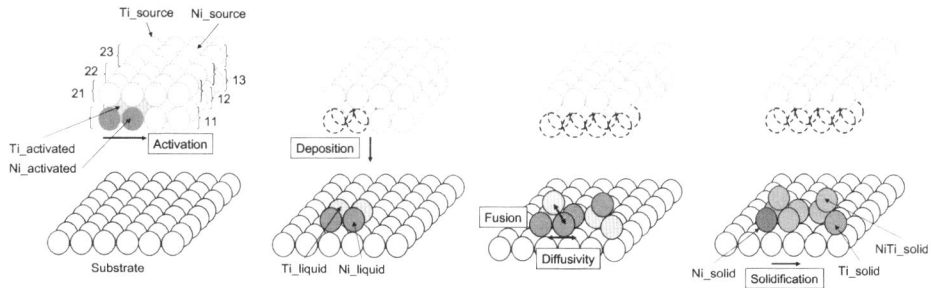

Fig. 3.11. Examples of events for single layer deposition.

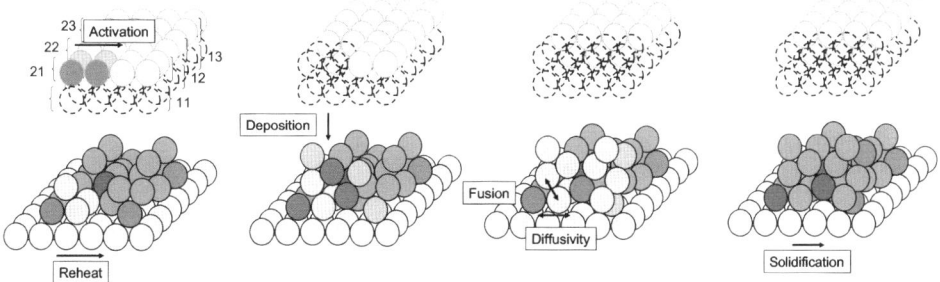

Fig. 3.12. Examples of events for multiple-layer deposition.

Table 3.3. Major events defined in DED process model.

Index	Events	Comments
(1)	Ni_src → Ni_actived (controlled species)"	activation
(2)	Ni_actived + vacuum → vacuum + Ni_actived (controlled events)	deposition
(3)	Ni_actived + substrate → Ni_liquid + substrate (controlled events)	deposition
(4)	Ni_actived + Ni_liquid → Ni_liquid + Ni_liquid (controlled events)	deposition
(5)	Ni_liquid + Ti_liquid → NiTi_liquid + NiTi_liquid	fusion
(6)	Ni_liquid + void + substrate → void + Ni_liquid + substrate	diffusion
(7)	Ni_liquid → Ni_solid (controlled species)	solidification
(8)	NiTi_liquid → NiTi_solid (controlled species)	solidification
(9)	Ni_solid → Ni_liquid (controlled species)	reheat
(10)	Ti_solid → Ti_liquid (controlled species)	reheat
(11)	NiTi_solid → NiTi_liquid (controlled species)	reheat

3.2.1.2 *Calibration of rates*

For each event, there is an associated rate. The values of rates are calibrated with sensitivity analysis. The rates involved in the modeled process are associated with deposition, fusion, diffusion, activation, reheat, and solidification. Activation and reheat rates are related to the laser scan speed. These rates can have a common effect on the simulation results such as conversion rate and porosity. Conversion rate is how much Ni and Ti is converted to NiTi. It is calculated by dividing the total number of NiTi particles at the final time step by the number of Ni and Ti powders at the first time step. Porosity is the ratio of void space to the volume of the built part, calculated from the number of solidified Ni, Ti, and NiTi particles divided by the total sites in the 3D part. For simplicity, sensitivity analysis is done on the diffusion rate while keeping other rates fixed. The results of sensitivity analysis are shown in Fig. 3.13. The calibration is done according to the values of porosity in Ref. 91.

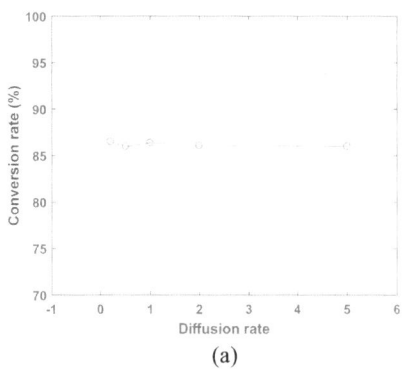 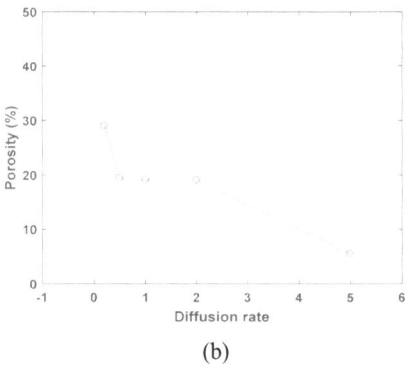

(a)   (b)

Fig. 3.13. Sensitivity analysis of diffusion rate with respect to (a) conversion rate and (b) porosity.

3.2.2 *Results and Discussions*

Laser condition is one of the most important factors that decide the quality of build in the DED process. After the model parameter calibration, the conversion rate and porosity of the NiTi were analyzed with different laser scan speeds and power levels. Results are shown in Fig. 3.14, which shows that higher laser power leads to slightly higher conversion rate. The conversion rates show the values between 85% and 90%. Conversion rate is more related to laser power than laser scan speed.

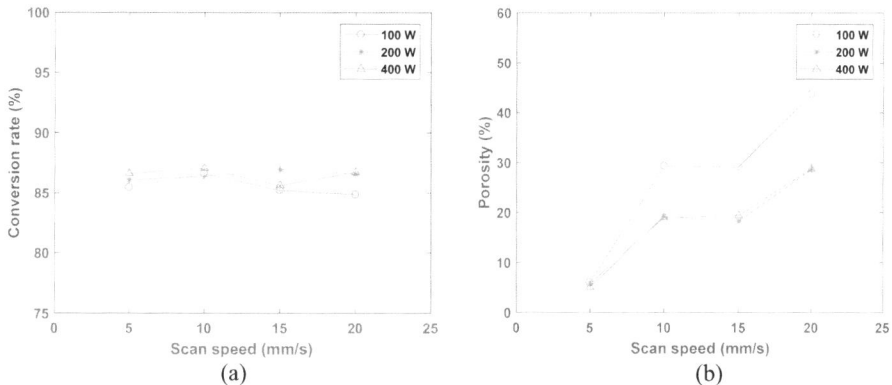

(a)   (b)

Fig. 3.14. Predicted effects of laser power and scanning speed. (a) Conversion rate with different laser scan speed and power, (b) Porosity with different laser scan speed and power.

Porosity increases with higher laser scan speed, especially with low laser power. In addition, laser condition of higher laser power results in lower porosity. The porosity of the structure varies from 5% to 45% depending on the different laser and deposition conditions. If the laser power is higher, the melting of powders can be more complete. There is a higher chance that the solid-state particles transform to the liquid state and moves around to fill the void space within the structure. The diffusion of the materials in the liquid state will be more effective. As a result, a denser structure is built. On the other hand, if the laser power level is low, there is less possibility for the solid state to completely liquefy and diffuse. Therefore, a structure with high porosity will be formed. However, there is no significant difference in porosity values between the laser conditions of 200 W and 400 W, compared to those of 100 W and 200 W, as shown in Fig. 3.14(b). There is a power level threshold for the materials to reach the melting point. A laser power level that is higher than the threshold will result in energy inefficiency or damage on the substrate. In addition, the scan speed also affects the porosity of the NiTi structure. If the scan speed is high, there is less time for the powder to fully absorb the laser energy. Therefore, porous structures will be generated. A low scanning speed gives enough time for laser energy to be transferred to the powders, and denser structures can be built. In particular, with the low scan speed of 5 mm/s, there is no difference in the porosity values among the different laser conditions. We can conclude that porosity is related to the amount of heat transferred to the powder within a time unit. Higher laser power and slower scan speed will result in lower porosity. The simulation results show a good agreement with the experimental results reported by other researchers. For example, Bimber et al.[93] showed that the values of porosity of the NiTi alloy fabricated by the DED process was between 7% and 33%, depending on the process conditions. In addition, Krishna et al.[94] reported that the porosity of NiTi alloy samples was 12% to 36% when the laser power was 150 W to 200 W and the scan speed was 10 mm/s to 20 mm/s. The value of recovered deformation, which is related to the conversion rate in the cKMC model, was also ranging from 75% to 100%.

Pores and defects formed during the DED process affect the properties and performance of the shape memory alloy. Therefore, it is important to select the process parameters that create dense structures with high-purity NiTi. The proposed reduced-order model can be used to construct the process-structure relationship efficiently for process optimization.

3.3 Summary

ICME simulation tools can accelerate the materials and process design. In this study, a multiscale modeling framework that integrates MD, PF-TLBM, FEA, and cKMC mechanisms was used to examine the DED process of Nitinol. The multiscale modeling techniques enable the predictions of the material structures and properties at different length and timescales, which provides the insight into the effects of Nitinol composition on material properties and process parameters on microstructures and morphology of the final build. MD at the atomic scale provides the predictions of material properties such as melting temperature and lattice thermal conductivity for different compositions of SMAs. These properties are important factors to be considered when device and processes are designed. Purely relying on experimental studies to decide material properties is time consuming.

Time scale is the major concern in simulating manufacturing processes. MD has difficulty simulating manufacturing processes such as solidification and machining. It is only appropriate for ultrafast processes at the scale of nanoseconds (e.g. nanoscale sintering[95]). Mesoscale simulations are needed to build process-structure relationships. Here a new PF-TLBM scheme for solidification and grain growth at mesoscale is demonstrated, where liquid-to-solid phase transition and the effects of thermal gradient and liquid flow are revealed. These external factors are directly related to process parameters such as laser power and scanning speed. It is shown that the composition of Nitinol has a significant influence on the grain growth patterns during solidification, and on the final phase distribution. The impact of process parameters on the dendritic growth and grain structures needs to be investigated thoroughly in order to establish the accurate process-structure relationship. Given the computational complexity of phase field simulation, larger scale simulations such as FEA are still valuable, even though material-related information cannot be predicted by them. Here, FEA is used to examine the geometry and dynamics of the melt pool by combining heat transfer, phase transition, and fluid flow models. The size and geometry of the melt pool and fluid velocity field can be predicted. High-fidelity PFM or FEA simulations are too expensive for extensive sampling in order to elucidate full P-S-P relationships and process optimization. Reduced-order modeling is a more viable approach. Here, a cKMC model is demonstrated to predict porosity and morphology of the build from different laser power levels and scanning speeds. It was shown that high laser power and slow scan speed help form a dense structure with reasonable conversion rate. This reduced-order model can reduce computational time from hours or days for FEA and PFM simulations to minutes for cKMC. Together, these multiscale modeling techniques enable a detailed study of the P-S-P relationships for additively manufactured Nitinol. Information exchange between these simulation tools is necessary for a comprehensive simulation-based process optimization framework. These methods are generic and can also be applied to investigate other materials and additive manufacturing systems.

Each of the individual simulation methods employed in this study provided some unique insight into the process-structure or structure-property relationships of Nitinol. However, to establish the complete P-S-P relationships, the integration of multiple modeling techniques is critical. Given the nature of information exchange between these models, the full integration requires that the simulations be synchronized while the input-output information update between them continue concurrently. A full automation can be computationally challenging in either a standalone or a distributed environment. The complexity of integrated simulation increases when more complex and detailed process models are introduced, e.g. including more physics, heterogeneous material models, and dynamic or adaptive process planning. In addition, these simulation techniques require significant initial development time to prepare and verify the simulation methods. The calibration and validation of these models themselves can be a daunting task. Once models are ready to predict, the requirement of generating massive amounts of data to examine the P-S-P relationships can be computationally prohibitive, because of the curse-of-dimensionality issue where many process parameters and factors need to be considered. When developing the models and examining the results, the developer also must ensure that the predictions are robust and accurate by carefully considering and quantifying uncertainty associated with these models. This is important to ensure validity of the simulation predictions.

In future work, these multiscale modeling techniques can be further enhanced to avoid some of the limitations of the current approaches. The models currently use certain assumptions to reduce computational complexity, which can hinder the accuracy and robustness of prediction. Each model balances the tradeoffs of fidelity to computational efficiency and can be adjusted to capture more information. For instance, the MD models can be expanded to coarse grained models that allow for the calculation of properties that require a longer simulation time period. More sensitivity analysis needs to be done for the PF-TLBM simulations. It has been shown that the PFM predictions are sensitive to the settings of process parameters and material properties.[96] More physical information can also be added into the PFM, FEA, and cKMC simulations for an improved accuracy while balancing the computational time for different physical phenomena. Physical phenomena or events can potentially be combined if strong correlations can be identified, which can reduce the computational draw during the computation of the models. Currently several simulations utilize experimentally-obtained properties. The availabilities of these properties can be limited because of the capabilities in experiment observations for small length scales or short time scales. Model calibration and validation can be generalized with the calculated properties from ab initio simulations. Information assimilation between simulations and experiments can overcome the limitations in experiments in order to generate accurate simulation models. For instance, the interatomic potentials in MD models can be calibrated from both density functional theory calculations and experimental observations. MD can be used to predict more physical properties required for PFM, FEA, and cKMC simulations.

Acknowledgements

This research was supported in part by the National Science Foundation with grants IGERT-1258425 and CMMI-1306996, and the Basic Research Lab Program through the National Research Foundation of Korea funded by the MSIT (2018R1A4A1059976).

References

1. R. Rahmanian, N. Shayesteh Moghaddam, C. Haberland, D. Dean, M. Miller, and M. Elahinia, Load bearing and stiffness tailored NiTi implants produced by additive manufacturing: A simulation study. *In Proceedings Volume 9058, Behavior and Mechanics of Multifunctional Materials and Composites*, p. 905814, San Diego, California, United States (March 2014).

2. S. Shiva, I. A. Palani, S. K. Mishra, C. P. Paul, and L. M. Kukreja, Investigations on the influence of composition in the development of Ni-Ti shape memory alloy using laser based additive manufacturing, *Opt. Laser Technol.* **69**(1), 44–51 (2015).

3. M. H. Elahinia, M. Hashemi, M. Tabesh, and S. B. Bhaduri, Manufacturing and processing of NiTi implants: A review, *Prog. Mater. Sci.* **57**(5), 911–946 (2012).

4. M. Elahinia, N. Shayesteh Moghaddam, M. Taheri Andani, A. Amerinatanzi, B. A. Bimber, and R. F. Hamilton, Fabrication of NiTi through additive manufacturing: A review, *Prog. Mater. Sci.* **83**, 630–663 (2016).

5. X. Wang, S. Xu, S. Zhou, W. Xu, M. Leary, P. Choong, M. Qian, M. Brandt, and Y. M. Xie, Topological design and additive manufacturing of porous metals for bone scaffolds and orthopaedic implants: A review, *Biomaterials* **83**, 127–141 (2016).

6. C. Haberland, M. Elahinia, J. M. Walker, H. Meier, and J. Frenzel, On the development of high quality NiTi shape memory and pseudoelastic parts by additive manufacturing, *Smart Mater. Struct.* **23**(10), 104002 (2014).

7. S. Saedi, A. S. Turabi, M. T. Andani, C. Haberland, H. Karaca, and M. Elahinia, The influence of heat treatment on the thermomechanical response of Ni-rich NiTi alloys manufactured by selective laser melting, *J. Alloys Compd.* **677**, 204–210 (2016).

8. B. V. Krishna, S. Bose, and A. Bandyopadhyay, Fabrication of porous NiTi shape memory alloy structures using laser engineered net shaping, *J. Biomed. Mater. Res. Part B Appl. Biomater.* **89**(2) 481–490 (2009).

9. M. Taheri Andani, C. Haberland, J. M. Walker, M. Karamooz, A. Sadi Turabi, S. Saedi, R. Rahmanian, H. Karaca, D. Dean, M. Kadkhodaei, and M. Elahinia, Achieving biocompatible stiffness in NiTi through additive manufacturing, *J. Intell. Mater. Syst. Struct.* **27**(19), 2661–2671 (2016).

10. Z.-Y. Zeng, C.-E. Hu, L.-C. Cai, X.-R. Chen, and F.-Q. Jing, Molecular dynamics study of the melting curve of NiTi alloy under pressure, *J. Appl. Phys.* **109**(4), 043503 (2011).

11. P. Chowdhury, L. Patriarca, G. Ren, and H. Sehitoglu, Molecular dynamics modeling of NiTi superelasticity in presence of nanoprecipitates, *Int. J. Plast.* **81**, 152–167 (2016).

12. Y. U. Wang, Y. M. Jin, and A. G. Khachaturyan, The effects of free surfaces on martensite microstructures: 3D phase field microelasticity simulation study, *Acta Mater.* **52**(4), 1039–1050 (2004).

13. H. M. Paranjape, S. Manchiraju, and P. M. Anderson, A phase field – Finite element approach to model the interaction between phase transformations and plasticity in shape memory alloys, *Int. J. Plast.* **80**, 1–18 (2016).

14. A. Baz and J. Ro, Thermo-dynamic characteristics of nitinol-reinforced composite beams, *Compos. Eng.* **2**(5–7), 527–542 (1992).

15. L. C. Brinson and R. Lammering, Finite element analysis of the behavior of shape memory alloys and their applications, *Int. J. Solids Struct.* **30**(23), 3261–3280 (1993).

16. P. Entel, M. E. Gruner, A. Dannenberg, M. Siewert, S. K. Nayak, H. C. Herper, and V. D. Buchelnikov, Fundamental aspects of magnetic shape memory alloys: Insights from ab initio and Monte Carlo studies, *Mater. Sci. Forum* **635**, 2–12 (2009).

17. V. D. Buchelnikov, P. Entel, S. V. Taskaev, V. V. Sokolovskiy, A. Hucht, M. Ogura, H. Akai, M. E. Gruner, and S. K. Nayak, Monte Carlo study of the influence of antiferromagnetic exchange interactions on the phase transitions of ferromagnetic Ni-Mn- X alloys (X = In, Sn, Sb), *Phys. Rev. B* **78**(18), 184427 (2008).

18. S. Yoo, X. C. Zeng, and S. S. Xantheas, On the phase diagram of water with density functional theory potentials: The melting temperature of ice Ih with the Perdew–Burke–Ernzerhof and Becke–Lee–Yang–Parr functionals, *J. Chem. Phys.* **130**(22), 221102 (2009).

19. S. Yoo, S. S. Xantheas, and X. C. Zeng, The melting temperature of bulk silicon from ab initio molecular dynamics simulations, *Chem. Phys. Lett.* **481**(1–3), 88–90 (2009).

20. E. Schwegler, M. Sharma, F. Gygi, and G. Galli, Melting of ice under pressure., *Proc. Natl. Acad. Sci. U.S.A.* **105**(39), 14779–14783 (2008).

21. S. W. Watt, J. A. Chisholm, W. Jones, and S. Motherwell, A molecular dynamics simulation of the melting points and glass transition temperatures of myo- and neo-inositol, *J. Chem. Phys.* **121**(19), 9565–9573 (2004).

22. J. R. Morris and X. Song, The melting lines of model systems calculated from coexistence simulations, *J. Chem. Phys.* **116**(21), 9352–9358 (2002).

23. A. B. Belonoshko, R. Ahuja, and B. Johansson, Quasi– Ab Initio Molecular Dynamic Study of Fe Melting, *Phys. Rev. Lett.* **84**(16) 3638–3641 (2000).

24. J. R. Morris, C. Z. Wang, K. M. Ho, and C. T. Chan, Melting line of aluminum from simulations of coexisting phases, *Phys. Rev. B* **49**(5), 3109–3115 (1994).

25. T. Farrell and D. Greig, The thermal conductivity of nickel and its alloys, *J. Phys. C Solid State Phys.* **2**(8), 1465 (1969).

26. S. Plimpton, Modeling Thermal Transport and Viscosity with Molecular Dynamics (Presentation), Sandia National Lab (SNL-NM) (2014).

27. X. Zheng, D. G. Cahill, P. Krasnochtchekov, R. S. Averback, and J.-C. Zhao, High-throughput thermal conductivity measurements of nickel solid solutions and the applicability of the Wiedemann–Franz law, *Acta Mater.* **55**(15), 5177–5185 (2007).

28. M. W. Ackerman, K. Y. Wu, and C. Y. Ho, Lattice thermal conductivity and Lorenz Function of Copper-Nickel and Silver-Palladium alloy systems. In *Thermal Conductivity 14*, P.G. Klemens and T.K. Chu (eds) Springer (1976).

29. M. G. Holland, Analysis of lattice thermal conductivity, *Phys. Rev.* **132**(6) 2461–2471 (1963).

30. I. Steinbach, Phase-field models in materials science, *Model. Simul. Mater. Sci. Eng.* **17**(7), 073001 (2009).

31. D. Liu and Y. Wang, Mesoscale multi-physics simulation of solidification in selective laser melting process using a phase field and thermal lattice boltzmann model. In *37th Computers and Information in Engineering Conference Volume 1* (2017).

32. D. Liu and Y. Wang, Mesoscale multi-physics simulation of rapid solidification of Ti-6Al-4V alloy, *Addit. Manuf.* **25**, 551–562 (2019).

33. B. B. Laird, The solid-liquid interfacial free energy of close-packed metals: Hard-spheres and the turnbull coefficient, *J. Chem. Phys.* **115**(7) 2887–2888 (2001).

34. P. Ternik and R. Rudolf, Numerical analysis of continuous casting of niti shape memory alloy, *Int. J. Simul. Model.* **15**(3), 522–531 (2016).

35. C. DellaCorte, S.V. Pepper, R. Noebe, D.R. Hull and G. Glenon, Intermetallic Nickel-Titanium alloys for oil-lubricated bearing applications, *Power Transm. Eng.* **8**, 26–35 (2009).

36. I. Loginova, G. Amberg, and J. Ågren, Phase-field simulations of non-isothermal binary alloy solidification, *Acta. Mater.* **49**(4), 573–581 (2001).

37. D. Hu and R. Kovacevic, Sensing, modeling and control for laser-based additive manufacturing, *Int. J. Mach. Tools Manuf.* **43**(1), 51–60 (2003).

38. A. J. Pinkerton, Advances in the modeling of laser direct metal deposition, *J. Laser Appl.* **27**(S1), S15001 (2015).

39. I. Tabernero, A. Lamikiz, E. Ukar, L. N. López de Lacalle, C. Angulo, and G. Urbikain, Numerical simulation and experimental validation of powder flux distribution in coaxial laser cladding, *J. Mater. Process. Technol.* **210**(15), 2125–2134 (2010).

40. G. Zhu, D. Li, A. Zhang, and Y. Tang, Numerical simulation of metallic powder flow in a coaxial nozzle in laser direct metal deposition, *Opt. Laser Technol.* **43**(1), 106–113 (2011).

41. P. Peyre, P. Aubry, R. Fabbro, R. Neveu, and A. Longuet, Analytical and numerical modelling of the direct metal deposition laser process, *J. Phys. D. Appl. Phys.* **41**(2), 025403 (2008).

42. S. Kumar, S. Roy, C. P. Paul, and A. K. Nath, Three-dimensional conduction heat transfer model for laser cladding process, *Numer. Heat Transf. Part B Fundam.* **53**(3), 271–287 (2008).

43. A. F. A. Hoadley and M. Rappaz, A thermal model of laser cladding by powder injection, *Metall. Trans. B* **23**(5), 631–642 (1992).

44. G. Zhu, A. Zhang, D. Li, Y. Tang, Z. Tong, and Q. Lu, Numerical simulation of thermal behavior during laser direct metal deposition, *Int. J. Adv. Manuf. Technol.* **55**(9–12), 945–954 (2011).

45. C. Cho, G. Zhao, S.-Y. Kwak, and C. B. Kim, Computational mechanics of laser cladding process, *J. Mater. Process. Technol.* **153–154**, 494–500 (2004).

46. M. F. Gouge, J. C. Heigel, P. Michaleris, and T. A. Palmer, Modeling forced convection in the thermal simulation of laser cladding processes, *Int. J. Adv. Manuf. Technol.* **79**(1–4), 307–320 (2015).

47. V. Manvatkar, A. De, and T. DebRoy, Heat transfer and material flow during laser assisted multi-layer additive manufacturing, *J. Appl. Phys.* **116**(12), 124905 (2014).

48. L. Han, K. M. Phatak, and F. W. Liou, Modeling of laser cladding with powder injection, *Metall. Mater. Trans. B* **35**(6), 1139–1150 (2004).

49. H. Qi, J. Mazumder, and H. Ki, Numerical simulation of heat transfer and fluid flow in coaxial laser cladding process for direct metal deposition, *J. Appl. Phys.* **100**(2), 024903 (2006).

50. X. He and J. Mazumder, Transport phenomena during direct metal deposition, *J. Appl. Phys.* **101**(5), 053113 (2007).

51. S. Wen and Y. C. Shin, Modeling of transport phenomena during the coaxial laser direct deposition process, *J. Appl. Phys.* **108**(4), 044908 (2010).

52. Y. Lee and D. F. Farson, Simulation of transport phenomena and melt pool shape for multiple layer additive manufacturing, *J. Laser Appl.* **28**(1), 012006 (2016).

53. B. Zheng, Y. Zhou, J. E. Smugeresky, J. M. Schoenung, and E. J. Lavernia, Thermal behavior and microstructure evolution during laser deposition with laser-engineered net shaping: Part II. Experimental investigation and discussion, *Metall. Mater. Trans. A* **39**(9), 2237–2245 (2008).

54. T. Amine, J. W. Newkirk, and F. Liou, An investigation of the effect of direct metal deposition parameters on the characteristics of the deposited layers, *Case Stud. Therm. Eng.* **3**, 21–34 (2014).

55. Y. Zhang, G. Yu, X. He, W. Ning, and C. Zheng, Numerical and experimental investigation of multilayer SS410 thin wall built by laser direct metal deposition, *J. Mater. Process. Technol.* **212**(1), 106–112 (2012).

56. S. Ghosh and J. Choi, Three-dimensional transient finite element analysis for residual stresses in the laser aided direct metal/material deposition process, *J. Laser Appl.* **17**(3), 144–158 (2005).

57. L. Costa, R. Vilar, T. Reti, and A. M. Deus, Rapid tooling by laser powder deposition: Process simulation using finite element analysis, *Acta Mater.* **53**(14), 3987–3999 (2005).

58. E. Toyserkani, A. Khajepour, and S. Corbin, 3-D finite element modeling of laser cladding by powder injection: effects of laser pulse shaping on the process, *Opt. Lasers Eng.* **41**(6), 849–867 (2004).

59. H. El Cheikh, B. Courant, S. Branchu, J.-Y. Hascoët, and R. Guillén, Analysis and prediction of single laser tracks geometrical characteristics in coaxial laser cladding process, *Opt. Lasers Eng.* **50**(3), 413–422 (2012).

60. M. Alimardani, E. Toyserkani, and J. P. Huissoon, A 3D dynamic numerical approach for temperature and thermal stress distributions in multilayer laser solid freeform fabrication process, *Opt. Lasers Eng.* **45**(12), 1115–1130 (2007).

61. F. Brückner, D. Lepski, and E. Beyer, Modeling the influence of process parameters and additional heat sources on residual stresses in laser cladding, *J. Therm. Spray Technol.* **16**(3), 355–373 (2007).

62. S. A. Khairallah and A. T. Anderson, Mesoscopic simulation model of selective laser melting of stainless steel powder, *J. Mater. Process. Technol.* **214**(11), 2627–2636 (2014).

63. S. A. Khairallah, A. Anderson, A. M. Rubenchik, J. Florando, S. Wu, and H. Lowdermilk, Simulation of the main physical processes in remote laser penetration with large laser spot size, *AIP Adv.* **5**(4), 047120 (2015).

64. N. Shen and K. Chou, Thermal modeling of electron beam additive manufacturing process: powder sintering effects. In *ASME 2012 Intl. Manufacturing Science and Engineering Conf.* (2012).

65. M. Jamshidinia, F. Kong, and R. Kovacevic, Temperature distribution and fluid flow modeling of Electron Beam Melting® (EBM). In *Volume 7: Fluids and Heat Transfer, Parts A, B, C, and D* (2012).

66. G. Bugeda Miguel Cervera and G. Lombera, Numerical prediction of temperature and density distributions in selective laser sintering processes, *Rapid Prototyp. J.* **5**(1), 21–26 (1999).

67. S. Kolossov, E. Boillat, R. Glardon, P. Fischer, and M. Locher, 3D FE simulation for temperature evolution in the selective laser sintering process, *Int. J. Mach. Tools Manuf.* **44**(2-3), 117–123 (2004).

68. I. A. Roberts, C. J. Wang, R. Esterlein, M. Stanford, and D. J. Mynors, A three-dimensional finite element analysis of the temperature field during laser melting of metal powders in additive layer manufacturing, *Int. J. Mach. Tools Manuf.* **49**(12–13), 916–923 (2009).

69. L. Dong, A. Makradi, S. Ahzi, and Y. Remond, Three-dimensional transient finite element analysis of the selective laser sintering process, *J. Mater. Process. Technol.* **209**(2), 700–706 (2009).

70. M. Matsumoto, M. Shiomi, K. Osakada, and F. Abe, Finite element analysis of single layer forming on metallic powder bed in rapid prototyping by selective laser processing, *Int. J. Mach. Tools. Manuf.* **42**(1), 61–67 (2002).

71. A. Hussein, L. Hao, C. Yan, and R. Everson, Finite element simulation of the temperature and stress fields in single layers built without-support in selective laser melting, *Mater. Des.* **52**, 638–647 (2013).

72. D. Riedlbauer, P. Steinmann, and J. Mergheim, Thermomechanical finite element simulations of selective electron beam melting processes: performance considerations, *Comput. Mech.* **54**(1), 109–122 (2014).

73. K. Dai and L. Shaw, Thermal and stress modeling of multi-material laser processing, *Acta Mater.* **49**(20), 4171–4181 (2001).

74. J. F. Li, L. Li, and F. H. Stott, A three-dimensional numerical model for a convection–diffusion phase change process during laser melting of ceramic materials, *Int. J. Heat Mass Transf.* **47**(25) 5523–5539 (2004).

75. M. Shiomi, A. Yoshidome, F. Abe, and K. Osakada, Finite element analysis of melting and solidifying processes in laser rapid prototyping of metallic powders, *Int. J. Mach. Tools Manuf.* **39**(2), 237–252 (1999).

76. F.-J. Gürtler, M. Karg, K.-H. Leitz, and M. Schmidt, Simulation of Laser Beam Melting of Steel Powders using the Three-Dimensional Volume of Fluid Method, *Phys. Procedia* **41**, 881–886 (2013).

77. K. Dai and L. Shaw, Thermal and mechanical finite element modeling of laser forming from metal and ceramic powders, *Acta Mater.* **52**(1), 69–80 (2004).

78. K. Dai and L. Shaw, Parametric studies of multi-material laser densification, *Mater. Sci. Eng. A* **430**(1-2), 221–229 (2006).

79. K. Dai and L. Shaw, Finite element analysis of the effect of Volume shrinkage during laser densification, *Acta Mater.* **53**(18), 4743–4754 (2005).

80. J. Tille and J. C. Kelly, The surface tension of liquid titanium, *Br. J. Appl. Phys.* **14**(10), 717–719 (1963).

81. V. E. Zinov'ev, *Thermal Properties of Metals at High Temperatures (in Russian).* Metallurgiya, Moscow (1989).

82. H. M. Lu and Q. Jiang, Surface tension and its temperature coefficient for liquid metals, *J. Phys. Chem. B.* **109**(32), 15463–15468 (2005).

83. H. Wang, S. Yang, and B. Wei, Density and structure of undercooled liquid titanium, *Chinese Sci. Bull.* **57**(7), 719–723 (2012).

84. Engineering ToolBox, Metals - as Liquids (Online). Available at https://www.engineeringtoolbox.com/liquid-metal-boiling-points-specific-heat-d_1893.html (Accessed June 27, 2018).

85. K. Zhou, H. P. Wang, J. Chang, and B. Wei, Experimental study of surface tension, specific heat and thermal diffusivity of liquid and solid titanium, *Chem. Phys. Lett.* **639**, 105–108 (2015).

86. R. E. Rozas, A. D. Demirağ, P. G. Toledo, and J. Horbach, Thermophysical properties of liquid Ni around the melting temperature from molecular dynamics simulation, *J. Chem. Phys.* **145**(6), 064515 (2016).

87. Z. Sparks, T. E., & Fan, Measurement of laser absorption coefficient of several alloys for diode laser, Unpubl. Rep. (2006).

88. Y. Wang, Controlled kinetic monte carlo simulation of nanomanufacturing processes. In *Volume 2: 31st Computers and Information in Engineering Conference, Parts A and B*, pp. 241–252 (2011).

89. Y. Wang, Controlled Kinetic Monte Carlo Simulation for Computer-Aided Nanomanufacturing, *J. Micro Nano Manuf.* **4**(1), 011001 (2015).

90. Y. Zhu and X. Pan, Kinetic Monte Carlo simulation of 3-D growth of NiTi alloy thin films, *Appl. Surf. Sci.* **321**, 24–29 (2014).

91. T. M. Rodgers, J. D. Madison, and V. Tikare, Simulation of metal additive manufacturing microstructures using kinetic Monte Carlo, *Comput. Mater. Sci.* **135**, 78–89 (2017).

92. J.-H. Song, K.-H. Choi, R. Dai, J.-O. Choi, S.-H. Ahn, and Y. Wang, Controlled kinetic Monte Carlo simulation of laser improved nano particle deposition process, *Powder Technol.* **325**, 651–658 (2018).

93. B. A. Bimber, R. F. Hamilton, J. Keist, and T. A. Palmer, Anisotropic microstructure and superelasticity of additive manufactured NiTi alloy bulk builds using laser directed energy deposition, *Mater. Sci. Eng. A* **674**, 125–134 (2016).

94. B. V. Krishna, S. Bose, and A. Bandyopadhyay, Fabrication of porous NiTi shape memory alloy structures using laser engineered net shaping, *J. Biomed. Mater. Res. Part B Appl. Biomater.* **89B**(2), 481–490 (2009).

95. J. M. Sestito, F. Abdeljawad, T. A. L. Harris, Y. Wang, and A. Roach, An atomistic simulation study of nanoscale sintering: The role of grain boundary misorientation, *Comput. Mater. Sci.* **165**(April), 180–189 (2019).

96. A. Tran, D. Liu, H. Tran, and Y. Wang, Quantifying uncertainty in the process-structure relationship for Al–Cu solidification, *Model. Simul. Mater. Sci. Eng.* **27**(6), 064005 (2019).

https://doi.org/10.1142/9789811222825_0004

Chapter 4

Towards Direct Deposition of Continuous-Fibers on Curved Surfaces

Chi-Chung Li, Chengkai Dai, Wei-Hsin Liao, and Charlie C.L. Wang

4.1 Introduction

Recent technological advancement has dramatically accelerated the application of *Three-Dimensional* (3D) printing in fabricating low-volume parts with high complexity economically. Although metallic parts can be 3D printed by using expensive machines with intensive energy consumption – e.g., by *Selective Laser Melting* (SLM)[1] or *Direct Metal Laser Sintering* (DMLS),[2] public and commercial sectors generally apply 3D printing on thermoplastics due to the consideration of accessibility and economic reasons. As a thermoplastic printing technique, *Fused Deposition Modeling* (FDM) gained tremendous popularity based on its stability, accuracy and number of materials that can be used in fabrication. However, weak mechanical properties of 3D printed thermoplastics in FDM limit their usage in applications requiring high mechanical strength.

Numerous attempts have been made to enhance the mechanical performance of 3D printed parts by means of making new materials,[3] optimizing structures[4] or changing the environment of fabrication.[5] On the other aspects, effort has been made by Belter and Dollar[6] to add epoxy (one kind of thermoset) to fill up the voids of thermoplastic part. Another method proposed by Compton and Lewis (2014)[7] is to add short fibers in the feedstock to enhance the overall mechanical properties of epoxy-based polymer. Similarly, 3D printed powders are used by Christ et al. and Tekinalp et al.[8,9] studied how to add short fibers into thermoplastics. Specifically, good alignments of short fibers can be achieved in 3D printed thermoplastics to have significantly enhanced tensile strength and modulus. *Carbon Fiber Reinforced Polymer* (CFRP) composite filaments were prepared from carbon fiber and ABS in extrusion processes by Ning et al.[10] The study of Llewellyn-Jones et al.[11] has demonstrated a method that can align the fibers in the desired 3D architecture by using ultrasonic forces.

Building upon the results of using short fibers in 3D printing, it was hypothesized that further enhancement in performance is possible by depositing continuous fiber instead of chopped short fibers. A design oriented method has been investigated by Brooks and Molony[12] to manually reinforce 3D printed thermoplastic parts by carbon fibers and epoxy at the regions with high stresses. In another manually reinforced case study taken on *Polylactic Acid* (PLA) molded plastic and printing filaments by Canela et al.,[13] significant improvement in both strength and stiffness can be observed. Shortly speaking, reinforcement with continuous-fiber is a very attractive means for further enhancing the mechanical properties of 3D printed thermoplastics.

Besides reinforcement provided by continuous-fiber, more and more research has been conducted to align the 3D printed filament according to the external loading to be added. Specifically, it has also been argued that making a structure (or printing paths) conformal to the axial surface or the curves of principal stresses can help to fabricate a model stronger in mechanical properties.[14–16]

The purpose of this chapter is to explore the possible methodology to realize the direct deposition of continuous-fibers in a sandwich structure on curved 3D surfaces. Preliminary tests have been conducted to demonstrate the performance improvement after

(1) converting a planar-layer-based filament deposition into a curved-layer-based fabrication, and
(2) further reinforcing continuous carbon fibers between the curved layers of plastic filaments.

Physical experiments are conducted on a hardware setup with 6 *degrees-of-freedom* (DOF) motion provided by a robotic arm. With the help of such a hardware platform, we are able to reinforce 3D printed parts by a process of continuous-fiber deposition between layers of PLA matrix in 3D printing.

4.2 Literature Review

In this section, we review the relevant research of high DOF 3D printing and the progress of using continuous-fiber in 3D printing. Research developed in these two areas provides the essential technology to realize the continuous-fiber reinforced 3D printing on curved surfaces.

4.2.1 *Robotic and Multi-axis 3D Printing*

Although called 3D printing, the practice of Additive Manufacturing (AM) is still mainly conducted in a manner of accumulating materials layer by layer in planes.

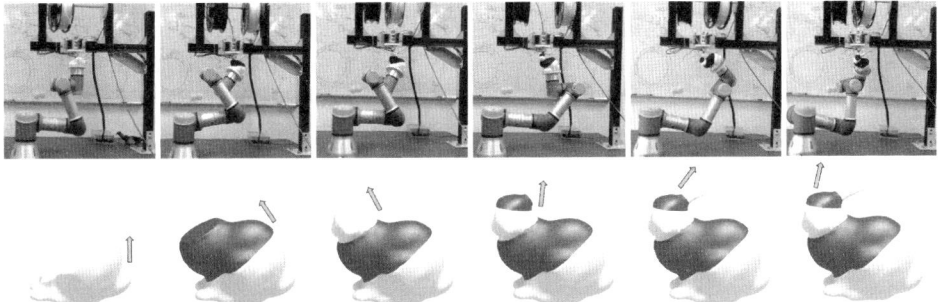

Fig. 4.1. The progress of using the multi-directional 3D printing system to fabricate a freeform model bunny. Different parts of the model is fabricated along different directions (see arrows in illustration), and filaments in different colors are used for making different parts.

Three major dimensions of AM technology development are towards the directions of multi-materials, faster printing and higher-flexibility of 3D printing. Specifically, more and more recent approaches try to introduce more DOFs in motion to overcome the limitation of plane based manufacturing.

Curved Layer Fused Deposition Modelling (CLFDM) developed in Chakraborty et al.[17] provides the function of FDM fabrication by using a tool-path with dynamically changed z-values within individual layers, which is different from conventional FDM having fixed z-values with each individual layers. Recent work along the same research direction can be found in the work of Huang and Singamneni (2015), Singamneni et al. and Allen and Trask,[18–20] which however can only deal with relatively simple shapes such as height fields. The recent development in Shembekar et al.[21] belongs to this category of technology but can handle more complicated surfaces.

Researchers have started to explore the new DOF of AM in motion to improve the process of fabrication. The concept of freeform AM without supporting structure is demonstrated in the work of Keating and Oxman[22] by using 6DOF provided by a robotic arm. There is no detail of tool-path generation provided in their paper. The hardware system conducted in our physical verification borrows their idea of fixing the nozzle of material deposition to obtain better quality of material adhesion with the help of gravity. A five-axis motion system similar to five-axis CNC machining is developed in Pan et al.[23] to accumulate materials onto an existing model, which is an extension of their prior work of building-around-inserts implemented on a three-axis motion platform.[24] Only small components with relative simpler planning tasks can be handled by their algorithm. Material accumulation is taken around a cubic component in Gao et al.,[25] which is hard to be generalized. A multi-directional 3D printing technique is recently developed in Wu et al.[26] to segment a given model into support-free regions so that they can be fabricated one

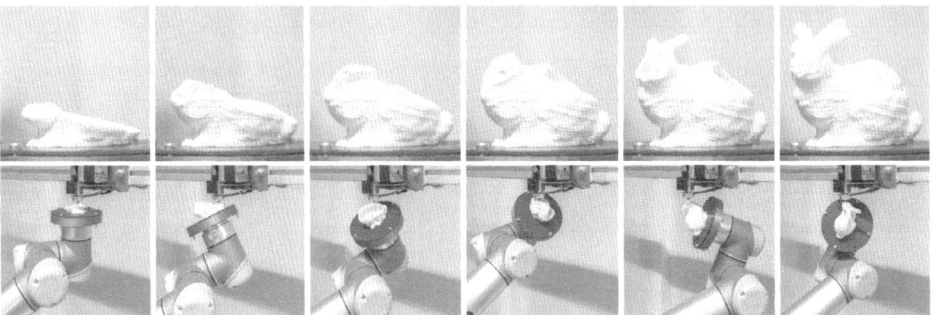

Fig. 4.2. By exploiting all 6DOF provided by a robotic arm, the material accumulation / placement along different orientation on a general curved surface becomes possible.

by one – each in a different direction with the help of a robotic arm (see Fig. 4.1). Planar layers are employed to fabricate each part, and it cannot handle the material accumulation on a general curved surface. In short, the capability of fully 6DOF motion has not been fully utilized in these approaches.

Recently, algorithms for using high DOF in 3D printing have been developed in the community of computer graphics Wu et al.[27] proposed an algorithm to compute collision-free tool paths to extrude materials for a given mesh model edge by edge. A global planning is taken on a directed graph Huang et al.[28] considers not only collision constraint but also stability constraint in the process of fabrication. A divide-and-conquer strategy is developed to compute collision-free and stable sequence of fabrication. Both these approaches have the step of collision detection included in the loop of computation, which is time-consuming and also prevent applying these algorithm to problems in large scales – i.e., with many primitives. Differently, the recent approach of Dai et al.[29] tackles the challenge of tool-path planning for multi-axis 3D printing by two successive decompositions: 1) volume-to-surfaces and 2) surfaces-to-curves. A convex-front is employed to generate the field of material accumulation sequence so that the collision detection between a printer head and the already printed model can be avoided. The curved working layers generated by this approach in fact enables the manufacturing process planning for continuous-fiber fabrication. Figure 4.2 shows the material accumulation conducted on different curved working surface. Moreover, continuity in different levels (positions, orientation and poses) are preserved during their 5DOF 3D printing, which can also be applied in 3D printing with continuous-fiber reinforcement.

4.2.2 *Continuous-Fiber in 3D Printing*

For the research of continuous-fiber reinforcement Prüß and Vietor,[30] categorized different production strategies of continuous-fibers into three types: I) pre-impregnation, II) in-nozzle impregnation and III) direct deposition. We review relevant work below by following these categories.

In aerospace industry, Automated Fiber Placement (AFP) has been developed to fabricate fiber composites. Continuously pre-impregnated (pre-preg) fibers can be heated and pressed against a model (or existing layers of the laminate) in thermoplastic AFP processes.[31] The matrix materials around fibers are fused together during the fabrication process. Required composite with high performance is usually manufactured tape-by-tape and then layer-by-layer. As a small-scale counterpart, Mark Two 3D printer invented by Markforged[32] is able to deposit filaments reinforced with continuous-fiber in FDM. A special extruder is developed in their system to print filaments with pre-preg carbon fibers and produce 3D models in the conventional line-by-line process of layered manufacturing. Currently, the cost per unit volume of pre-preg carbon fiber[33] is USD\$3/cm^3. It is about 10 to 20 times higher than that of typical carbon fibers.[34] The fact that pre-preg carbon fibers are proprietary[35] and relatively expensive blocks the wide usage of this technique.

For the second type of continuous-fiber embedding, recently attempts have been made to achieve it in FDM by inserting a narrow strand of fiber into the back opening of a specially designed nozzle while the filament is extruded along with the molten thermoplastic. When cooling down, fibers are embedded in the strand of thermoplastic and thereby reinforces a printed part. The whole process is an integration of impregnation and 3D printing; therefore, the approach is referred as in-nozzle impregnation Matsuzaki et al.[36] recently developed a method that impregnates fibers with filament within the heated nozzle of 3D printers, where thermoplastic filaments and continuous fibers are supplied separately to the nozzle. Tian et al.[37] conducts a similar strategy to reinforce fibers into printed models and the maximum flexural strength of 335MPa and the flexural modulus of 30GPa can be observed on their printed composite specimens with 27% fiber content. In the experiments of Li et al.,[38] further enhanced mechanical property is found by using the preprocessed carbon fiber with polylactic acid sizing agent. All these approaches using very thin carbon fiber bundles (e.g., the one made up of 1000 single carbon fibers in Li et al.[38]).

In the above two types of continuous-fiber reinforcement, the width of fibers to be used is constrained by the mechanical design of nozzle in printing head. As a result, the strand based deposition could be very slow. If carbon fiber could be directly deposited on the printing part, the fiber deposition speed could be much higher (see the comparison listed in Table 1). In the rest of this chapter, we will

Table 4.1. Comparison of different approaches for reinforcing continuous-fibers in additive manufacturing

Strategy	Pre-impregnation	In-nozzle Impregnation	Direct Deposition
Width	*Narrow (Strand)*	*Narrow (Strand)*	*Wide (Strand/Tape)*
Cost	*Higher*	*Lower*	*Lower*
Availability	*Proprietary*	*Non-proprietary*	*Non-proprietary*
Dep. Speed	*Slower*	*Slower*	*Faster*

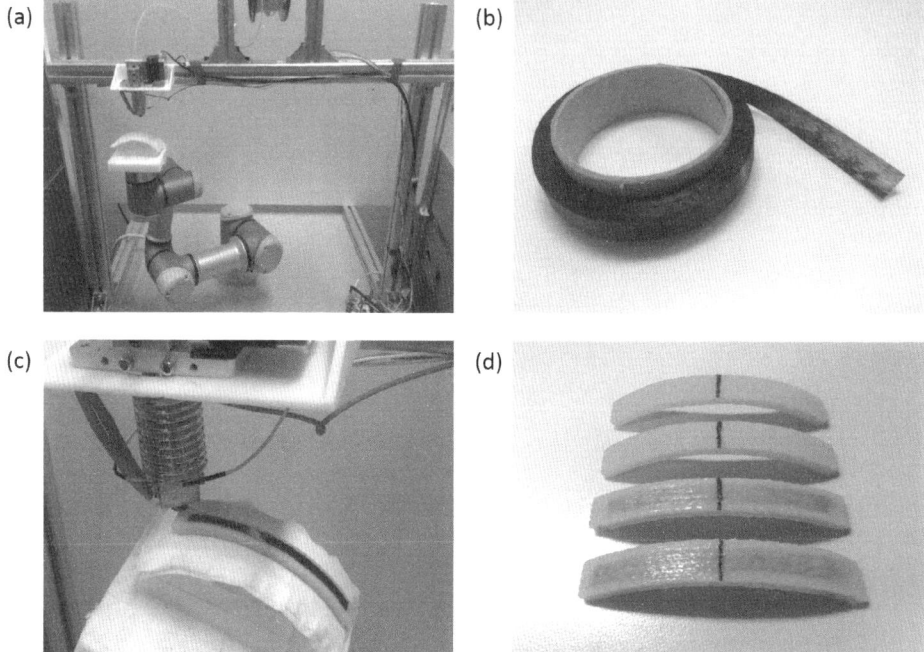

Fig. 4.3. The hardware setup in our research for continuous-fiber reinforced 3D printing with 6DOF motion: (a) 6DOF platform used for direct deposition, (b) tapes of carbon fibers reinforcements used for 3D printing, (c) accumulating PLA on top of ironed fibers, and (d) samples fabricated by curved deposition with and without continuous-fiber reinforcement.

introduce our preliminary research towards this direction of direct carbon fiber deposition.

Along the other thread of research, it has been observed that the mechanical strength of parts made by FDM is strongly affected by the arrangement of thermoplastic strands. The adhesive strength across plastic strands is much weaker than the longitudinal strength along the strands. It has been advised to keep plastic strands as continuous as possible during the printing and orient filaments along the loading direction to maximize mechanical performance. In planar layered AM, choosing the right printing direction can let strands be continuous for simple objects but it does not apply to parts with freeform curved surface. Chakraborty et al.[17] proposed curved layer FDM (CLFDM) algorithms to generate non-planar paths for printing curved objects; however, only simulation is given in their results. Singamneni et al.[19] applied the curved layer FDM algorithms to fabricate physical specimen to verify this conclusion. Recently, Huang and Singamneni[18] further considered the factor of adaptive slice thickness into the fabrication of curved layers. Testing specimens have been made using CLFDM, and significant enhancement of mechanical performance was observed on these specimens. Besides, the surface finish of a model made by CLFDM can be improved due to absence of stair-case effect. Differently, we consider how to incorporate the material properties of continuous-fiber into the curved layers of PLA with the help of a robotic system.

4.3   Direct Deposition: Our Study

In this chapter, we introduce a direct deposition method to place tapes of continuous carbon fibers between matrix of thermoplastics (e.g., PLA) so that the mechanical strength of a printed model can be reinforced. We wish more work will be motivated towards the direct deposition of continuous-fibers in 3D printing. Fig. 4.3(a) shows our hardware setup with 6DOF motion used for the direct deposition of continuous-fiber. Note that when planar layers are utilized for the continuous-fiber reinforcement, the direct deposition can also be conducted on the planar motion platform of conventional FDM 3D printer (e.g., the one shown in Fig.4.5(a)).

4.3.1   *Processing Method*

Our approach can be considered as a variant of AFP process in 3D printing. Continuous tapes of carbon fibers are placed between layers of thermoplastic matrix – i.e., continuous-fibers and PLA used in our experiments are fused in a sandwich way. A layer of thermoplastic must be deposited as the substrate before the direct deposition of fibers. When a layer of thermoplastic matrix is ready, head of fibers (in tape) is first placed on top of the fused thermoplastics at the desired location and with an orientation normal to the surface. Heat is then applied on the interface between fibers and thermoplastics in a conductive, convective or radiant manner. When the heated region is cooled down (preferably under pressure), the placed fiber will be locally bonded with the thermoplastic. The deposition mechanism, which applies heat and pressure, travels along the length of a tape to continuously bond fibers to the previous layer of thermoplastic substrate.

With the help of 6DOF motion platform, the head of deposition would trace a planned path on the working surface and keep applying perpendicular pressure to

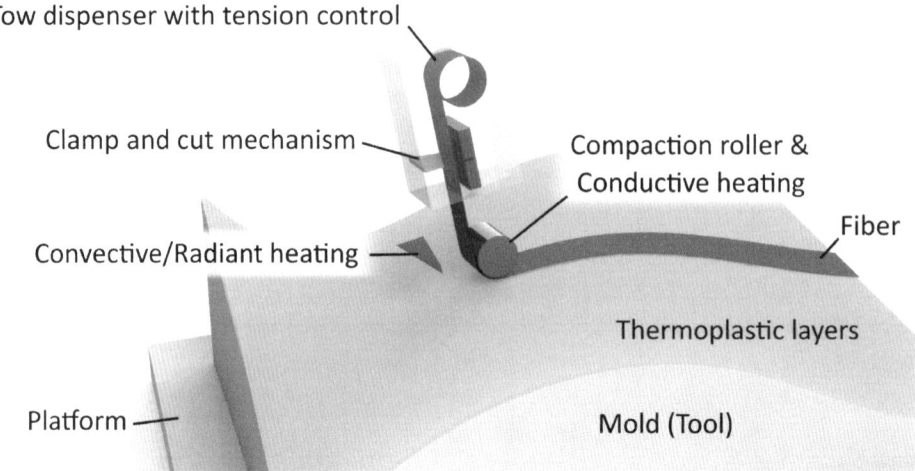

Fig. 4.4. Illustration of heated deposition mechanism for continuous-fiber reinforcement in 3D printing.

the deposited fibers. At the end of a path, chopper in the mechanism is applied to cut the continuous-fiber into segments (see Fig. 4.4 for an illustration of the mechanism). This step of direct deposition is similar to the AFP technique used in the aerospace industry. Differently, AFP only uses pre-preg tow whereas in this approach raw spread tow (that only contains the reinforcing fibers) are directly bonded onto thermoplastic substrate.

To complete the process of direct deposition, a layer of thermoplastic is fused on top of the placed fibers in a way similar to conventional FDM but again with 6DOF motion (see Fig. 4.3(c)). It is noted that height or extrusion compensation has to be considered for the volume taken by additional fiber. In our practice, the thickness of thermoplastic layer (in the range of 0.3mm to 0.6mm) is much larger than that of fibers (about 0.08mm). No compensation is taken for the fiber layer, and the height increase between thermoplastic layers is conducted by using the uniform offsetting method.[39] After the upper layer of thermoplastic matrix is fused, continuous tapes of fibers have been securely embedded into the fabricated part (see Fig. 4.3(d) for the results of fabrication).

4.3.2 *Detail of Experiments*

Materials employed in our experimental fabrication are PLA supplied by Dazzle-light and spread tow of carbon fiber supplied by Easy Composite Ltd. PLA thermoplastic is chosen for its popularity and ease of printing. The spread tow made

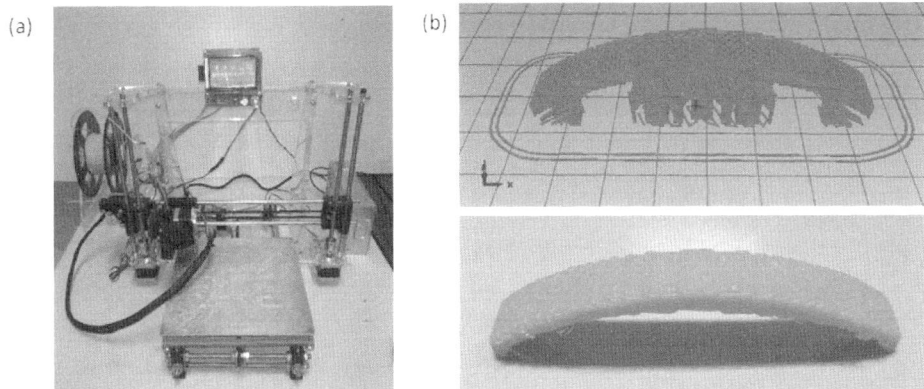

Fig. 4.5. Comparison to the specimen fabricated by a conventional 3D printing system with planar motion and (b) the model with curved surfaces fabricated from planar tool-paths.

from Gradfil TR50S 15K carbon fiber tow and contain 6% Polyamide by weight is utilized. In particular, spread tow, an organized and thin bundle of fiber (0.08mm thick), is used as its thickness facilitates the infiltration of plastic and heat conduction in the direct deposition process. The 15mm-wide spread tow is first split into three 5mm-wide ones before application.

In our hardware setup for fabrication, a Universal Robot UR3 robot arm is employed to move the building platform so that 6DOF relative motion can be realized between the platform (and also the working envelope) and a stationary nozzle. When fabricating models with curved layers, a mold is added onto the building platform to provide a curved surface where the extruded plastic could adhere to. In our tests, the mold is a 3D printed rectangular block with cylindrical surface on the top (80° arc at 60mm radius), and it is covered with masking tape. PLA matrix is printed at 210°. The diameter of nozzle is 1mm and the thickness of PLA layer is controlled at 0.6mm. Note that, our expendable setup actually allows multiple nozzles to be installed on the frame and multi-material printing can be realized. For the comparison taken with conventional 3D printing, Reprap Prusa i3 was used as a FDM platform to fabricate testing samples. The machine accumulates PLA with a hotbed at 60°. The temperature of nozzle is set as 230° for planar samples for tensile and bending tests. To be consistent with the samples fabricated by 6DOF motion platform, 210° is used for PU samples in curved shape (i.e., the one shown in Fig.4.5(b)).

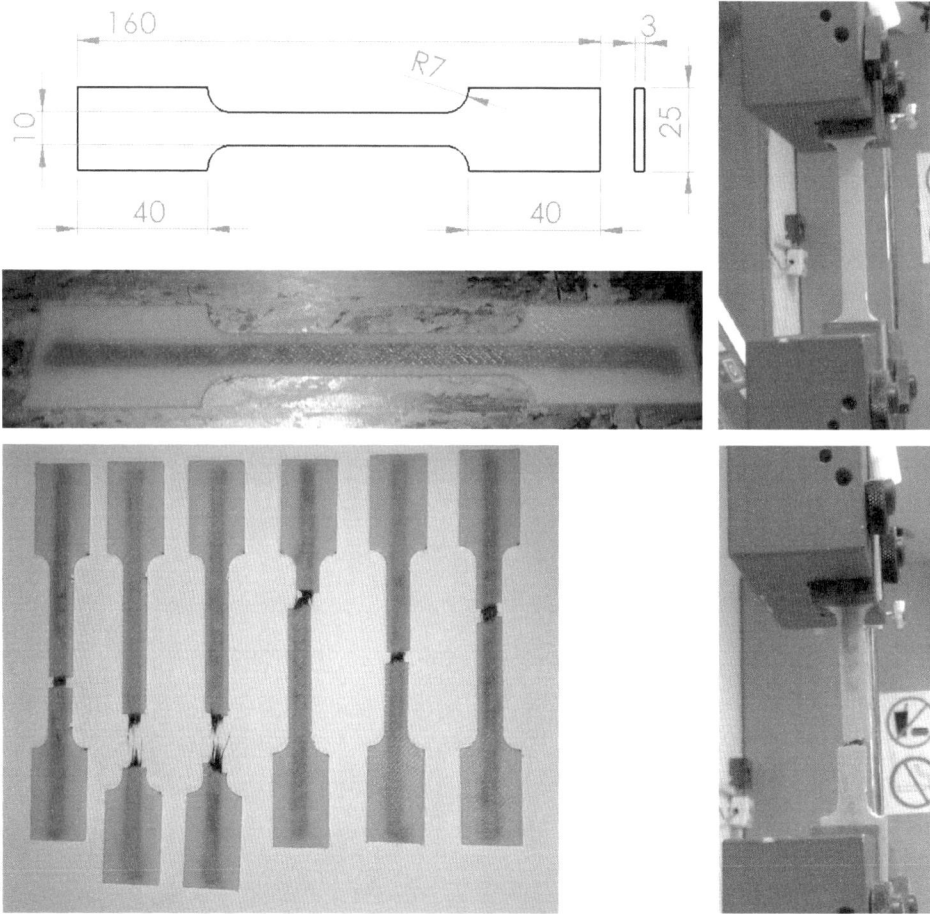

Fig. 4.6. Tensile tests taken on reinforced models with planar layers.

4.4   Results

In this section, we present the results of experimental tests for comparing the mechanical properties of 3D printed specimen in different aspects. Planar and curved models are fabricated to test their mechanical properties with and without reinforcement.

4.4.1   *Mechanical Properties of Continuous-fiber Reinforcement by Planar Layers*

The first type of tests are taken on models printed by reinforced and non-reinforced planar layers, which are denoted by PR (i.e., planar with carbon-fiber reinforcement) and PU (i.e., planar unreinforced specimen) in the charts of testing results.

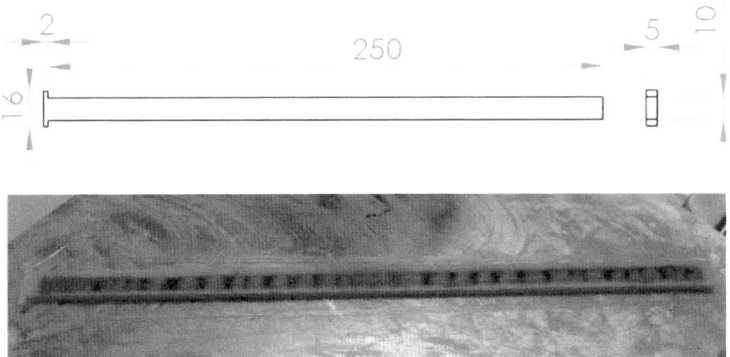

Fig. 4.7. Planar specimen (with carbon-fiber reinforced) that are used in the bending tests.

Both the tensile and the bending tests are conducted.

Twelve dumbbell-shape samples are fabricated and six of them are reinforced with five layers of 150mm × 5mm carbon fiber spread tow (as shown in Fig. 4.6). Tensile tests for these specimen are conducted on a H5KS Benchtop Materials Testing Machine made by Tinius Olsen. All samples are applied with a preload of 15N and stretched in the speed of 10mm/min. Average ultimate tensile strengths of 1.700kN and 3.535kN are observed on PU and PR samples respectively. The corresponding stress-strain curves generated in these tests are shown in Fig. 4.8(a). The estimated Young's modulus are $E_{PU} = 1.25$GPa and $E_{PR} = 2.32$GPa – i.e., about 85.6% improvement has been observed on reinforced samples for its stiffness.

For bending tests, eight rectangular bars with dimension 250mm × 10mm × 5mm (thickness) are fabricated with planar layers. Four of the samples are reinforced with nine layers of 240mm × 5mm carbon fiber spread tow (as shown in Fig. 4.7). All samples are subjected to three-point flexural tests with a maximal loading of 7.27N applied at the middle. By using the simple beam theory, flexural rigidity (EI) of a beam can be obtained by $EI = P \cdot L^3/(48d)$, where P denotes the force applied, L is the distance between the supports and d is measured central deflection. According to the measurements shown in Fig. 4.8(b), the flexural rigidity of unreinforced and reinforced bars are 0.351N · m^2 and 0.671N · m^2 respectively. On average, about 91.1% of enhancement in flexural rigidity is achieved on the reinforced specimen.

4.4.2 *Mechanical Properties of Continuous-fiber Reinforcement by Curved Layers*

After studying the mechanical reinforcement on models fabricated by planar layers, it is more interesting to study the bending behavior on specimen made by curved

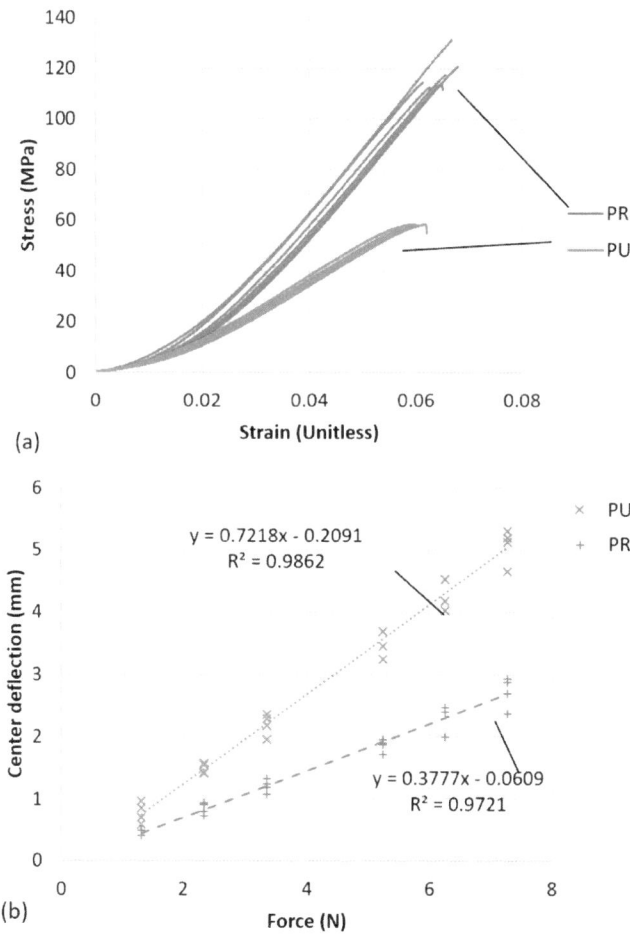

Fig. 4.8. Curves of mechanical property generated by physical tests on reinforced planar specimen (PR) vs. unreinforced planar specimen (PU). (a) Stress-strain curves of tensile tests on samples as shown in Fig. 4.6. (b) Deflection-force curves of three-points bending tests generated on planar rectangular sample bars as shown in Fig. 4.7.

layers. The tests are taken on curved bars which have an arc in $75°$ with 60mm radius. Width of the bars is set as 10mm. 6DOF motion platform is employed to fabricate 12 such models that each has seven PLA curved layers in the thickness of 0.6mm. Six out of the 12 specimen are reinforced by direct deposition with 6 layers of 68mm $\times$ 5mm spread tow of carbon fibers. Moreover, to compare with the models fabricated by conventional 3D printing, we also made 6 models in the same shape by planar layers as shown in Fig. 4.5. Specifically, three different types of fabrication are conducted as explained in Fig. 4.9(b).

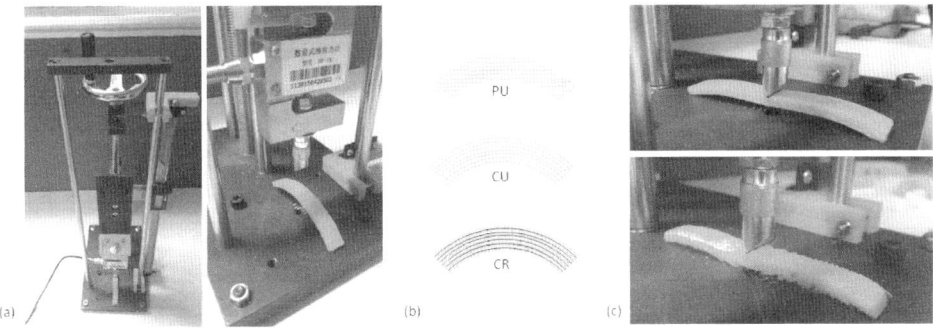

Fig. 4.9. (a) The hardware setup of bending test applied to the specimen. (b) Specimen in curved shape fabricated by different strategies: PU – by planar unreinforced layers, CU – by curve unreinforced layers, and CR – by curve reinforced layers. (c) Under the same loading, delamination can be found on the model fabricated from planar layers (bottom).

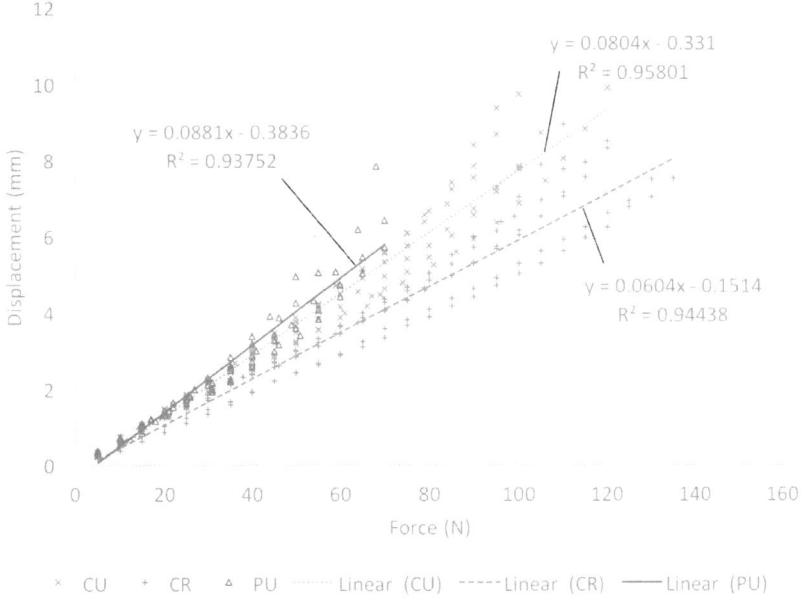

Fig. 4.10. Displacement-force diagram generated by bending tests on PU, CU, and CR specimen.

During the bending tests, compression force is applied on samples with a sharp-edge contact (see the black marks in Fig. 4.3(d). Handpi 1000N force gauge is used to measure the force applied and an attached digital caliper is used to measure displacement in the tests. A preload of 10N is applied before setting both the force gauge and caliper reading to be zero. The preload is used to ensure close and stable contact between the sharp edge and the sample.

Our experimental tests also help to generate the displacement-force diagram of bending as shown in Fig. 4.10. A straight line is used to estimate the displacement-force relationship of each type of specimen using least-square regression. Here the reciprocal of slope of fitting result, in terms of required force to create unit displacement, is conducted as a measurement of bending resistance. The observed average reciprocals of PU, CU, and CR specimen are $11.35N/mm$, $12.43N/mm$ and $16.54N/mm$ respectively. The reinforced samples are 33.06% more resistant to bending over the unreinforced ones. The bending stiffness has been significantly improved by adding continuous-fiber reinforcement. On the other aspect, our experimental results also prove that with an appropriate orientation of filament the mechanical property of a model fabricated by 3D printing can be enhanced – i.e., on both specimen without fiber reinforcement, the ones printed with curved layers (CU samples) shows 9.50% more resistance to deformation than PU samples.

4.4.3 *Fracture Tests Under Large Deformation*

Fracture in the form of delamination between layers is a phenomenon that can always be observed on the curved models fabricated by planar-layer-based AM. For example, such delamination can be observed PU samples when around $80N$ loadings are applied – see the bottom of Fig. 4.9(c). For the samples of CU and CR, fracture observed only when much larger loadings are applied – i.e., $170N$ and $198N$ for CU and CR specimen respectively.

Unlike PU specimen, fracture cannot be observed on CU and CR samples even after bending them into a flat shape. It only occurs after further deforming them into an inverse arc shape. This nice mechanical property is a benefit from the fabrication of layers conformal to the neutral surface of a beam, which also shows the effectiveness of continuous-fiber reinforcement with the help of 6DOF 3D printing.

4.5 Electrical Conductivity

As carbon fiber is electrically conductive, it could potentially be used as wires and form part of an electronic circuit. We successful were in attempting to fabricate a simple LED circuit in a reinforced hook using the direct deposition method proposed in this chapter (see Fig. 4.11). This suggests that the embedded fiber could have functions other than structural reinforcements and closer inspection should be given on embedding functional elements in FDM plastics. More discussion about 3D printing electronics can be found in Song et al.[40]

Fig. 4.11. Reinforced hook with a LED circuit – continuous carbon-fibers are served as wires to form an electronic circuit.

4.6 Discussion

In our current practice, the fiber-matrix combination, the type of fiber and the parameters of process are not optimized. Therefore, there is a lot of space to further improve the performance of 3D printed models reinforced with continuous-fiber. Future work could focus on developing a mathematical model for performance prediction and process optimization. Algorithms for computing and controlling the height of layers and the speed of material extrusion could also be developed to enable the fabrication of parts with higher percentage of fibers. Also, surface processing procedures could be taken on fibers before deposition to further enhance adhesion strength of the fusion bonding.

A major limitation of our current tests is that the process planning of embedding fibers and other functional component (such as electronics) are realized through a manual operation. With the increase of part's complexity, automated process planning for non-planar placement of continuous-fiber types is required. Future work could be done on developing such a printer head with advanced control system to automate and optimize the process of direct deposition.

In all our experimental tests, we fuse tapes of carbon fibers onto the surface layer of PLA substrate manually. Heat is delivered on top of the fiber in a conductive way by soldering iron with series K blade tip. The blade shape in iron's tip helps to deliver heat and pressure evenly across the width of the fiber, which is analogous to the household iron for clothes. Taking references from AFP (such as August et al.[31] and Pitchumani et al.[41]), the temperature controlled at the interface and the pressure applied throughout the heating and cooling process are major factors to determine the quality of bonding. If they are further optimized, the mechanical properties of reinforcement achieved by direct deposition of continuous-fibers could be even stronger than what we reported here.

Direct deposition of carbon fiber tapes is still taken manually in our experimental system. We plan to develop an automatic rolling mechanism to work together with the robotic system for automating the process of carbon-fiber deposition. Another possible future research is to study the pattern to fill a given freeform surface by using tapes of carbon-fibers. A similar work in computational architecture can be found in the work of Barton et al.[42] The computational methods to optimize a structure with continuous-fiber reinforcement need to be investigated in the future.

4.7 Summary

In this chapter, we present the new trend of direct deposition for 3D printing objects with continuous-fiber reinforced by a robot-assisted 3D printing platform, which can provide 6DOF motion to generate curved layers of matrix made by polylactic acid. We have realized a preliminary system for direct deposition of continuous-fibers although there are some steps conducted manually. Notable structure enhancement has been observed in our experimental results of curved layer fabrication with and without fiber reinforcement. In summary, spatial alignment of polylactic acid filaments can help structurally strengthen a model printed in 6DOF motion, which can be further improved by reinforcing continuous-fibers between curved layers.

In the preliminary work reported in this chapter, only specimens with simple shape and topology are fabricated and tested. It will be challenging and of course more interesting if a general method can be developed to generate the optimized tool-paths for the reinforcement of 3D printed parts. This is a work to be conducted in our near future research.

Acknowledgment

The project taken in this work is supported by Research Grants Council of the Hong Kong Special Administrative Region, China (Project No.: CUHK/14207414, 14202016, and 14202219).

References

1. I. Gibson, D. W. Rosen, and B. Stucker, *Additive Manufacturing Technologies: 3D Printing, Rapid Prototyping, and Direct Digital Manufacturing*, 2nd Edition, Springer (2015).

2. EOS. EOS e-manufacturing solutions: Industrial 3D printing. `http://www.eos.info/en` (Accessed May 2010).

3. Stratasys. 3D printing with FDM and Nylon 12. `http://www.stratasys.com/materials/fdm/nylon` (Accessed January 2015).

4. Z. Qin, B. G. Compton, J. A. Lewis, and M. J. Buehler, Structural optimization of 3D-printed synthetic spider webs for high strength, *Nature Communications* **6**, 7038 (2015). 10.1038/ncomms8038.

5. F. Lederle, F. Meyer, G.-P. Brunotte, C. Kaldun, and E. G. Hübner, Improved mechanical properties of 3D-printed parts by fused deposition modeling processed under the exclusion of oxygen, *Progress in Additive Manufacturing* **1**(1), 3–7 (2016). 10.1007/s40964-016-0010-y.

6. J. T. Belter and A. M. Dollar, Strengthening of 3D printed fused deposition manufactured parts using the fill compositing technique, *PLoS ONE*, **10**(4), 1–19 (04, 2015). 10.1371/journal.pone.0122915.

7. B. G. Compton and J. A. Lewis, 3D-printing of lightweight cellular composites, *Advanced Materials* **26**(34), 5930–5935 (2014). 10.1002/adma.201401804.

8. S. Christ, M. Schnabel, E. Vorndran, J. Groll, and U. Gbureck, Fiber reinforcement during 3D printing, *Materials Letters* **139**, 165–168 (2015). http://dx.doi.org/10.1016/j.matlet.2014.10.065.

9. H. L. Tekinalp, V. Kunc, G. M. Velez-Garcia, C. E. Duty, L. J. Love, A. K. Naskar, C. A. Blue, and S. Ozcan, Highly oriented carbon fiber–polymer composites via additive manufacturing, *Composites Science and Technology* **105**, 144–150 (2014). http://dx.doi.org/10.1016/j.compscitech.2014.10.009.

10. F. Ning, W. Cong, J. Qiu, J. Wei, and S. Wang, Additive manufacturing of carbon fiber reinforced thermoplastic composites using fused deposition modeling, *Composites Part B: Engineering* **80**, 369–378 (2015). 10.1016/j.compositesb.2015.06.013.

11. T. M. Llewellyn-Jones, B. W. Drinkwater, and R. S. Trask, 3D printed components with ultrasonically arranged microscale structure, *Smart Materials and Structures* **25**(2), 02LT01 (2016). 10.1088/0964-1726/25/2/02LT01.

12. H. Brooks and S. Molony, Design and evaluation of additively manufactured parts with three dimensional continuous fibre reinforcement, *Materials & Design* **90**, 276–283 (2016). 10.1016/j.matdes.2015.10.123.

13. J. M. Canela, I. B. Corral, L. Bade, P. Hackney, I. Shyha, and M. Birkett, Investigation into the development of an additive manufacturing technique for the production of fibre composite products, *Procedia Engineering* **132**, 86–93 (2015). 10.1016/j.proeng.2015.12.483.

14. T.-H. Kwok, Y. Li, and Y. Chen, A structural topology design method based on principal stress line, *Computer-Aided Design* **80**, 19–31 (2016). https://doi.org/10.1016/j.cad.2016.07.005.

15. K.-M. M. Tam and C. T. Mueller, Additive manufacturing along principal stress lines, *3D Printing and Additive Manufacturing* **4**(2) (2017). http://doi.org/10.1089/3dp.2017.0001.

16. Y. Chen, C. Zhou, and J. Lao, A layerless additive manufacturing process based on CNC accumulation, *Rapid Prototyping Journal* **17**(3), 218–227 (2011).

17. D. Chakraborty, B. Aneesh Reddy, and A. Roy Choudhury, Extruder path generation for curved layer fused deposition modeling, *Computer-Aided Design* **40**(2), 235–243 (2008). 10.1016/j.cad.2007.10.014.

18. B. Huang and S. B. Singamneni, Curved layer adaptive slicing (CLAS) for fused deposition modelling, *Rapid Prototyping Journal* **21**(4), 354–367 (2015). 10.1108/RPJ-06-2013-0059.

19. S. Singamneni, A. Roychoudhury, O. Diegel, and B. Huang, Modeling and evaluation of curved layer fused deposition, *Journal of Materials Processing Technology* **212**(1), 27–35 (2012). 10.1016/j.jmatprotec.2011.08.001.

20. R. J. Allen and R. S. Trask, An experimental demonstration of effective curved layer fused filament fabrication utilising a parallel deposition robot, *Additive Manufacturing* **8**, 78–87 (2015). 10.1016/j.addma.2015.09.001.

21. A. Shembekar, Y. Jung, A. Kanyuck, and S. K. Gupta. Trajectory planning for conformal 3d printing using non-planar layers. In *Proceedings of ASME International Design Engineering Technical Conferences & Computers and Information in Engineering Conference* (2018).

22. S. Keating and N. Oxman, Compound fabrication: A multi-functional robotic platform for digital design and fabrication, *Robotics and Computer-Integrated Manufacturing* **29**(6), 439–448 (2013). http://dx.doi.org/10.1016/j.rcim.2013.05.001.

23. Y. Pan, C. Zhou, Y. Chen, and J. Partanen, Multitool and multi-axis computer numerically controlled accumulation for fabricating conformal features on curved surfaces, *ASME Journal of Manufacturing Science and Engineering* **136**(3), 031007 (2014).

24. X. Zhao, Y. Pan, C. Zhou, Y. Chen, and C. C. L. Wang, An integrated CNC accumulation system for automatic building-around-inserts, *Journal of Manufacturing Processes* **15**(4), 432–443 (2013).

25. W. Gao, Y. Zhang, D. C. Nazzetta, K. Ramani, and R. J. Cipra. RevoMaker: Enabling multi-directional and functionally-embedded 3D printing using a rotational cuboidal platform. In *Proceedings of the 28th Annual ACM Symposium on User Interface Software and Technology*, pp. 437–446 (2015). 10.1145/2807442.2807476.

26. C. Wu, C. Dai, G. Fang, Y.-J. Liu, and C. C. L. Wang. RoboFDM: A robotic system for support-free fabrication using FDM. In *Proceedings of IEEE International Conference on Robotics and Automation* (2017).

27. R. Wu, H. Peng, F. Guimbretière, and S. Marschner, Printing arbitrary meshes with a 5DOF wireframe printer, *ACM Trans. Graph* **35**(4), 101:1–101:9 (2016). 10.1145/2897824.2925966.

28. Y. Huang, J. Zhang, X. Hu, G. Song, Z. Liu, L. Yu, and L. Liu, Framefab: Robotic fabrication of frame shapes, *ACM Trans. Graph* **35**(6), 224:1–224:11 (2016). 10.1145/2980179.2982401.

29. C. Dai, C. C. L. Wang, C. Wu, S. Lefebvre, G. Fang, and Y.-J. Liu, Support-free volume printing by multi-axis motion, *ACM Trans. Graph* **37**(4), 134:1–134:14 (2018). 10.1145/3197517.3201342.

30. H. Prüß and T. Vietor, Design for fiber-reinforced additive manufacturing, *Journal of Mechanical Design* **137**(11), MD–15–1106 (2015). 10.1115/1.4030993.

31. Z. August, G. Ostrander, J. Michasiow, and D. Hauber, Recent developments in automated fiber placement of thermoplastic composites, *SAMPE Journal* **50**(2) (2014).

32. Markforged. The Mark Two industrial strength 3D printer. https://markforged.com/mark-two/ (Accessed June 2016).

33. Markforged Material. Materials offered by Markforged. http://markforged.com/order-materials/ (Accessed June 2016).

34. EasyComposite. http://www.easycomposites.co.uk/ (Accessed December 2015).

35. G. T. Mark and A. S. Gozdz. Three Dimensional Printer with Composite Filament Fabrication. US Patent 9156205 B2 (2015).

36. R. Matsuzaki, M. Ueda, M. Namiki, T.-K. Jeong, H. Asahara, K. Horiguchi, T. Nakamura, A. Todoroki, and Y. Hirano, Three-dimensional printing of continuous-fiber composites by in-nozzle impregnation, *Scientific Reports* **6** (2016). 10.1038/srep23058.

37. X. Tian, T. Liu, C. Yang, Q. Wang, and D. Li, Interface and performance of 3D printed continuous carbon fiber reinforced PLA composites, *Composites Part A: Applied Science and Manufacturing* **88**, 198–205 (2016). 10.1016/j.compositesa.2016.05.032.

38. N. Li, Y. Li, and S. Liu, Rapid prototyping of continuous carbon fiber reinforced polylactic acid composites by 3D printing, *Journal of Materials Processing Technology* **238**, 218–225 (2016). 10.1016/j.jmatprotec.2016.07.025.

39. Y. Chen and C. C. L. Wang, Uniform offsetting of polygonal model based on layered depth-normal images, *Computer-Aided Design* **43**(1), 31–46 (2011). 10.1016/j.cad.2010.09.002.

40. Y. Song, R. A. Boekraad, L. Roussos, A. Kooijman, C. C. Wang, and J. M. Geraedts. 3D printed electronics: opportunities and challenges from case studies. In *Proceedings of ASME International Design Engineering Technical Conferences & Computers and Information in Engineering Conference* (2017).

41. R. Pitchumani, J. W. Gillespie Jr., and M. A. Lamontia, Design and optimization of a thermoplastic tow-placement process with in-situ consolidation, *Journal of Composite Materials* **31**(3), 244–275 (1997). 10.1177/002199839703100302.

42. M. Bartoň, H. Pottmann, and J. Wallner, Detection and reconstruction of freeform sweeps, *Computer Graphics Forum* **33**(2), 23–32 (2014). 10.1111/cgf.12287. Proc. Eurographics.

© 2020 World Scientific Publishing Company
https://doi.org/10.1142/9789811222825_0005

Chapter 5

Additive Manufacturing of Magnetic Particle-Polymer Composites

Lu Lu, Erina Baynojir Joyee, and Yayue Pan

5.1. Introduction

The synthesis of magnetic particle-polymer composites derives from the needs of novel intelligent materials whose physical properties can be varied by the presence of a magnetic field. This new generation of composite contains micro- or nano-sized magnetic functional fillers dispersing in a polymeric matrix. As a result, the magnetic particle-polymer composite demonstrates a combination of intense magnetic and elastic properties mainly depending on the functional filler and the host polymer. The dispersed fillers are magnetic field controllable materials that include magnetic metallic and alloy particles, ferrofluids, magneto-rheological fluids, and various magnetized particles. The compatibility of the host polymer with corresponding filler is indispensable. The magnetic particles can be incorporated into the polymer matrix either uniformly resulting in an isotropic composite, or in uniaxial direction and specific alignment leading to an anisotropic behavior.

The conventional manufacturing method, such as molding,[1-3] hinders the development of the magnetic particle-polymer composites with complicated shapes or anisotropic properties. Recently, great efforts have been made to develop multi-material and multi-functional AM technologies, specifically for polymer composite fabrication.[4-10] Composite materials such as magnetic particle-polymer composites have played a crucial role in development of AM technology due to their enhanced functional properties combined with the advantages of AM such as high design freedom and geometric flexibility. This combination of advanced composite materials and modern AM processes have enabled fabrication of structurally and functionally optimized magnetic-particle polymer composites for a wide range of practical applications.

In this chapter, an overview of recent advances in the development of magnetic particle-polymer composite with AM technologies is provided. This review is organized into five sections. The first section outlines the diversity of functional magnetic fillers and polymeric base materials. Additionally, two general techniques of feedstock preparation for additive manufacturing are described. The second section introduces reported categories of additive manufacturing processes for magnetic particle-polymer composites. For each category of processes, the benefits and drawbacks are summarized. This is followed by the discussion of material properties of isotropic composites with uniform and random particle distribution, and anisotropic composites characterized by directional filler alignment. Magnetic, mechanical, and thermal enhancements are discussed. Afterward, applications with respect to various functions are presented and elucidated. At the end, a short summary and future works are given.

5.2. Material

5.2.1. *Filler Material*

Magnetic particle-polymer composites contain different types of micron and nano sized magnetic filler particles dispersed in polymer materials. The magnetic filler materials not only reinforce the polymer composites but often can also enhance several material properties.[5] Magnetic field responsive particles and associated magnetic particle-polymer composites have garnered much attention from researchers due to their potential application in sensing and actuating, integrated memory devices, and diagnostic medicine.[5,11] In a proper environment, the particles act as a single magnetic domain and show superparamagnetic behavior in response to the applied magnetic field. The following subsections briefly discuss varied magnetic and non-magnetic filler materials used to develop various types of magnetic field responsive particle-polymer composites.

5.2.1.1. *Magnetic filler material*

The most common method to develop a magnetic particle-polymer composite is to use a magnetic filler material and disperse it into a polymer matrix. Magnetic fillers are usually iron based particles and alloys either directly used as powders or mixed with aqua or oil based solutions. A commonly used and one of the most extensively studied magnetic filler materials is magnetorheological (MR) fluid. MR fluids are usually prepared by dispersing micro-size magnetic particles such as carbonyl iron (CI) into a carrier fluid (i.e. silicone or hydrocarbon oil), often with stabilizer additives. Bastola et al.[12] developed a hybrid MR elastomer (MRE) using a 3D printing method and used MR fluid as the magnetic component, made with CI particles and silicone grease. Hua-jin et al. proposed an MR fluid by dispersing gelatin- carbonyl iron composite in silicone oil.[13] Xu-feng et al.[14] attempted to improve the magnetorheological properties of the MR fluid by dispersing $Fe_{76}Cr_2Mo_2Sn_2P_{10}B_2C_2Si_4$ amorphous alloy particles into silicon oil. Another common fluid based magnetic field responsive material is ferrofluid, an aqua or oil based solution of magnetite (Fe_3O_4) particles. Zhu et al. and Wang et al. demonstrated magnetic field assisted additive manufacturing techniques by mixing base material with water based ferrofluids.[15, 16] Varga et al.[17] also used ferrofluid magnetic solution (17.2% volume fraction of magnetite) to fabricate smart magnetic particle-polymer composites with controlled anisotropy. Xia et al.[18] fabricated micro-nanomachines using ferrofluids comprising of synthesized Fe_3O_4 nanoparticles, chemically modified by 3-(trimethoxysilyl)propyl methacrylate (MPS). One major drawback of the fluid based magnetic filler materials is the sedimentation problem. Either additive stabilizers are added or different combination of magnetic particles with varying volume fractions are mixed with the feedstock solution to tackle this issue and increase the stability.

Another solution is to use magnetic particle powders (Fig. 5.1 (a)) directly, instead of magnetic filler fluid, and mix them with the base polymer materials. Bollig et al.[19] for example, used magnetic iron filament as a magnetic material. Some researchers used highly pure iron powder, also called carbonyl iron powder (CI) (Fig. 5.1 (b)) with different polymer matrices to develop micro-nano particle-polymer composites.[3, 20-25] Boczkowska et al.[21] used two types of commercially available CI powder and mixed them with polyurethane polymer to test the rheological properties of the composites. In Ref. 20, CI particles were physically embedded in polymer matrix for adding magnetic function to shape memory polymers. Tsumori et al.[22] developed chainlike clusters of CI particles and polydimethylsiloxane (PDMS) polymer matrix. Varga et al.[3] used a similar combination of PDMS and CI powder and tested the effect of magnetic field on elastic modulus of the particle-polymer composites. Chung et al.[26] demonstrated magnetic polymer microstructures using magnetite nanoparticles mixed with PDMS while Zhang et al.[27] developed mesoporous bioactive glass (MBG)-polycaprolactone (PCL)-Fe_3O_4 polymer composite. Tian et al.[28] used stabilized magnetite with Oleic acid and Au et al.[29] used synthesized magnetite dispersed in SU-8 polymer matrix. Apart from the above mentioned magnetic materials, many studies have experimented with different types of magnetic particles such as different forms of ferrite powder,[30] NdFeB magnetic powder,[31, 32] stainless steel micro-powder,[33] Co-based nanoparticles,[34] and different Iron-Nickel based magnetic alloys.[35, 36]

5.2.1.2. *Non-magnetic filler material*

Generally, the nonmagnetic particles or platelets are chosen for customizable reinforcement architectures within the particle-polymer composite and desired material properties. Libanori et al. demonstrated such customized platelet reinforced composites, fabricated through a directed assembly of inorganic alumina platelets (Fig. 5.1 (e)) using combined mechanical and magnetic stimuli.[37] The alumina platelets were magnetized by the electrostatic adsorption of superparamagnetic iron oxide nanoparticles (SPIONs) in deionized water. Kokkinis et al.[38] also used similar SPION-coated alumina platelets and explored particle orientation control using additive manufacturing process. Remarkably, SPION-coated micro-platelets demonstrated ultrahigh magnetic response (UHMR), with very low magnetic field (~ 0.8 mT). A study conducted by Erb et al.[39] used a relatively lower magnetic field as well (1 to 10 mT) to align micrometer scale reinforcing fillers coated with a small concentration of superparamagnetic nanoparticles. They used alumina platelets and calcium sulfate hemihydrate rods as reinforcements.

Martin et al.[40] further studied the use of SPION coated reinforced alumina particles in magnetic printing and fabricated bioinspired magnetic particle-polymer composite architectures. A different application of SPION coated nanocomposite was demonstrated by Sommer et al.,[42] who reported a technique to create oriented porosity in different injectable materials with SPION coated calcium sulfate rods as magnetic filler material. Niebel et al.[43] proposed an alternative non-aqueous sol–gel approach which involved directly magnetizing and chemically modifying the surface of the inorganic micro-platelets and non-magnetic polymer fibers. A dense and homogenous coating of Fe_3O_4 on alumina allows magnetically responsive Pickering emulsions. In some other studies,[44, 45] magnetic particle-polymer composites were developed with poly(methyl methacrylate) (PMMA) spheres coated with Fe_2O_3 magnetic nanoparticles (Fig. 5.1 (c)).[44]

Fig. 5.1. SEM picture of the (a) iron oxide particles, (b) carbonyl iron (CI) particles, (c) CI-PMMA core-shell structured particles, (d) Ni-Mn-Ga powder, and (e) alumina platelets. Adapted with permission from Filipesei et al., Choi et al., Mostafaei et al., and Libanori et al.[5, 46, 47, 37]

In addition, extensive research has been done recently on utilizing Fe_2O_3 and Fe_3O_4 to synthesize carbon nanotubes (CNT) and multi-walled CNTs (MWCNTs) for developing magnetic carbon nanomaterials. CNTs have potential application in electrical devices, magnetic data storage and heterogenic catalysis[48-50] and are of much scientific and analytical interest. Zhou et al.[51] and Jia et al.[52] used Fe_3O_4 to synthesize MWCNT and developed Fe_3O_4-CNT nanocomposites. The Fe_3O_4 nanoparticles were grafted on to the multi walled CNTs using the in situ hydrothermal method in a Teflon-lined stainless-steel autoclave (50 mL capacity). The observed size of the Fe_3O_4 nanoparticles were 20 ~ 50 nm and the magnetic CNT composite exhibited high magnetic saturation value. Cao et al.,[53] Jiang et al.,[47] and Correa et al.[55] used precipitation technique, in-situ solvothermal technique and layer-by-layer polymer wrapping method to coat iron oxide on MWCNTs respectively and the resulting nanocomposite exhibited typical ferromagnetic characteristics at room temperature. Zhang et al.[54] presented an approach to graft Fe_3O_4 nanoparticles on MWCNTs modified by polyethyleneimine (PEI) and demonstrated enhanced electrical and magnetic properties of the nanocomposite. Some other combinations of CNT-magnetic nanocomposites are $CNT/P_{1-x}Q_xFe_2O_4$ composites (P, Q = Co, Zn, Ni, Mn) which have potential use as electronic nanodevices.[54] Some sample SEM images showing the influence of external magnetic field on iron oxide coated CNTs are depicted in Fig. 5.2 (adapted with permission from Ref. 55).

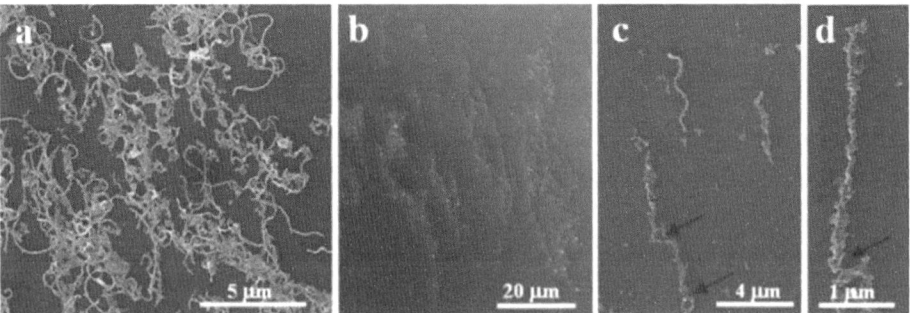

Fig. 5.2. SEM images of carbon nanotubes coated with magnetite, (a) in absence, and (b, c, d) presence, of an external magnetic field.

5.2.2. Base Material

Different additive manufacturing techniques for fabricating magnetic particle-polymer composites utilize different types of polymer resins as host materials, based on the requirement of printing process itself and the printing setup.

In extrusion based printing systems like the direct ink writing (DIW) system, the ideal base material should have appropriate viscosity values and a surface tension of at least 420 mN/m with low permeability and permittivity.[30] Silicon based polymer materials are one of the most common types of base resins used in extrusion based 3D printing research. Robles et al.[30] used SS-3045 silicone due to its low surface tension, high viscosity and good chemical compatibility with magnetic ferrite powders. Bastola et al.[12] used a commercially available UV-curable silicon sealant resin while Wang et al.[16] used silicon based polydimethylsiloxane (PDMS) with silicone oil. Kokkinis et al.[38] also proposed polyurethane acrylate (PUA) oligomers as base resin for 3D printing ink. Few commonly used resins in extrusion based printing are commercial grey photopolymer,[31] polycaprolactone (PCL),[27] and UV-sensitive monomer (Pentaerythritol tetraacrylate) mixed with photo-initiators.[56] Another popular extrusion based printing method is the fused deposition modelling (FDM) method which commonly uses commercially available polylactic acid (PLA)[19, 57] or acrylonitrile butadiene styrene (ABS) pellets as the base material.

Compared to FDM techniques, stereolithography apparatus (SLA) based printing techniques provide higher resolution and requires the polymers to be UV-sensitive and generally mixed with photo-initiators. Martin et al.[40] made the UV-sensitive resin by mixing some common polymer base resins such as Ebecryl 230 and isobornyl acrylate in a 1:3 ratio. Hassan et al.[20] used a monomer (Benzyl methacrylate) and mixed it with a crosslinker poly(ethylene glycol) dimethacrylate (PEGDA). Domingo-Roca et al.[58] also used monomers and created different polymer matrices by adding PEGDA and bisphenol-A ethoxylate dimethacrylate (BEMA) to improve 3D-printing resolution. In Ref. 28, two-photon photopolymerization printing was conducted using butyl methacrylate as monomer and PO3-TMPTA (56 wt. %) as crosslinker. Chung et al.[26] on the other hand used an optofluidic maskless lithography (OFML) system with PDMS as the base resin. Some other commercially available photopolymers such as 3DM-ABS from 3D Materials, G + from MakerJuice Inc, have also been tested as polymer base material.[62]

In selective laser melting method, silicon powders have been investigated as the polymer base material.[59-61] Urethane has been tested as the polymer base material in binder jet printing.[32]

5.3. Additive Manufacturing Processes for Magnetic Particle-Polymer Composites Fabrication

5.3.1. *Fused Deposition Modeling*

As the most widely used AM process, extrusion based printing technology has been applied to fabricate magnetic polymer composites. Fused Deposition Modeling (FDM) method utilizes a nozzle to deposit molten filament along the tool path to build 3D structures. As the printing feedstock, the solid filament is prepared by mixing magnetic fillers with polymer before starting the printing process. The uniform mixing results in the homogeneous composite filament for use as the feedstock of FDM machine.

Researchers have successfully combined various magnetic functional fillers with the polymer at varied loading fraction[19,33,36,62-68] to prepare the composite filament feedstock. For instance, Nickel-Iron ($Ni_{81}Fe_{19}$) alloy particles were reported to be mixed into polyethylene (PE) matrix at different volume fractions up to 40 vol. % (Fig. 5.3(a)).[36] The preparation of this homogeneous magnetic composite filament was realized by using a propeller mixer or a twin screw extruder. FDM technique was utilized to shape PE/NiFe composites into a parallelepiped magnetic material block. Similarly, Khatri et al.[33] prepared four different filaments with acrylonitrile butadiene styrene (ABS) as the host polymer and stainless steel micro powder at 10, 20, 30, and 40 vol. % as the functional filler (Fig. 1 (b)). Test samples were printed using a commercial FDM desktop printer (MakerBot Replicator 2X, MakerBot, New York, NY, USA).

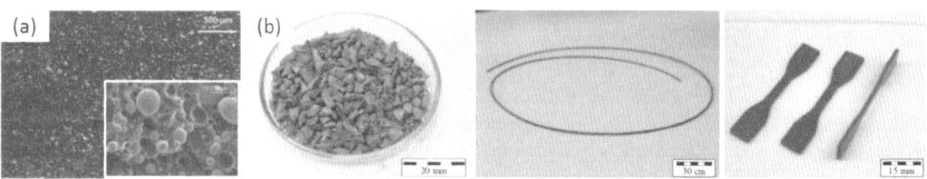

Fig. 5.3. (a) SEM micrographs of PE/NiFe composites (30 vol. % NiFe) at zoom out view and 25× zoom in view,[36] (b) the ABS–Steel composite: the feedstock, the extruded composite filament, and printed dog bone samples.[33]

5.3.2. *Direct Ink Writing*

Numerous researches have been conducted on the development of Direct Ink Writing (DIW) technique in magnetic particle-polymer composite production. DIW process uses liquid ink as the feedstock. After depositing through the disperser, liquid material will be solidified to construct the 3D parts, usually by UV light exposure or heat treatment.

In general, the ink is a mixture of magnetic particles and polymer for composite fabrication. For instance, Liu et al.[56] prepared UV-curable ferrite paste consisted of NiZn ferrite powder as the magnetic filler. A commercial multi-extruder direct-extrusion 3D printer (System 30M, Hyrel 3-D) was customized by affixing a UV-LED module to one extruder as shown in Fig. 5.4(a). After printing each layer, the UV-LED module was used to solidify the print to avoid slumping problem. Shen et al.[31] reported a novel "UV-assisted Direct Write (UADW)" fabrication technique with magnetic inks composed of NdFeB particles and UV-curable polymer. In this study, instead of mounting the UV-LED to one extruder, the UV light source was placed under the transparent substrate. For each layer, the micro-dispensing system deposited the ink onto this transparent substrate followed by the UV light exposure.

Without premixing process, magnetic particles and polymer can be deposited separately. Bastola et al.[12] employed a multi-material 3D printer (BioFactory, RegenHU, Switzerland) to fabricate the hybrid fluid-elastomeric MR elastomers (H-MRE). Because of the high viscosity of the MR fluid and the elastomeric resin, the extrusion-based time-pressure print heads were chosen for dispensing. A plastic container used as the printing substrate defined the size and shape of the sample. As illustrated in Fig. 5.4(b), the printing process started with depositing the MR fluid to form different patterns, and then dispensing polymer resin to cover the formed pattern. UV curing unit solidified each layer once the resin being spread out to achieve the desired layer thickness. By repeating these steps, a multiple-layer sample can be produced. Dot and line patterns of MR fluids were printed with accurately controlled printing parameters including extrusion pressure, feed rate, printing nozzle diameter, and initial height.

To control the orientation of the magnetic particles, the external magnetic field are usually integrated into the DIW printer. Kokkinis et al.[38] developed a Multi-material Magnetically assisted 3D printing (MM-3D printing) platform to enable 5D programmability: local control of composition, particle orientation, and 3D shaping capability. The ink was prepared by dispersing magnetic responsive anisotropic particles in a photocurable liquid resin. A commercial DIW 3D printer (3D Discovery, regenHV Ltd) was equipped with a rotating permanent magnet for particle alignment and a spot curing lamp for material solidification as demonstrated by the schematic drawing in Fig. 5.4(c). The object was programmed to move between the printing, alignment and curing stations during the printing process. Dual-component dispenser (Dreeflow EcoPuo450, ViscoTec, Germany) containing inks with different particle loading fractions in the individual syringes allowed localized filler concentration within one printed part. The MM-3D printing technique provides a pathway for manufacturing complex functional heterogeneous composite architectures.

Apart from aligning the magnetic fillers after the ink deposition, the reorientation can be achieved during the dispensing. By means of applying a magnetic field to the dispensing nozzle via a permanent magnet or an electromagnetic coil, the magnetized NdFeB particles were reoriented along the field direction as illustrated in Fig. 5.4(d).[69] Consequently, the deposited ink was imparted with patterned magnetic polarity. Additionally, changing polarity of the applied field or depositing direction can program ferromagnetic domains in printed composites. In this way, complex geometric 3D structures with tunable magnetic properties can be produced.

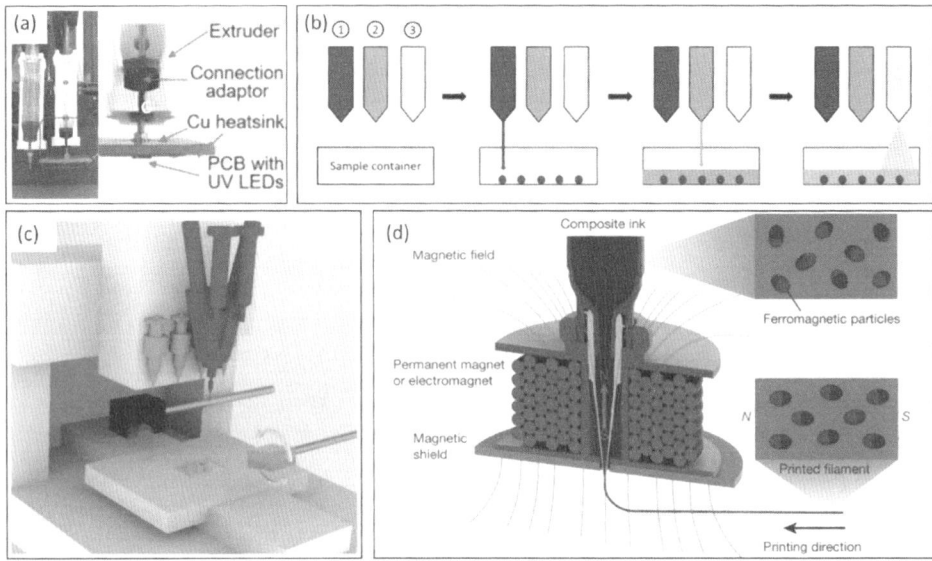

Fig. 5.4. (a) Extruder equipped with a UV LED module,[56] (b) schematic diagram of the H-MRE fabrication process,[12] (c) schematics of the MM-3D printing setup,[38] (d) schematics of the printing process with magnetic particles reorientation.[69]

5.3.3. *Polymerization-based Process*

With the multiscale printing capability and relatively fast printing speed, polymerization related process is a promising method to handle mixtures of particles and polymer. Standard stereolithography (SLA) printing processes involve curing a photosensitive polymer resin layer with a certain thickness and shape predefined by the 2D slice of the digital model. Commercial SLA 3D printers were investigated to produce homogenous composites with magnetic particles uniformly distributed in photosensitive polymer resin. For instance, Domingo-Roca et al.[58] produced magnetic composites by a UV-based SLA 3D printer (ASIGA PicoPlus27 machine, ASIGA, Anaheim Hills, California, USA) with a mixture of Fe_3O_4 nanoparticles and UV-light photocurable polymer resin (Fig. 5.5(a)). The printed composites were characterized to be soft permanent magnets with randomly distributed magnetic domains.

The conventional method of preparing the feedstock for SLA is mixing the particles and polymer due to the facile processability. However, the homogeneous particle distribution confines the development of the anisotropic property. To address this challenge, magnetic field assistance was introduced to precisely control the orientation of magnetic fillers during the printing process. Martin et al.[40] created a SLA-based 3D magnetic printing process to fabricate dense particle-polymer composites with precisely coordinated particle orientation for each voxel of printed material as shown in Fig. 5.5(b). To capacitate the magnetically guided particle reorientation, the traditionally nonmagnetic ceramic particles were treated by coating with iron oxide nanoparticles. Each layer was divided into several groups of voxels according to the designed reinforcement orientation. A rotating magnetic field was applied to a group of active voxels to align the long axes of the reinforcing microparticles. The orientation of the reinforcement was fixed in the polymer matrix after polymerization by a digital light projector (DLP). These two steps were repeated for each unique reinforcement orientation that is to be incorporated into the composite layer, leading to a complex macroscopic composite architecture.

Furthermore, with the assistance of a small-scaled external magnetic field, it is possible to control local particle distribution and alignment direction in liquid photopolymer. In the Magnetic-Field-Assisted Projection SLA process, a certain amount of magneto-rheological (MR) fluid was deposited through a programmable micro-deposition nozzle into the resin vat (Fig. 5.5(c)).[62,64] A small permanent magnet or electromagnet can be placed under the resin vat to direct the magnetic particle to form a designed pattern with unidirectional orientation at the predefined desired location. After projection of a digital mask image, the particle distribution pattern was locked in the cured photopolymer. Hence, material intelligence can be realized by localized particle distribution in the polymer matrix.

The two-photon photopolymerization (TPP) process can produce nanocomposite with a high resolution, which is specifically desired for micro-nanomachines fabrication.[18,28] Xia et al.[18] uniformly dispersed Fe_3O_4 nanoparticles into photoresist to prepare homogeneous photopolymerizable resin for TPP process. Figure 5.5(d) illustrates the fabrication process for the remotely controllable micromachine model by TPP with this stable ferrofluids resin. By incorporating Fe_3O_4 nanoparticles in the composite structures, the printed functional machines can respond to external magnetic fields.

5.3.4. *Ink Jetting*

As an example for Ink Jetting process, the drop-on-demand inkjet print head equipped with an electromagnet was used to fabricate novel magnetic composite materials involving aligned magnetic particles as shown in Fig. 5.6(a).[34] The printing ink contains ferromagnetic Co-based nanoparticles with a volume loading

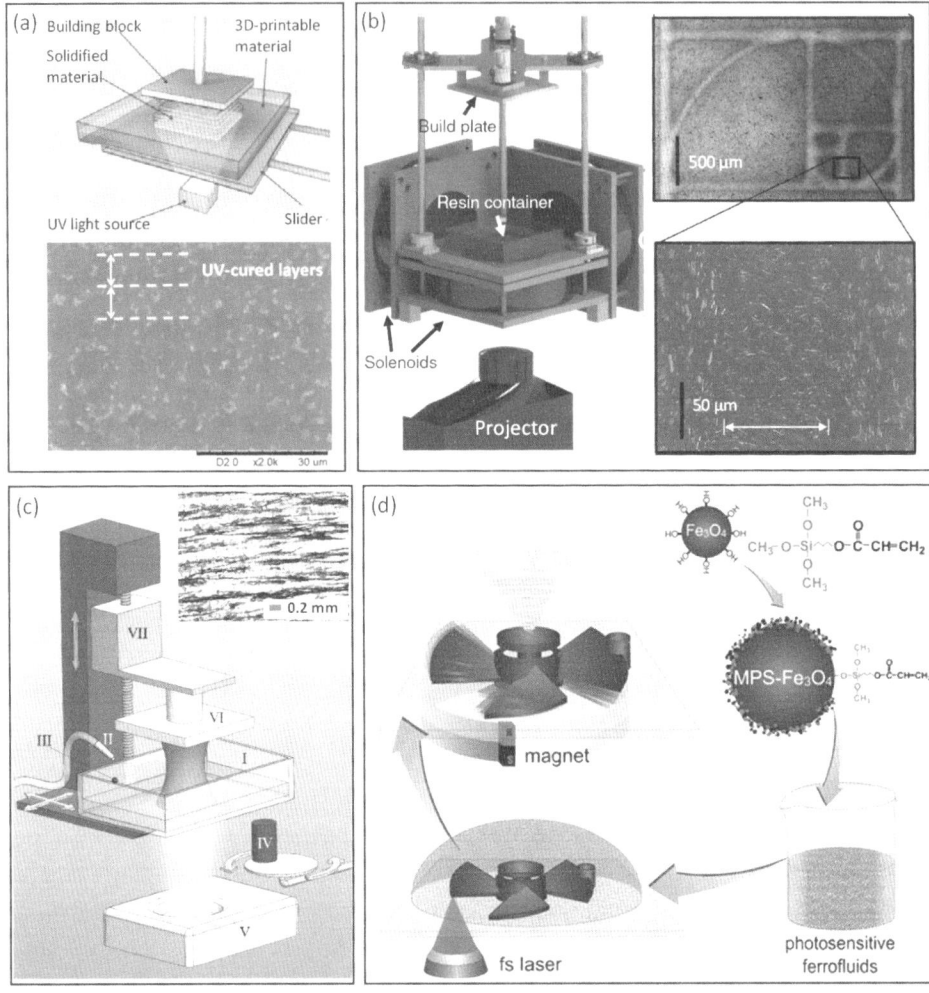

Fig. 5.5. (a) Schematic drawing of the commercial SLA printer (top) and the SEM images of the 3D printed magnetic composites with randomly distributed magnetic particles (bottom),[58] (b) the 3D magnetic-printer setup,[40] (c) a schematic diagram of the Magnetic-Field-Assisted Projection SLA system and the micro image of the printed composited with unidirectional particle chains (top),[63] (d) fabrication procedure of remotely controllable micromachine by TPP.[18]

fraction at 40%. Linear alignment and radial alignment were achieved by applying the external magnetic field during the ink jetting process (Fig. 5.6(b)). The printed single layer samples were air-dried. By investigating the printed unaligned and aligned sample, it was found that the particle alignment leads to high frequency and low hysteresis losses in the alignment-induced hard axis direction.

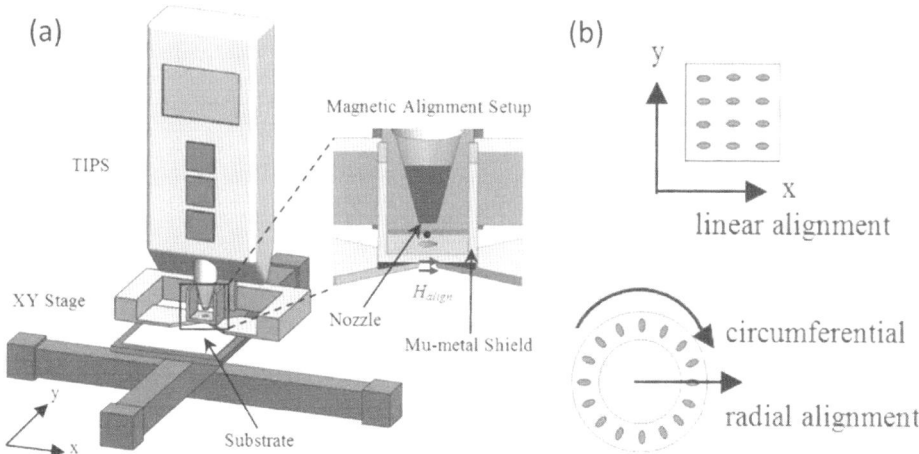

Fig. 5.6. (a) Schematic of the drop-on-demand Ink Jetting set up with an electromagnet mounted on the print head, (b) schematics of linear alignment (top) and radial alignment (bottom).

5.3.5. *Binder Jetting*

In general, Binder Jetting process has two steps: spreading the powder and depositing the binder, as illustrated in Fig. 5.7(a). Paranthaman et al.[32] provided a new pathway for fabricating near-net-shape isotropic bonded magnets by utilizing an inkjet print-head to deposit a polymer binding agent to a bed of NdFeB powder. The selection of deposition region is according to the shape of cross-section. Once a layer of powder was bounded together, a heat lamp was used to solidify the binding agent. Then, a new layer of loosen powder was spread on the top for the next layer printing. After all the layers were printed, the "green" part will be post-sintered in an oven. The resultant composite contains evenly spread isotropic NdFeB magnet powder particles (Fig. 5.7(b)).

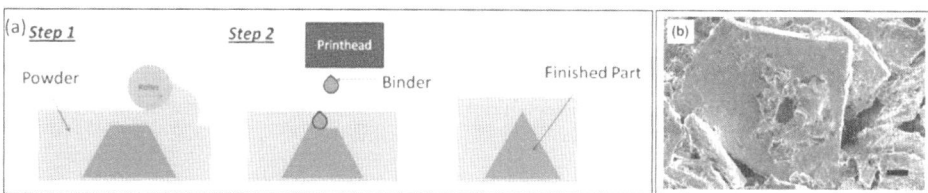

Fig. 5.7. (a) Schematic of the Binder Jetting process,[70] (b) SEM image of the printed magnet part (scale bar: 50μ).[32]

5.3.6. *Slip Casting*

In Slip Casting process, the first step is to deposit particle-polymer suspension into a dry porous mold with a pre-defined geometry.[71] Through the pores of the mold wall, capillary forces or vacuum removes liquid from the suspension. Thereby, a layer of hammed particles is built against the mold wall. With the presence of an external magnetic field, the orientation of dispersed magnetic-responsive particles can be controlled during the printing process as demonstrated in Fig. 5.8. The printed composite exhibited heterogeneous properties with localized texture and composition.

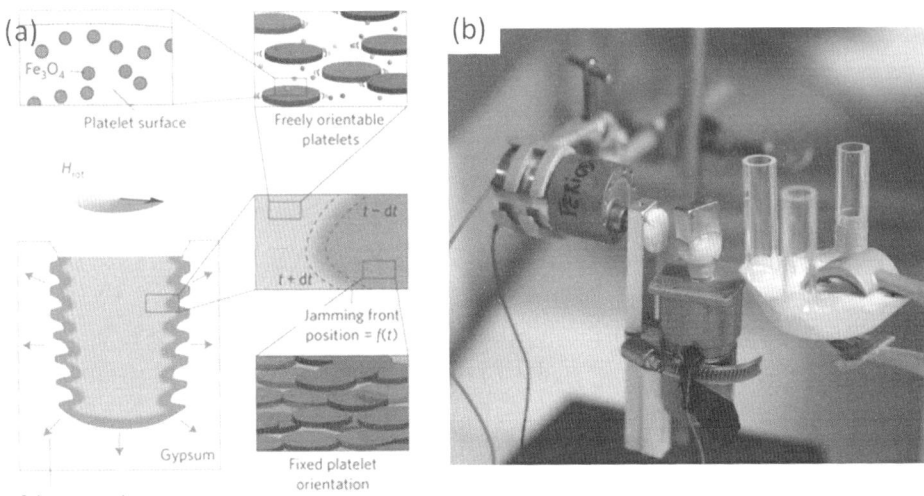

Fig. 5.8. (a) Schematics of the slip casting process under an external rotating magnetic field, (b) picture of the testbed.[71]

5.3.7. *Process Benefits and Drawbacks*

The genetic benefits and limitations from these additive manufacturing technologies remain in the production of the magnetic polymer composite. The popularity of FDM is attributed to the low cost and simple setup. However, there will be shape distortion caused by the nozzle tip shape reflecting in the rounding of all corners and edges. High part accuracy and fine surface finish are the two main advantages of photopolymerization technology. This flexible process supports different setup configurations in a wide range of printing scales from 100 nm to 1.5 m. However, the material selection is restricted to photopolymers, and the increased filler loading hindered the penetration of UV into the photopolymer.[31] Direct ink writing process works with various liquid phase materials but has limited ink viscosity ranges. Heavy filler loading often causes nozzle clogging issue. Material jetting has the advantage of high printing speed and multiple material production. It suffers from the similar limitation as the DIW method, as the superabundant filler content usually causes jetting orifice clogging issues. Binder jetting enables composite materials that are difficult to obtain by a direct method on account of the combination or the high loading fraction. However, the printed part has poor accuracy and surface finish. Slip casting process allows for various constituents with a wide range in type and size, but the usage of mole confines the geometric flexibility.

5.4. Properties of Magnetic Particle-Polymer Composites

5.4.1. *Magnetic Property*

The unique magnetic properties stemming from the magnetic fillers have attracted research interests in the past few years. Since the polymer is not magnetic responsive, the inclusion of various magnetic particles as the functional filler enables the magnetic related functionalities of the polymer composites.

The concentration of magnetic functional fillers plays an essential role in determining the composite magnetic property. Higher filler ratios resulted in an increased magnetic response.[31,33,36] Functional magnetic hysteresis characterization of the developed steel-ABS composite showed a remarkable increase in ferromagnetism at higher filler ratios.[33] A microwave characterization of NiFe–PE composite samples was carried out using a coaxial line method,[36] and found that the NiFe particle volume fraction determined the composite's electromagnetic properties. As a result, the composite's microwave properties can be predicted as a function of the initial component concentrations. As expected, the increased load rate caused substantial enhancement in the composite permittivity and permeabilities. The printed NdFeB-polymer composites possessed the highest intrinsic coercivity with one of the highest magnetic remanence values, indicating the preservation of the raw NdFeB powder's magnetic property.[31] The relative magnetic remanence was growing in a linear trend with the NdFeB volume fraction as plotted in Fig. 5.9(a).

Part design, printed geometry, and manufacturing process settings can also affect the magnetic performance of printed parts. A study of structural printing parameters, specifically for the FDM process, was conducted on printed iron-PLA composite parts.[19,57] Effects of printing settings including fill factor and fill pattern on composites magnetic property were investigated. According to the vibrating sample magnetometer (VSM) characterization results, all of the printed parts exhibited soft magnetically hysteresis loops with low coercivity and reached saturation below 1000 mT. It was found that a high fill factor along with a high iron particle concentration gave rise to a higher saturation moment (Fig. 5.9(b)). Nevertheless, the fill pattern did not affect the magnetic property significantly.

Anisotropic magnetic properties can be achieved by controlling the polarity,[69] orientation,[62] and distribution[63,64] of the embedded magnetic fillers. As a demonstration, a simple straight composite wire was printed with an opposite magnetization on each side as shown in Fig. 5.9(c).[69] It was realized by switching the applied field direction on the nozzle during the printing. With the presence of a uniform magnetic field, the straight composite wire deformed into an "m" shape. This transformation was reversible once the magnetic field was removed. The orientation of the magnetic particle alignment was proved to affect the magnetic field responsive property of printed composite regarding trigger condition under magnetic actuation.[64] When parallel to the magnetic field induction line, the particle chain exerted larger magnetic force than in perpendicular orientation, resulting in a longer trigger distance as shown in Fig. 5.9(d).

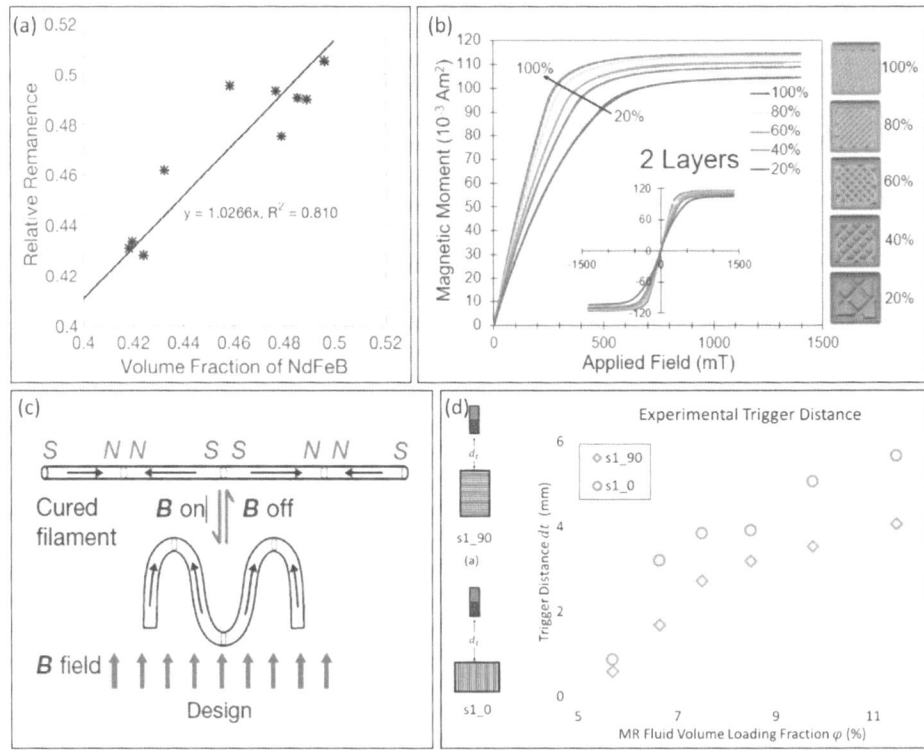

Fig. 5.9. (a) The relative magnetic remanence of various NdFeB volume fraction,[31] (b) effect of fill factor on magnetic moment measured under varied applied field for cubes with two layers,[57] (c) 3D-printed soft composite wire,[69] (d) experimental trigger distance measured with different chain orientations.[63]

5.4.2. *Mechanical Property*

The existence of metallic fillers inside the polymer matrix changes the mechanical properties of the printed composites compared to the pure polymer in terms of tensile strength, Young's modulus, and stiffness, etc. Intensive studies have been conducted to investigate the impact of filler loading on the resultant composite mechanical property. To name a few, Khatri et al. did tensile and flexural tests of magnetic ABS composite specimens printed by FDM, as shown in Fig. 5.10(a).[33] Compared to the pure ABS polymer, the ultimate tensile strength of the printed stainless steel-ABS composites dropped proportionally with increasing filler content, due to the exceedingly brittle characteristic of the stainless steel filler. Accordingly, Young's modulus and the flexural strength decreased dramatically with higher filler ratios. As a result of the stronger resistance created by the increased metal particle content, a considerable enhancement was observed in the secant modulus of elasticity. The elastomeric matrix and the presence of highly viscous MR fluid lead to a greater stiffness and a higher damping capacity performed by the printed H-MRE comparing to conventional MREs.[12] Furthermore, the dynamic stiffness and damping capacity of the printed H-MREs were tunable by applying a moderate magnetic field. It opens the possibility for fabricating a tunable spring-damper element with the printed H-MRE materials.

Additionally, post-processing of green parts can further improve the mechanical properties of the printed composites.[31,32] For example, the printed dumbbell-shaped composites with 90 wt% NdFeB loading exhibited a tensile strength increased by approximately four times and Young's modulus increased by around nine times after post-curing and heat treatment, as shown in Fig. 5.10(b).[31] Meanwhile, the elongation at failure reduced from 1.7% to 0.6%, resulting from the stronger but more brittle properties of the printed material.

In addition to the particle loading fraction, by adjusting the particle orientation, alignment, and local distribution, the mechanical property of the heterogeneous magnetic-particle polymer composites can be precisely programmed. For example, Martin et al.[40] fabricated a block with unidirectional reinforcement orientation by a SLA-based 3D magnetic printing process as demonstrated in Fig. 5.10(c). Based on the tensile test results, axes along the reinforcement orientation expressed improved stiffness by 29%, hardness by 23%, and rupture strain by 100%, compared to the perpendicular axes. Similarly, composite with different particle alignment was produced by MM-3D printing system and was tested with a load applied parallel and vertical to the orientated magnetic platelets. In comparison with perpendicular alignment, elastic modulus, strength, and strain-at-rupture were increased by 150%, 150%, and 110% with in-plane particle alignment.

A similar result was shown by Joyee et al.,[41] with particle-polymer composites printed with M-PSL process. The authors did both analytical modeling and experimental characterization, showing that a higher Young's modulus can be achieved when particle chains are aligned parallel to the external magnetic force direction, while a quite smaller modulus, almost same as the modulus of pure polymer, is observed when particle chains are perpendicular to the external force direction (Fig. 5.9 (d)).

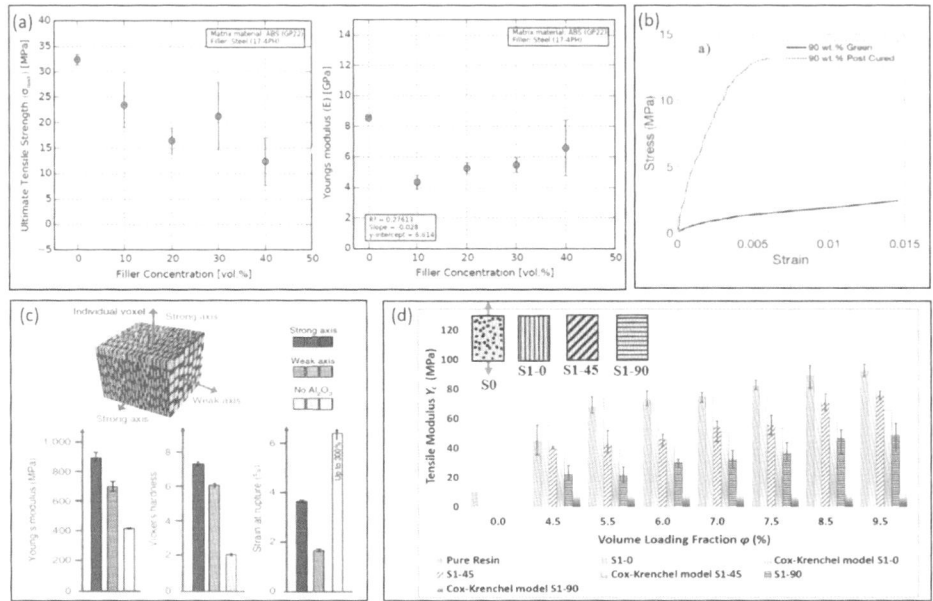

Fig. 5.10. (a) Ultimate tensile strength and Young's modulus for composites with different filler concentration,[33] (b) stress-strain curves of the printed sample before and after post-curing,[31] (c) schematic of a block with unidirectional reinforcement orientation (left); tensile tests results for strong and weak axes (right),[40] (d) comparison of analytical and experimental Young's modulus of iron oxide particle-polymer composites with different particle chain alignments.[41]

5.4.3. *Thermal Property*

Additionally, the thermal property of the magnetic polymer composite was studied. Copper particles were mixed separately with the ABS thermoplastic to develop new metal-polymer composite filament for FDM processes.[68] To investigate the impact of the filler concentration on the composite thermal properties, metal powder loading fractions ranging from 10 to 50 wt. % were tested. As shown in Fig. 5.11, the thermal expansion coefficient of the specimens declined with higher copper contents. Specifically, the thermal expansion coefficient for composites with 50 wt. % of copper particles reduced by 29.5% compared to the pure ABS. The reduction in the thermal expansion coefficient provides a potential solution to the distortion problem of the FDM printed product. In contrast to the decreased thermal expansion coefficient, a rise in thermal conductivity was observed with high loading of the metal powder.

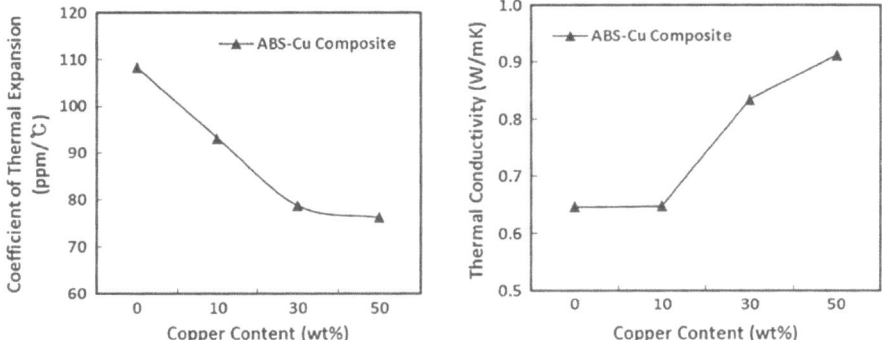

Fig. 5.11. Thermal expansion coefficient (left) and thermal conductivity (right) of the composites with different copper content.[67]

5.5. Applications

As is evidenced by the abounded examples, AM has been demonstrated as an effective pathway to fabricate the magnetic particle-polymer composites. Geometric flexibility controlled anisotropic properties, and advanced functionality of the printed composites open up opportunities for different applications. Some examples are presented in this section.

5.5.1. *Polymer-based Permanent Magnet*

AM shows the enormous potential of fabricating polymer-bonded magnets with arbitrary shapes at a reduced cost.[32,58] The 3D-printable permanent polymer-based magnet shown in Fig. 5.12(a) was developed using the DLP-SLA 3D-printing approach.[58] After applying a poling process, the magnetic domain within the printed magnet was reoriented to follow a well-defined polarity of a NdFeB permanent magnet as characterized in Fig. 5.12(b). By adjusting the type, size and concentration of the functional filler, magnetic properties of the printed magnet can be finely tuned expanding the applicability range. The high degree of property tunability allows the implementation in desired fields including medical science, biotechnology, and biomedicine. Through additive manufacturing, systems in micro- and nano-scale can be developed and controlled with external magnetic field.

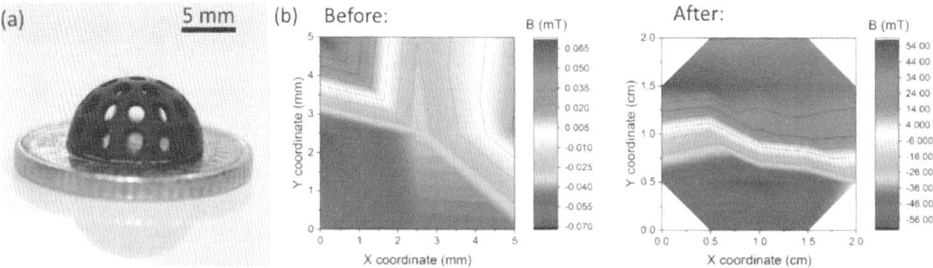

Fig. 5.12. (a) 3D-printed hollow half-sphere permanent magnet, (b) the magnetic domain of the printed magnet before and after the poling process.[58]

5.5.2. *Magnetic Transformer*

Bollig et al.[19, 57] demonstrated the possibility of 3D printing a magnetic transformer core by using FDM method. Filament consisting of polylactic acid (PLA) and iron (40 wt. %) was used to fabricate toroidal geometries with ferromagnetic properties. With the well-planned combination of printing settings, the most desirable transformers with entirely customizable designs can be successfully printed, as illustrated in Fig. 5.13 (a). In SEM images, iron particles appearing brighter color were randomly distributed in the composite (Fig. 5.13(b)). Circuit hysteresis loops revealed that magnetic moment contained within the transformer core was proportional to the fill factor (Fig. 5.13(c)). The development of high-resolution customized transformers allows for many applications in information storage, computer science, and pressure sensors with lower production costs.

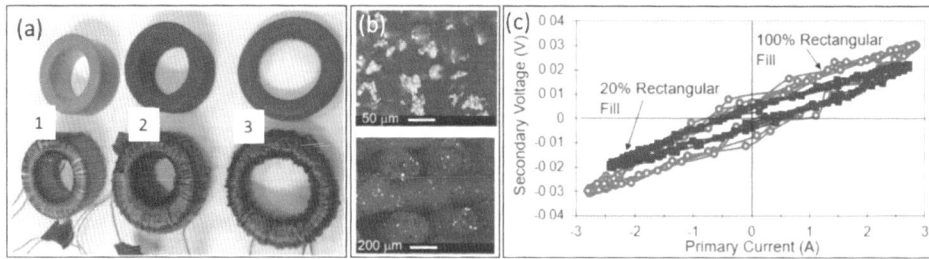

Fig. 5.13. (a) Picture of a commercial core (1), a 3D printed transformer core (2) and a toroidal shaped transformer core (3), (b) SEM images of 45° cross-section (top) and cross-section (bottom) of a printed transformer core, (c) circuit hysteresis loops for core printed with different fill factors.[19]

5.5.3. *Shape Memory Polymer*

Shape memory polymers (SMPs) are smart materials capable of switching between different shapes in response to external stimuli, such as temperature, magnetic and electric field. A SLA 3D printing technique was employed to embed the carbonyl iron particles (CIPs) in the polymer matrix for the fabrication of SMPs that exhibit shape recovery behavior triggered by temperature.[20] To illustrate the shape recovering behaviors of the printed SMPs, a Y-shaped sample was first stretched in an oven at a temperature higher than the glass transition temperature and then was put in a freezer to settle the temporary shape. At room temperature, the sample retained the temporary shape even after the release of the tension. The shape recovery happened when the sample was reheated above the glass transition temperature. This shape memory recovery process was shown in Fig.5.14. With the addition of embedded CIPs, the glass transition temperature and rubbery modulus decrease in the graded structure. Therefore, by adjusting the content of functional filler, the glass transition temperature can be controlled. With the shape recoverability, the SMPs can be applied in numerous fields such as wearable devices, artificial muscles, biomedical devices, and soft robotics.

Fig. 5.14. Shape recovery behaviors of the printed Y-shaped composite sample: (a) original shape, (b) deformed shape, (c) recovered shape.[20]

5.5.4. *Remote Control Machine*

Magnetic force drive technique for remote control has the advantages of simplicity, safe operation, and non-contact control. AM techniques address the challenge of appending magnetic components to machines during the fabrication process. Lu et al. developed a magnetic field-assisted projection stereolithography (M-PSL) process,[62] which successfully fabricated magnetic field-responsive smart polymer composite machines. By embedding magnetic particles in selected regions of the parts, the printed composite machine can respond to external magnetic stimuli and perform desired functions. Three examples, a two-wheel roller following a moving magnet, an impeller spinning with a rotating magnetic field, and a flexible film deforming with the presence of magnetic field, were printed for demonstration of potential applications, such as sensing and actuation in soft robotics, biomedical devices, and autonomous systems (Fig. 5.15). Additionally, Joyee et al.[64,65] developed a fully 3D printed bio-inspired multi-material monolithic soft robot using the M-PSL process. The robot can be remotely controlled using an external magnetic field and demonstrated superior locomotion capabilities (Fig. 5.15 (d)).

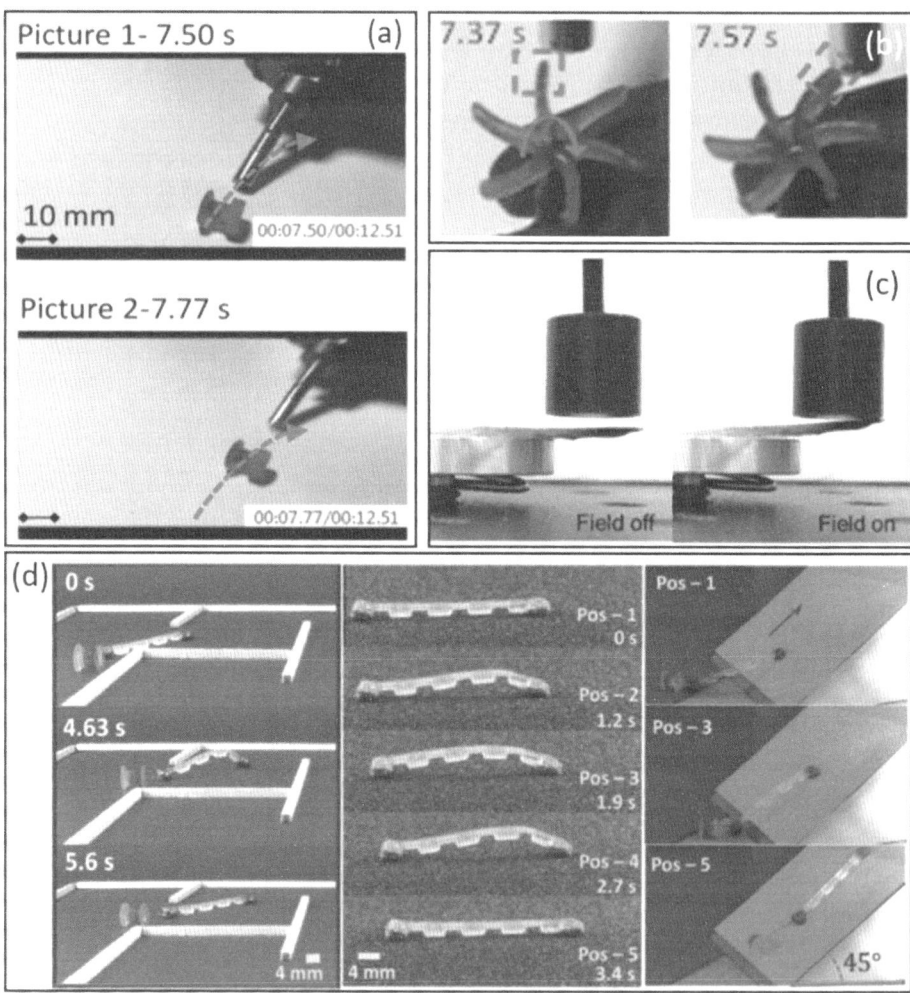

Fig. 5.15. Magnetic field-responsive smart polymer composite machines: (a) a two-wheel roller, (b) an impeller, (c) a flexible film,[62] (d) an untethered soft robot.[64]

In addition, micro-sized turbine and spring printed uniform ferrofluid resin were developed using the two-photon technology for remote control under external magnetic field.[18] As illustrated in Fig. 5.16, A piece of magnet was used to trigger the micro-turbine spinning and sway movement of the micro-spring in acetone. Magnetic force remote control was successfully demonstrated with the smart micromachines performing desired tasks. Easy fabrication and precise remote control of intelligent micromachines were enabled by the additive manufacturing technology and the photopolymerizable ferrofluid. The multiform manipulation allows for a broad range of application, such as blood vessels for health care, and micro-mixer in microfluidic field.

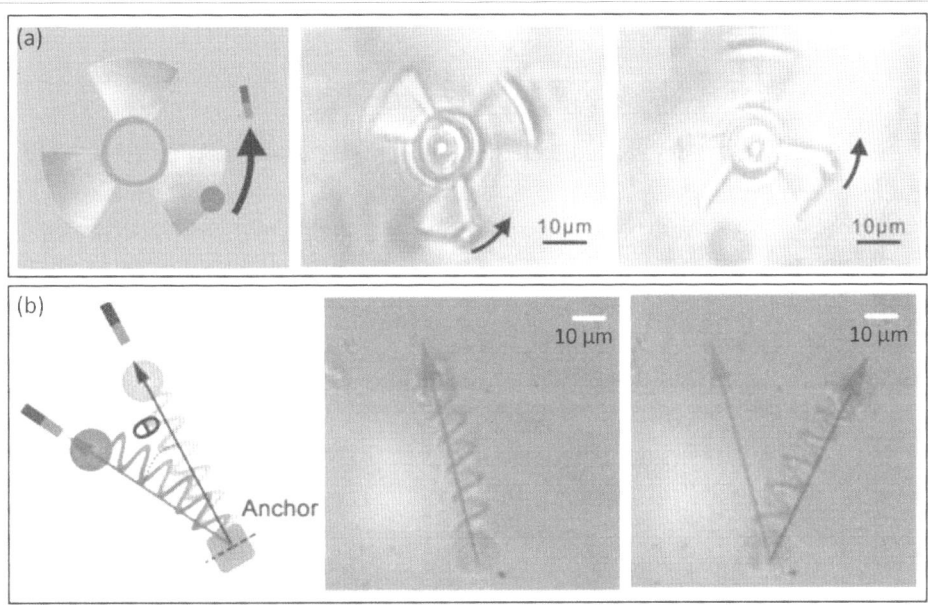

Fig. 5.16. Remote control of the (a) micro-turbine, and (b) micro-spring in acetone.[18]

5.5.5. *Magnetic Multi-material Scaffold*

As shown in Fig. 5.17, Zhang et al.[27] fabricated 3D magnetic composite scaffolds containing Fe_3O_4 nanoparticles, mesoporous bioactive glass (MBG), and polycaprolactone (PCL) by the 4th 3D Bioplotter™ (EnvisionTEC GmbH, Germany). The printed scaffolds exhibited excellent compressive strength, apatite-forming bioactivity, sustained anticancer drug delivery, magnetic heating properties, and significantly stimulated proliferation. There is considerable potential for using the printed scaffolds in the treatment and regeneration of bone defects caused by bone tumors, local anticancer drug delivery, and magnetic hyperthermia.

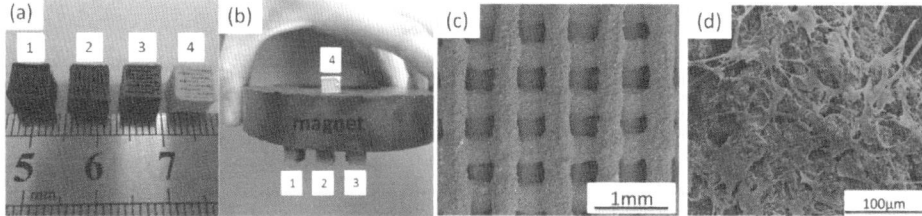

Fig. 5.17. (a) Picture of the printed scaffold samples containing 9.3 (Sample 1), 6.2 (Sample 2), 3.1 (Sample 3) and 0 (Sample 4) wt% Fe3O4, (b) responses of the printed samples to a magnet, (c) SEM image of Sample 1, (d) SEM image of the attachment of h-BMSCs cells on the printed scaffold samples after culturing for seven days.[27]

A hybrid manufacturing process combining FDM and SLA was also reported to fabricate complex magnetic multi-material scaffolds suitable for the regeneration of complex tissues in the form of coaxial and bilayer structures.[66, 67] The region of the scaffold committed to bone regeneration was manufactured through FDM by reinforcing iron oxide particles (20% wt) into Poly(e-caprolactone) (Fig. 5.18(a)), whereas the region of the scaffold devoted to cartilage regeneration was manufactured through SLA with a mixture containing poly(ethylene glycol) diacrylate (PEGDA) and magnetic nanoparticles (MNPs) at 6 wt% (Fig. 5.18(b)). The compressive mechanical behavior of these scaffolds was in a wide range covering those spanning hard and soft tissues. The finely structured multi-material scaffold with tailored mechanical and degradation property performs the load-bearing function of the bone. Furthermore, the MNPs embedded region of the scaffold is able to switch an external magnetic field on and off, providing a unique mechanism to trigger the sequential biological events that occur during tissue regeneration on demand.

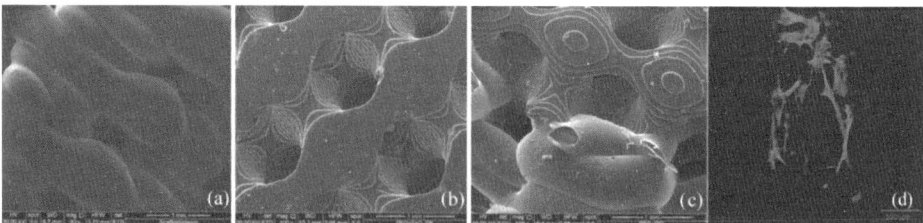

Fig. 5.18. SEM images of the structures printed by (a) FDM, and (b) SLA, and (c) the interface between those two regions, (d) microscopy images of cell adhesion and spreading on the printed composite scaffold.[23]

5.6. Summary

Overall, recent advances in the development of AM technologies for fabrication of magnetic particle-polymer composites are introduced and discussed in this chapter.

Relevant host material and functional fillers, along with the material preparation methods were presented. Different combination of filler and base material can cause a wide array of enhanced material properties. Examples of studies were covered across seven categories of AM processes. Advantages and drawbacks of each AM category are summarized and compared. The printed composites display impressive enhancements in magnetic, mechanical and thermal properties. Applications including magnetic transformer, shape changing polymer, remote control machine, and scaffold have been demonstrated.

Despite remarkable progress in the fabrication of magnetic polymer composites and structures as well as applications, there are still many challenges that need to be addressed. To name a few, high filler content may cause manufacturing failures, such as nozzle clogging, insufficient bonding, etc. Moreover, the materials used as functional fillers are usually restricted to metal and alloy. Considering the exciting properties can be added into the functional composite by involving advanced materials such as CNT, magnetizing treatment methods for these non-magnetic fillers is needed.

Future direction of the magnetic particle polymer composites fabricated with AM technology will be focused on both industrial and bio-applications. For industrial applications, a lot of research is going on fabricating electric nano-devices, integrated sensors, and 3D printed energy storage devices. The diversity of magnetic particle and chain distribution, their enhanced mechanical, electrical, magnetic and thermal properties with the added advantage of different 3D printing techniques, make it attractive for the future fabrication of different integrated devices. In addition, the AM technology will also see a wide range of bio-applications in the near future. Additive manufacturing of magnetic particle-polymer composites enable monolithic fabrication and can produce locally engineered material stiffness and graded functionalities. In the coming years, fabrication challenges presented in magnetic particle-polymer composite will be addressed by AM technologies, specifically focusing on functionality, high spatial resolution, geometric flexibility, multi-material production, and multi-scale production. More practical applications will be expanded to myriad fields in the near future beyond doubt.

References

1. T. Hamemann and D.V. Szabó, Polymer nanoparticle composites: From synthesis to modern applications, *Materials* **3**(6), 3468–3517 (2010).
2. Z. Varga, G. Filipesei, A. Szilágyi, and M. Zrínyi, Electric and magnetic field-structured smart composites, *Macromolecular Symposia* **227**(1), 123–134 (2005).

3. Z. Varga, G. Filipesei, and M. Zrínyi, Magnetic field sensitive functional elastomers with tunable elastic modulus, *Polymer* **47**(1), 227–233 (2006).

4. L. Lu, X. Tang, S. Hu, and Y. Pan, Acoustic field assisted particle patterning for smart polymer composite fabrication in stereolithographyl, *3D Printing and Additive Manufacturing* **5**(2), 151–159 (2018).

5. G. Filipesei, I. Csetneki, A. Szilagyi, and M. Zrinyi, Magnetic field responsive smart polymer composites. In *Oligomers-Polymer Composites- Molecular Imprinting*. Advances in Polymer Science Vol 206. Springer (2007).

6. C. Zhou, Y. Chen, Z.G. Yang, and B. Khoshnevis, Development of multi-material mask-image-projection-based stereolithography for the fabrication of digital materials. In *Annual Solid Freeform Fabrication Symposium*, Austin, Texas (2011).

7. B. Jackson, K. Wood, and J. Beaman, Discrete multi-material selective laser sintering (M2 SLS): Development for an application in complex sand casting core arrays. In *Proc Solid Freeform Fabr*, pp. 176–182 (2000).

8. P.J. Bartolo and J. Gaspar, Metal filled resin for stereolithography metal part, *CIRP Annals Manufacturing Technology* **57**(1), 235–238 (2008).

9. S. Kumar and J.P. Kruth, Composites by rapid prototyping technology, *Materials & Design* **31**(2), 850–856 (2010).

10. G. Wurm, B. Tomaneok, K. Holl and J. Trenkler, Prospective study on eranioplasty with individual carbon fiber reinforced polymer (CFRP) implants produced by means of stereolithography, *Surgical Neurology* **62**(6), 510–521 (2004).

11. A.H. Lu, E.E. Salabas and F. Schüth, Magnetic nanoparticles: Synthesis, protection, functionalization, and application, *Angewandte Chemie International Edition* **46**(8), 1222–1244 (2007).

12. A.K. Bastola, V.T. Hoang and L. Li, A novel hybrid magnetorheological elastomer developed by 3D printing, *Materials & Design* **114**, 391–397 (2017).

13. H.J. Pan, H.J. Huang, L.Z. Zhang, J.Y. Qi and S.K. Cao, Rheological properties of magnetorheological fluid prepared by gelatin-carbonyl iron composite particles, *Journal of Central South University of Technology* **12**(4), 411–415 (2005).

14. X.F. Dong, M.A. Ning, Q.I. Min, J.H. Li, X.C. Guan and J.P. Ou, Properties of magneto-rheological fluids based on amorphous micro-particles, *Transactions of Nonferrous Metals Society of China* **22**(12), 2979–2983 (2012).

15. T. Zhu, R. Cheng, G.R. Sheppard, J. Locklin and L. Mao, Magnetic-field-assisted fabrication and manipulation of nonspherical polymer particles in ferrofluid-based droplet microfluids, *Langmuir* **31**(31), 8531–8534 (2015).

16. L. Wang, F. Li, M. Kuang, M. Gao, J. Wang, Y. Huang, L. Jiang and Y. Song, Interface manipulation for printing three-dimensional microstructures under magnetic guiding, *Small* 11(16), 1900–1904 (2015).

17. Z. Varga, G. Filipcsei and M. Zrinyi, Smart composites with controlled anisotropy, *Polymer* **46**(18), 7779–7787 (2005).

18. H. Xia, J. Wang, Y. Tian, Q.D. Chen, X.B. Du, Y.L. Zhang, Y. He and H.B. Sun, Ferrofluids for fabrication of remotely controllable micro-nanomachines by two-photon polymerization, *Advanced Materials* **22**(29), 3204–3207 (2010).

19. L.M. Bollig, P.J. Hilpiseh, G.S. Mowry and B.B. Nelson-Cheeseman, 3D printed magnetic polymer composite transformers, *Journal of Magnetism and Magnetic Materials* **442**, 97–191 (2017).

20. R.U. Hassan, S. Jo and J. Seok, Fabrication of a functionally graded and magnetically responsive shape memory polymer using a 3D printing technique and its characterization, *Journal of Applied Polymer Science* **135**(11), 45997 (2018).

21. A. Boczkowska, S.F. Awietjan, S. Pietrzko and K.J. Kurzydlowski, Mechanical properties of magnetorheological elastomers under shear deformation, *Composites Part B: Engineering* **43**(2), 636–640 (2012).

22. F. Tsumori, H. Kawanishi, K. Kudo, T. Osada and H. Miura, Development of three-dimensional printing system for magnetic elastomer with control of magnetic anisotropy in the structure, *Japanese Journal of Applied Physics* **55**(6S1), 06GP18 (2016).

23. V. Kumar and D.J. Lee, Iron particle and anisotropic effects on mechanical properties of magneto-sensitive elastomers, *Journal of Magnetism and Magnetic Materials* **441**, 105–112 (2017).

24. M. Farshad and A. Benine, Magnetoactive elastomer composites, *Polymer Testing* **23**(3), 347–353 (2004).

25. M.R. Jolly, J.D. Carlson, B.C. Munoz and T.A. Bullions, The magnetoviscoelastic response of elastomer composites consisting of ferrous particles embedded in a polymer matrix, *Journal of Intelligent Material Systems and Structures* **7**(6), 613–622 (1996).

26. S.E. Chung, J. Kim, S.E. Choi, L.N. Kim and S. Kwon, In situ fabrication and actuation of polymer magnetic microstructures, *Journal of Microelectromechanical Systems* **20**(4), 785–787 (2011).

27. J. Zhang, S. Zhao, M. Zhu, Y. Zhu, Y. Zhang, Z. Liu and C. Zhang, 3D-printed magnetic Fe3O4/MBG/PCL composite scaffolds with multifunctionality of bone regeneration, local anticancer drug delivery and hyperthermia, *Journal of Materials Chemistry B* **2**(43), 7583–7595 (2014).

28. Y. Tian, Y.L. Zhang, J.F. Ku, Y. He, B.B. Xu, Q.D. Chen, H. Xia and H.B. Sun, High performance magnetically controllable microturbines, *Lab on a Chip* **10**(21), 2902–2905 (2010).

29. T.H. Au, D.T Trinh, D.B. Do, D.P. Hguyen, Q.C. Tong and N.D. Lai, Free-floating magnetic microstructures by mask photolithography, *Physica B: Condensed Matter* **532**, 59–63 (2018).

30. U. Robles, J. Kasemodel, J. Avila, T. Benitez, R.C. Rumpf, 3D printed structures by microdispensing materials loaded with dielectric and magnetic powders, *IEEE Transactions on Components, Packaging and Manufacturing Technology* **8**(3), 492–498 (2018).

31. A. Shen, C.P. Bailey, A.W. Ma and S. Dardona, UV-assisted direct write of polymer-bonded magnets, *Journal of Magnetism and Magnetic Materials* **462**, 220–225 (2018).

32. M.P. Paranthaman, C.S. Shafer, A.M. Elliott, D.H. Siddel, M.A. McGuire, R.M. Springfield, J. Martin, R. Fredette and J. Ormerod, Binder jetting: A novel NdFeB bonded magnet fabrication process, *JOM* **68**(7) 1978–1982 (2016).

33. B. Khatri, K. Lappe, D. Noetzel, K. Pursche and T. Hanemann, A 3D-printable polymer-metal soft-magnetic functional composite -- Development and characterization, *Materials* **11**(2), 189 (2018).

34. H. Song, J. Spencer, A. Jander, J. Nielsen, J. Stasiak, V. Kasperchik and P. Dhagat, Inkjet printing of magnetic materials with aligned anisotropy, *Journal of Applied Physics* **115**(17) 17E308 (2014).

35. C.V. Mikler, V. Chaudhary, T. Borkar, V. Soni, D. Jaeger, X. Chen, R. Contieri, R.V. Ramanujan and R. Banerjee, Laser additive manufacturing of magnetic materials, *JOM* **69**(3) 532–543 (2017).

36. Y. Arbaoui, P. Agaciak, A. Chevalier, V. Laur, A. Maalouf, J. Ville, P. Roquefort, T. Aubry and P. Queffelee, 3D printed ferromagnetic composites for microwave applications, *Journal of Materials Science* **52**(9), 4988–4996 (2017).

37. R. Libanori, R.M. Erb and A.R. Studart, Mechanics of platelet-reinforced composites assembled using mechanical and magnetic stimula, *ACS Applied Materials & Interfaces* **5**(21), 10794–10805 (2013).

38. D. Kokkinis, M. Schaffner and A.R. Studart, Multimaterial magnetically assisted 3D printing of composite materials, *Nature Communications* **6**, 8643 (2015).

39. R.M. Erb, R. Libanori, N. Rothfuehs and A.R. Studart, Composites reinforced in three dimensions by using low magnetic fields, *Science* **335**(6065), 199–204 (2012).

40. J.J. Martin, B.E. Fiore and R.M. Erb, Designing bioinspired composite reinforcement architectures via 3D magnetic printing, *Nature Communications* **6**, 8641 (2015).

41. E.B. Joyee, L. Lu and Y. Pan, Analysis of mechanical behavior of 3D printed heterogeneous particle- polymer composites, *Composites Part B: Engineering* **173**, 106840 (2019).

42. M.R. Sommer, R.M. Erb and A.R. Studart, Injectable materials and magnetically controlled anisotropic porosity, *ACS Applied Materials & Interfaces* **4**(10), 5086–5091 (2012).

43. T.P. Niebel, F.J. Heiligtag, J. Kind, M. Zanini, A. Lauria, M. Niederberger and A.R. Studart, Multifunctional microparticles with uniform magnetic coatings and tunable surface chemistry, *RSC Advances* **4**(107), 62483–62491 (2014).

44. W.J. Ahn, H.S. Jung, H.J. Choi, Pickering emulsion polymerized smart magnetic poly (methyl methacrylate)/Fe2O3 composite particles and their stimulus-response, *RSC Advances* **5**(29), 23094–23100 (2015).

45. J. Gass, P. Poddar, J. Almand, S. Srinath and H. Srikanth, Superparamagnetic polymer nanocomposites with uniform Fe3O4 nanoparticle dispersions, *Advanced Functional Materials* **16**(1), 71–75 (2006).

46. J.S. Choi, B.J. Park, M.S. Cho, H.J. Choi, Preparation and magnetorheological characteristics of polymer coated carbonyl iron suspensions, *Journal of Magnetism and Magnetic Materials* **304**(1), e374–e376 (2006).

47. A. Mostafaei, K.A. Kimes, E.L. Stevens, J. Toman, Y.L. Krimer, K. Ullakko and M. Chmielus, *Acta Materialia* **131**, 482–490 (2017).

48. L. Jiang and L. Gao, Carbon nanotube-magnetite nanocomposites from solvothermal processes: Formation, characterization, and enhanced electrical properties, *Chemistry of Materials* **15**(14), 2848–2853 (2003).

49. X. Teng and H. Yang, Effects of surfactants and synthetic conditions on the sizes and self-assembly of monodisperse iron oxide nanoparticles, *Journal of Materials Chemistry* **14**(4), 774–779 (2004).

50. Y.F. Chong, K.L. Pey, A.T.S. Wee, M.O. Thompson, C.H. Tung and A. See, Laser-induced amorphization of silicon during pulsed-laser irradiation of TiN/Ti/polycrystallinesilicon/SiO2/silicon, *Applied Physics Letters* **81**(20), 3786–3788 (2002).

51. X. Zhou, C. Fang, Y. Li, N. An and W. Lei, Preparation and characterization of Fe3O4-CNTs magnetic nanocomposites for potential application in functional magnetic printing ink, *Composites Part B: Engineering* **89**, 295–302 (2016).

52. B. Jia, L. Gao and J. Sun, Self-assembly of magnetite beads along multiwalled carbon nanotubes via a simple hydrothermal process, *Carbon* **45**(7), 1476–1481 (2007).

53. C. Huiqun, Z. Meifang, L. Yaogang, Decoration of carbon nanotubes with iron oxide, *Journal of Solid State Chemistry* **179**(4), 1208–1213 (2006).

54. Q. Zhang, M. Zhu, Q. Zhang, Y. Li and H. Wang, The formation of magnetite nanoparticles on the sidewalls of multi-walled carbon nanotubes, *Composites Science and Technology* **69**(5), 633–638 (2009).

55. M.A. Correa-Duarte, M. Grzelczak, V. Salgueirino-Maceira, M. Giersig, L.M. Liz-Marzan, M. Farle, K. Sierazdki and R. Diaz, Alignment of carbon nanotubes under low magnetic fields through attachment of magnetic nanoparticles, *Journal of Physical Chemistry B* **109**(41), 19060–19063 (2005).

56. L. Liu, T. Ge, K.D. Ngo, Y. Mei and G.Q. Lu, Ferrite paste cured with ultraviolet light for additive manufacturing of magnetic components for power electronics, *IEEE Magnetics Letters* **9**, 1–5 (2018).

57. L.M. Bollig, M.V. Patton, G.S. Mowry, B.B. Nelson-Cheeseman, Effects of 3D printed structural characteristics on magnetic properties, *IEEE Transactions on Magnetics* **53**(11), 1–6 (2017).

58. R. Domingo-Roca, J.C. Jackson, J.F.C. Windmill, 3D-printing polymer-based permanent magnets, *Materials & Design* **153**, 120–128 (2018).

59. M. Garibaldi, I. Ashcroft, J.N. Lemke, M. Simonelli and R. Hague, Effect of annealing on the microstructure and magnetic properties of soft magnetic Fe-Si produced via laser additive manufacturing, *Scripta Materialia* **142**, 121–125 (2018).

60. J.N. Lemke, M. Simonelli, M. Garibaldi, I. Ashcroft, R. Hague, M. Vedani, R. Wildman and C. Tuck, Calorimetric study and microstructure analysis of the order-disorder phase transformation in silicon steel built by SLM, *Journal of Alloys and Compounds* **722**, 293–301 (2017).

61. M. Garibaldi, I. Ashcroft, N. Hillier, S.A.C. Harmon and R. Hague, Relationship between laser energy input, microstructures and magnetic properties of selective laser melted Fe-6.9%wt Si soft magnets, *Materials Characterization* **143**, 144–151(2018).

62. L. Lu, P. Guo and Y. Pan, Magnetic-field-assisted projection stereolithography for three-dimensional printing of smart structures, *Journal of Manufacturing Science and Engineering* **139**(7), 071008 (2017).

63. L. Lu, E.B. Joyce and Y. Pan, Correlation between microscale magnetic particle distribution and magnetic-field-responsive performance of three-dimensional printed composites, *Journal of Micro and Nano-Manufacturing* **6**(1), 010904 (2018).

64. E.B. Joyce and Y. Pan, A fully three-dimensional printed inchworm-inspired soft robot with magnetic actuation, *Soft Robotics* **6**(3), 333–345 (2019).

65. E.B. Joyce and Y. Pan, Multi-material additive manufacturing of functional soft robot, *Procedia Manufacturing* **34**, 566–573 (2019).

66. R. De Santis, U. D'Amora, T. Russo, A. Ronca, A. Gloria and L. Ambrosio, 3D fibre deposition and stereolithography techniques for the design of multifunctional nanocomposite magnetic scaffolds, *Journal of Materials Science: Materials in Medicine* **26**(10), 260 (2015).

67. U. D'Amora, T. Russo, A. GLoria, V. Rivi0065ccio, V. D'Anto, G. Negri, L. Ambrosio and R. De Santis, 3D additive-manufactured nanocomposite magnetic scaffolds: Effect of the application mode of a time-dependent magnetic field on hMSCs behavior, *Bioactive Materials* **2**(3), 138–145 (2017).

68. S. Hwang, E.I. Reyes, K.S. Moon, R.C. Rumpf and N.S. Kim, Thermo-mechanical characterization of metal/polymer composite filaments and printing parameter study for fused deposition modeling in the 3D printing process, *Journal of Electronic Materials* **44**(3), 771–777 (2015).

69. Y. Kim, H. Yuk, R. Zhao, S.A. Chester and X. Zhao, Printing ferromagnetic domains for untethered fast-transforming soft materials, *Nature* **558**(7709), 274 (2018).
70. D.A. Snelling, C.B. Williams, C.T.A. Suchicital and A.P. Druschitz, Binder jetting advanced ceramics for metal-ceramic composite structures, *The International Journal of Advanced Manufacturing Technology* **92**, 531–545 (2017).
71. H. Ferrand, F. Bouville, T.P. Niebel and A.R. Studart, Magnetically assisted slip casting of bioinspired heterogeneous composites, *Nature Materials* **14**(11), 1172 (2015).

Chapter 6

Additive Manufacturing of Bio-Inspired Structures via Nanocomposite 3D Printing

Yang Yang, Xiangjia Li, and Yong Chen

6.1 Introduction

Living organisms are composed of natural materials built by hard and soft phases arranged in complex hierarchical architectures with dimensions spanning from the nanoscale to the macroscale.[1,2] The creatures in nature possess almost perfect structures and functions after millions of years of evolution.[3,4] For example, nacre exhibits high fracture toughness and energy absorption properties due to its brick-and-mortar structure (Fig. 6.1(a)). The horn of bighorn sheep (*Ovis canadensis*) is a prime example where such high performance is generated through expert mechanical designs (Fig. 6.1(b)).[5,6] Tubules with an elliptically shaped cross section extend along the longitudinal direction of the horn and are dispersed within a laminated structure formed by keratinous cells. The Arapaima gigas fish scale with Bouligand type collagen is a natural body armor possessing excellent mechanical properties as well as superior flexibility (Fig. 6.1(c)).[7] Finally the hierarchical structure of a porcupine fish spine provides the protection of its body (Fig. 6.1(d)). These biological skills and attributes represent capabilities that are far beyond conventional engineering systems.[8]

In addition to mechanical properties, other physical properties are critical to survival for numerous species. For example, from a hydrodynamic perspective, the microstructure of shark skin significantly reduces the drag force in water, whereas the eggbeater microstructure of the surface of Salvinia Molesta, an aquatic fern native to south-eastern Brazil, achieves a degree of superhydrophobicity allowing this free-floating plant to remain buoyant in water.[9–11]

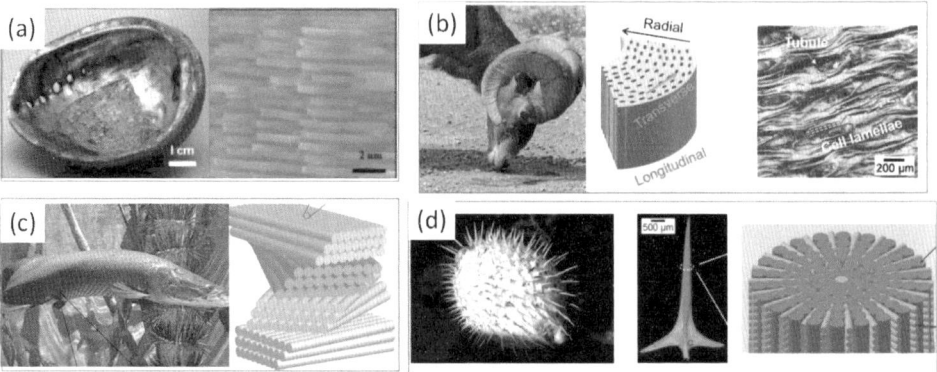

Fig. 6.1. The hierarchical structures in nature. (a) Layered structure in nacre; (b) the structure of bighorn sheep horn; (c) Bouligand structure in Arapaima gigas scales; and (d) schematic diagram of the hierarchical structure of a porcupine fish spine.

Water is the headspring of life. However, more and more places are experiencing severe water shortage due to the worsening environmental problem. To survive in extreme arid region, creatures in nature have evolved special mechanism to collect water from surroundings at ambient temperature.[12–15] For example, one kind of beetles, named as *Onymacris unguicularis*, can live in dry desert due to the special micro-scale textures on their back (refer to Fig. 6.2(a)).[16] The textures divide the desert beetle's back into hydrophobic and hydrophilic regions. The water droplet firstly condenses in the flat hydrophobic region, and the water droplet will roll off quickly from the hydrophobic region to the hydrophilic region when the gathered drop is big enough.[16] Inspired by this special texture, several devices were developed with the special arrangement of hydrophobic and hydrophilic materials for water harvest and droplet transportation.[17–23] Spider silks have the capability of water collection with unique micro-structures of spindle-knots and joints, which generate surface energy gradient with different Laplace pressures (refer to Fig. 6.2(b)). These features enlighten researchers to develop artificial fibers by using dip-coating and co-axial electrospinning process for high efficiency fog harvest and directional transportation of water droplet.[24,25] Besides, special micro-structures on the skin enable moisture-harvesting lizards, such as *Phrynosoma cornutum* and *Moloch horridus,* to live in arid places on earth. These micro-structures consist of capillary channels (shown in Fig. 6.2(d)), which can collect water and further transport it to lizards' mouth.[26,27]

In addition to animals, plants also show remarkable mechanism of water harvest. For example, one type of desert moss, called *Syntrichia caninervis,* flourishes in drought land thanks to its hair shaped micro structures (Fig. 6.2(c)).[28] Similarly, as shown in Fig. 6.2(e), *Cotula fallax* shows excellent fog harvesting behavior due to unique hierarchical structure consisting of macro leaves and micro hairs, and such features play critical roles in water collection and retention for a long persistent period.[29] Also, cactii can live in harsh environments with dry air for extensively long time. It has been found that the apex angle of 10-15 degrees at their spines is adequate to generate large Laplace pressure difference, which in turn greatly enhances the water collection and movement.[19,30] Inspired by the fog collection principle of the cactus, many kinds of water collectors with micro-scale spine shaped tips have been studied.[31–34]

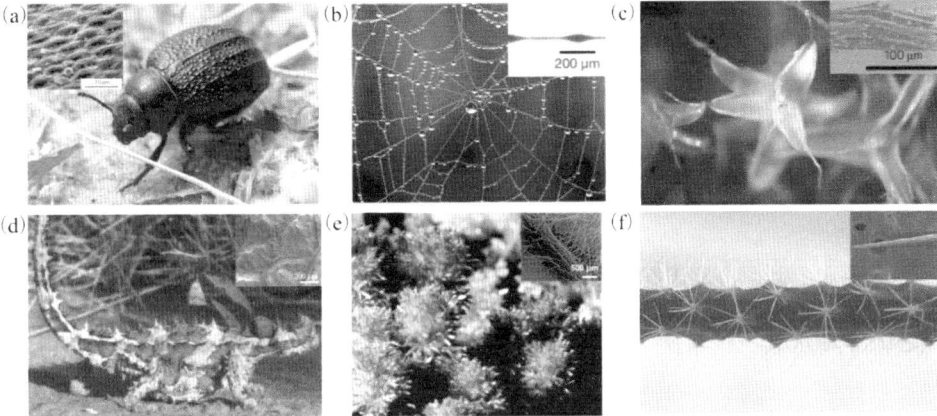

Fig. 6.2. Water collection phenomena of animals and plants in nature. (a) Image of desert beetle and its micro-structures on the back; (b) image of spider web and its special spindle-knots; (c) image of water collection *Syntrichia caninervis* and micro-structures on its leaf; (d) image of moisture harvesting *Moloch horridus* and micro-structures on its skin; (e) image of *Cotula fallax* and micro-structure of its leaf; and (f) image of cactus and the micro image of its spine.

Given the variety of excellent mechanical, hydrodynamic, optical, and electrical properties, nature has developed high-performance materials and structures that provide valuable sources of inspiration for the design of next-generation structural materials. Biomimicry, by learning from nature's concepts and design principles, is driving a paradigm shift in modern materials science and technology. Of particular interest, biological constructs comprise a significant source of inspiration for the design of next-generation structural materials, especially the variety of excellent mechanical properties found in evolved organic structures.[35,36] Biomaterials and biostructures exhibiting unexpected properties have been studied with several inspirational examples that have fueled a recent escalation of interest in smart biological structures, based on the study of several notable examples.[37,38] However, rather than simply duplicating natural materials and structures, a great challenge is to understand the design principles and physical/chemical mechanisms that determine optimized structural organization in biological systems and its relationship to function. In addition to material and structural research, another important consideration is what are the synthesis and manufacturing pathways for the biomimetic materials and structures based on the identified physical/chemical principles? The complicated architectures in nature far exceed the capability of traditional design and fabrication technologies, which hinders the progress of biomimetic study and its use in engineering systems.

Additive manufacturing (a.k.a., 3D printing) has created new opportunities for manipulating and mimicking the intrinsically multi-scale, multi-material, and multi-functional structures. Three-dimensional (3D) printing has been demonstrated as an effective pathway to fabricate customized products with complicated 3D structures for a wide variety of applications in industry, academia, and daily usages.[39,40] During the past 30 years, many novel additive manufacturing (AM) processes have been successfully developed.[41] The AM processes are generally flexible in fabricating a computer-aided design (CAD) model with good control on the resulting geometric shape. Recent advances in material, process, and machine developments have enabled AM processes to evolve from prototyping (rapid prototyping) to manufacturing (rapid manufacturing).[42] From craftsmanship to mass production, many believe the future of manufacturing lies in mass customization to which AM may present an effective solution.[43] Here, we present an introduction of additive manufacturing of bio-inspired structures via nanocomposite 3D printing. The inspirations are from various creatures such as nacre, lobster claw, cactus, and salvinia molesta. In the chapter, various 3D-printing technologies are discussed: the first part is focused on the bio-inspired mechanically reinforced structures by 3D printing; and the second part is focused on the bio-inspired hydrodynamic (superhydrophobic and water collection) structures by 3D printing.

6.2 3D Printing of Biomimetic Mechanically Reinforced Structures

In this section, we will introduce the bioinspired mechanical reinforced structures through 3D printing. The reinforced mechanism and some new 3D printing technology will be discussed. In natural structures, the mineralized composites play an important role in biological structures for strong mechanical properties. These composites consist of a mineral reinforcement phase, such as hydroxyapatite, calcium carbonate, or silica, embedded in a biopolymer matrix, such as collagen or chitin. Besides, the material in nature is not a single material but with controllable alignment of collagen fibers in organic matrix. With the inspiration, additive manufacturing is evolving from single and multiple materials for structural purpose to composites and multi-functional 3D printing. The addition of microfillers (e.g., ceramic platelets and microfibers) and nanofillers (e.g., carbon nanotubes, graphene, etc.) have been developed to reinforce the 3D-printed structures.[44,45] In order to mimic biological structures, the combination of shear force, electrical-field, and magnetic-field-assisted methods with 3D printing technologies have been developed to get the anisotropic mechanical properties by controlled filler alignments (refer to Fig. 6.3). We define these as the *"field-assisted 3D printing"* and will discuss them in the following sub-sections.

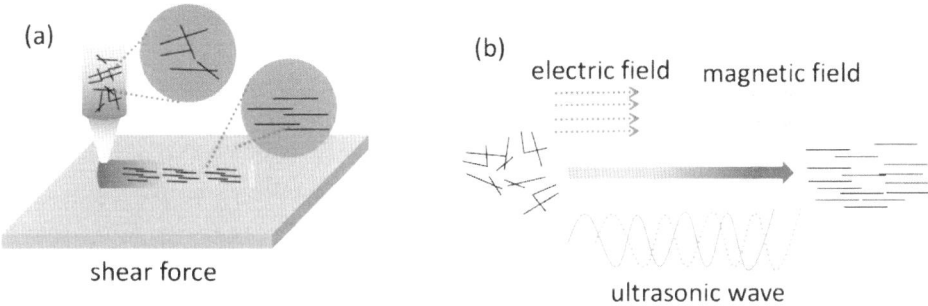

Fig. 6.3. Schematic diagram shows the alignment of fillers during 3D printing by (a) shear force, and (b) electric field, magnetic field, and ultrasonic wave.

6.3 Shear Force Assisted Bioinspired 3D Printing

Two key structural features in nature are the low-weight cellular architecture of the overall object reinforced with stiff fibers and the low weight of balsa by the honeycomb-type porous structure. In order to mimic these structures, the shear-force-assisted direct ink writing (DIW) process was developed to get the aligned silicon carbide whiskers and carbon fibers in polymer resins (Fig. 6.4(a)).[44] A new epoxy-based ink that enables 3D printing of cellular composites with controlled alignment of multi-scale, high aspect ratio fiber reinforcement to create hierarchical structures inspired by balsa wood was studied. Reinforced walls of the printed cellular structures were accomplished through the force alignment due to the shear stress during the extrusion. Reinforcements could be aligned around geometric stress concentrators or stiffness could be graded near fixture points to minimize damage. This approach is ideally suited to fabricate a wide range of bio-inspired composite structures with controlled architecture and mechanical properties. Similarly, a class of aligned carbon fiber reinforced composite was fabricated by the DIW technology. Carbon fibers were aligned in epoxy or aromatic thermoset resin via controlled micro-extrusion and the extruded mixture is subsequently cured into complex geometries.[46] The composites with carbon fiber alignment outperform equivalently filled composites with randomly oriented fibers.[47]

Fig. 6.4. Shear force assisted 3D printing technology. (a) Optical image of 3D printing of a triangular honeycomb composite aligned SiC/C-fiber. Reproduced with permission.[44] Copyright 2014, Wiley-VCH. (b) Schematic of the shear-induced alignment of cellulose fibrils and subsequent effects on anisotropic stiffness and swelling strain. Reproduced with permission[48]. Copyright 2016, Nature Publishing Group.

Special biomimetic structures with shape changing properties were 3D printed by aligning cellulose fibrils in hydrogel through shear force (Fig. 6.4(b)).[48] As a result, anisotropic modulus was produced, i.e., the filament expands easily in the radial direction (40%) but not in the longitudinal direction (10%) when immersed in water. The different expansion rates leads to the programmable folding behavior of designed artificial flowers. This four-dimensional (4D) printing method relies on a combination of materials and geometry that can be controlled in space and time. This technique has potential as a platform technology, where the hydrogel composite ink design can be extended to a broad range of matrices. Various shapes of morphology can be designed and printed by controlling the printing parameters, such as filament size, orientation and interfilament spacing. Another 3D printing process with shear force alignment is through the stereolithography process. Aluminum oxide nanowires were aligned by the shear flow generated by the lateral oscillation and the image patterned SLA printing.[49] The tensile strength was enhanced by 28% with 5wt% of aligned aluminum oxide nanowires. Shear force assisted 3D printing of composite materials offers an opportunity to combine the desired anisotropic mechanical, electrical and thermal properties with the shape flexibility of 3D printing.[50] This helps engineers to optimize their designs where composition and stiffness in a 3D structure need to be digitally adjusted.

6.4   Magnetic Field Assisted Bioinspired 3D Printing

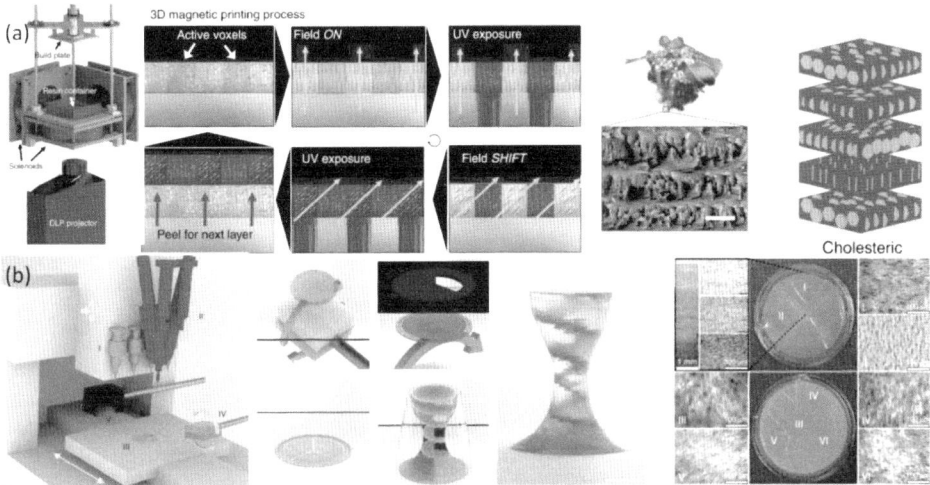

Fig. 6.5. Magnetic field assisted 3D printing technology. (a) Bioinspired composites with microstructured architectures created by 3D magnetic printing, the inspiration from Abalone shell, peacock mantis shrimp and Mammalian cortical bone, respectively. Reproduced with permission.[53] Copyright 2015, Nature Publishing Group. (b) Schematics of the multimaterial magnetic 3D printing of heterogeneous composites. Reproduced with permission.[56] Copyright 2015, Nature Publishing Group.

Magnetic field has been widely used in manufacturing processes because of its flexibility in controlling the alignment of fillers in polymer resins. For example, the magnetic-field-assisted freeze casting process can control the alignments to achieve enhanced strength and stiffness by two times[51] as well as the rotated alignments to fabricate bioinspired spiraling ceramics.[52] The combination of magnetic field with 3D printing was developed as 3D magnetic printing, which was successfully used to build bioinspired architectures inspired by bone, mollusk shells and mantis shrimp (Fig. 6.5(a)).[53] The alignments of magnetic response particles (decorated with superparamagnetic iron oxide nanoparticles) in the reactive resin were controlled by magnetic field.[54,55] A digital light projector (DLP) polymerizes the active grouping of voxels simultaneously in 10 s consolidating the structure and fixing the orientation of the reinforcement with UV light. The particle orientation can either be strengthened or weakened in individual voxels depending on whether the alignment is parallel (local hardening) or perpendicular (crack steering) to the loading direction. These 3D printed objects also show new mechanical properties such as programmable fracture toughness, that is not accessible using the homogeneous monoliths or conventional fabrication technology.[53]

In the multi-material magnetic-assisted 3D (MM-3D) printing, multiple materials can be printed by simply loading distinct syringes with inks containing different monomer compositions and ultrahigh magnetic response (UHMR) particle concentrations, which are aligned by a magnet or electromagnetic coils (Fig. 6.5(b)).[56] The magnetic alignment of 15wt% (4.4 vol%) platelets in the tensile loading direction increases the strength and elastic modulus by 49% and 52%, respectively. Compared with the shear-induced alignment method that has limited alignment directions, the magnetic alignment method can ensure deliberate texture control to eliminate the local swelling response.

6.5 Electrically Assisted 3D Printing of Bioinspired Structures

Dielectrophoresis (DEP) has been used to align fillers including ceramic (PZT short fiber), carbon nanotubes, graphite, and glass fibers in liquid resins using both alternating current (AC) and direct current (DC) fields.[57] The electric field can be applied to produce composites with uniform oriented structure or locally modified surfaces in selected areas. Biological architectures offer inspiration for the design of next-generation structural materials due to their low density, high strength and toughness through specially evolved structures. One of the inspiration originates from superior mechanical properties of naturally evolved composites featured with different orientations of reinforcing fibers or particles (known as Bouligand or twisted plywood structure). For example, the dactyl clubs of peacock mantis shrimp and gigas fish scales, Beetle wings, the claws of crab and lobster.[58] The Bouligand structure with ordered collagen or chitin fibers in one layer, yet heterogeneous between different layers is widely studied and proven to make a great contribution to the reinforcement of crack stopping.[59] The crack cannot follow a straight path, thereby increasing energy dissipation and impact resistance. However, the complicated architectures in natural materials far exceeds the design technology, which hinders the progress of study in reinforcement architectures. Here we present an electrically assisted additive manufacturing technology for the fabrication of reinforcement architecture with anisotropic layers of aligned surface-modified Multi-walled Carbon Nanotubes (MWCNT-S). The direction of MWCNT-S alignment in each layer is accurately controlled by a rotation stage, and how to use such control to induce improved mechanical properties is studied (Fig. 6.6).

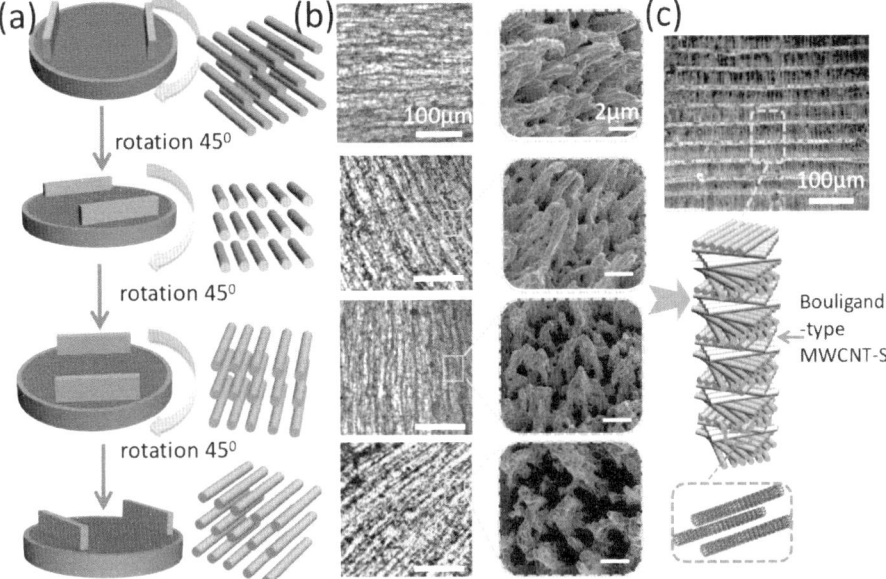

Fig. 6.6. Biomimetic architectures with Bouligand-type MWCNT-S can be recreated by electrically assisted 3D printing. (a) Schematic diagram of different alignment of carbon nanotubes by the rotation of the electrodes; (b) surface optical microscopy images and SEM images of fracture surface for different alignment of MWCNT-S; and (c) schematic diagram of layer by layer bioinspired Bouligand-type MWCNT-S fabricated by the electrically assisted nanocomposite 3D printing.

Figure 6.7(a) shows the setup of the electrically assisted 3D printing platform, in which photocurable Polymer A (PA)/MWCNT-S nanocomposites were deposited in a transparent glass tank. The photo curable resin (PA) is cured after mask images are projected upwards onto the bottom of the substrate by the digital micro-mirror device (DMD) based projection system.[60–62] A mask-image-projection-based stereolithography (MIP-SL) process is used due to its high quality surface finish, dimensional accuracy, high fabrication speed and low machine cost. Different from the laser-based stereolithography (SL) process, a DMD is used in the MIP-SL process to dynamically define the mask images that are projected on a photo curable resin surface (Fig. 6.7(b)). Compared with other methods (mechanical forces, shear flows and magnetic field), DC voltage is preferred for its easy processability and high efficiency in the alignment of carbon nanotubes.[63] Two parallel plate electrodes were used with DC voltages to get the parallel alignment. The alignment relaxation time is determined by $\tau^{-1} = (F(D)/3\eta)G$, $G = \varepsilon_0 \varepsilon E^2 / 2$, where ε_0 is the electric permittivity of vacuum, η is the matrix viscosity, G is rotational torque, ε is anisotropic dielectric constant, and $F(D)$ is the shape factor including aspect ratio D. The relaxation time is in proportional relation with the matrix viscosity. In order to reduce the time for alignment, a photo curable resin with low viscosity (90cp, 20 degrees) was chosen. The alignment of carbon nanotubes in PA/MWCNT-S with 1.5wt% filler loading takes 60s. The three key forces that dominate the rotation of carbon nanotubes are torque, Coulombic and electrophoresis forces that act on each nanotube due to the electric field (Fig. 6.7(d)). The polarization of CNT generated by the electric field leads to a torque force. Coulombic attraction is generated among oppositely charged ends of different CNTs. The electrophoresis force is induced by the presence of charged surface.[64] After the first layer is cured, the container with nanocomposites and electrodes is rotated by a stepper motor and the base is moved up for a certain distance (Fig. 6.7(b)). Hence, when the base is moved down to the container, the alignment of MWCNT-S has been changed due to the controlled rotation of nanocomposites relative to the base to achieve a Bouligand-type structure.[65]

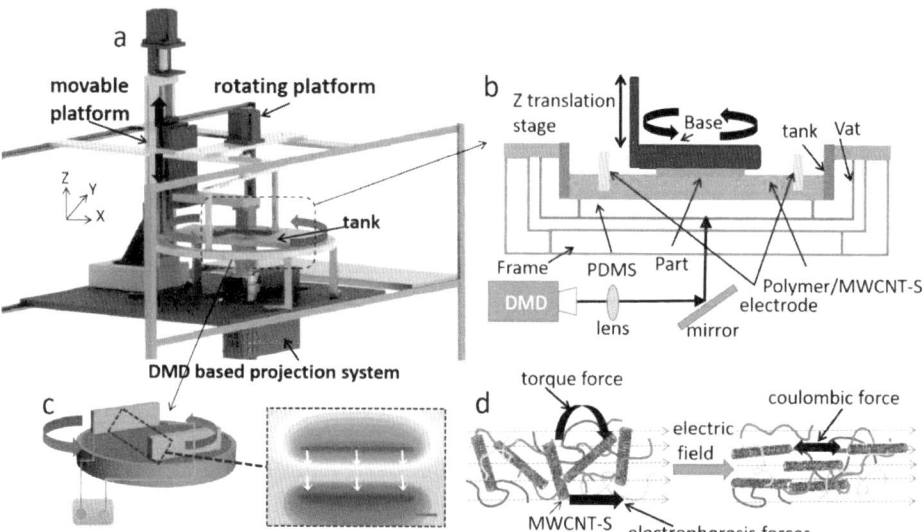

Fig. 6.7. Schematic diagram of the electrically assisted 3D printing platform for the creation of reinforcement anisotropic composites. (a) Diagram of electrically assisted 3D printing device, the rotation of electrodes is controlled by the platform; (b) a bottom-up projection process; (c) two parallel electrodes with applied DC electric field and the electrical potential simulated by *Comsol Multiphysics*; and (d) schematic diagram shows rotation of CNT in polymer resin under the application of electric field.

The Bouligand structure of fibers is shown to enhance the impact resistance under static loading conditions.[66] Here a static compression force was acted on the printed Menger structure (a smaller Menger model, 2.4 *mm*×2.4 *mm*, with layer thickness of 50 *μm*) by Instron-5942 to test the ability of impact resistance with different angles of Bouligand-type MWCNT-S. Figure 8(a) shows schematic diagram of layered structures for different values of N ($\alpha N = 180°$). For unidirectional alignment $N = 1$, the rotation angle is $180°$. For $N = 4$, the aligned CNT rotates $\alpha = 45°$ to form the second layer, $90°$ to form the third layer until it reaches the 5th layer to complete a $180°$ rotation. For $N = 15$, the rotation angle is $\alpha = 12°$, $\alpha = 6°$ for $N = 30$ and $\alpha = 2°$ for $N = 90$.

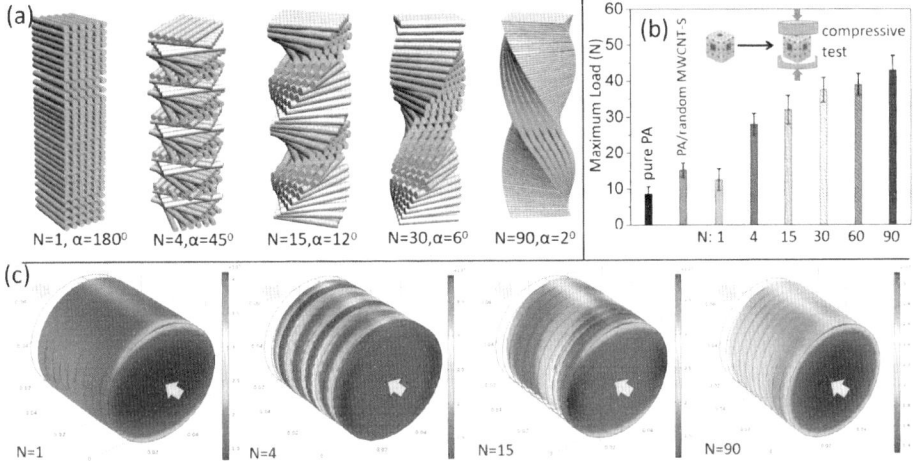

Fig. 6.8. Impact resistance test for Menger models with different rotation angles. (a) Schematic diagram of different types of layered pitch; (b) comparison of load of fracture for the models printed by pure resin, random MWCNT-S and aligned MWCNT-S with different N values; and (c) simulations by *Comsol Multiphysics* show the stress distribution for different values of N under the same compression (200 k Pa)-arrows show the direction of the force.

The comparison of impact resistance under the same load for different N shows an increased maximum load (the load for initial fracture) with the increment of N (Fig. 6.8(b)). The sample will deform in the XY plane under the compression in the Z direction. For example, for $N = 1$, if a crack in the matrix occurs in the $0°$ direction on the bottom layer, the stress concentration will affect the directionality of the damage zone in the second and subsequent layers (see Fig. 6.8(c), the stress is concentrated on the red area with the maximum stress higher than 400 kPa). A larger pitch length indicates a smaller rotation angle and a larger number of aligned MWCNT-S layers, which result in increased energy dissipation (Fig. 6.8(g)). This observation is consistent with previous studies that show a wide in-plane spread of damage and an increase in stiffness for a smaller fiber rotation angle.[67] Compared with the concentrated stress in $N = 1$, the stress is distributed in various layers ($N = 4$ and $N = 15$) until all through the sample ($N = 90$, Fig. 6.8g). Under a compression of 200 kPa, the maximum stress is 400 kPa for $N = 1$ and decrease to 300 kPa for $N = 90$. Additionally, as elastic properties are a function of fiber angle, the graded design with a larger value of N results in a smooth change in the in-plane stiffness and will reduce the interlaminar shear stresses (a key source of delamination). Thus a smaller rotation angle will enhance the ability to withstand greater deflections. Meanwhile, the strain is increased for a large N. This results in a rotating crack front, which yields a large surface area per unit crack length in the direction of crack propagation.[68]

6.6 3D Printing of Biomimetic Hydrodynamic Structures

6.6.1 *3D Printing of Biomimetic Water Collection Structures*

Currently, man-made bioinspired water harvest devices are commonly fabricated by the processes such as chemical or electrochemical erosion of metal wire,[32] lithography,[21] replica molding method,[69] mechanical punching,[70] etc. Typically, biomimicking cactus spine arrays (see in Fig. 6.9(a)), where polydimethylsiloxane (PDMS) prepolymer and cobalt magnetic particles (Co MPs) were used, were fabricated by a hybrid process including mechanical punching and template dissolving.[70] Artificial cones swing back and forth under magnetic fields, the swing of artificial cones facilitates the water transportation along the artificial cone, resulting in higher water collection efficiency. To study the wettability effect on the water collection of artificial cactus spine, a set of conical copper wires were fabricated by gradient electrochemical corrosion and chemical modification (refer to Fig. 6.9(b)), so that the gradient hydrophilic behavior is achieved. Although the aforementioned methods have successfully prepared the micro-structures for water collection, it is still a big challenge to faithfully replicate the natural structures due to the fabrication limitation of the traditional manufacturing methods.

For this purpose, AM technology has been introduced to cover the fabrication needs of biomimicking, since AM shows advantages in building multi-material and multi-scale 3D objects, and superiority in fabricating freeform objects with complex geometric shape.[6,9,72] By progressively depositing material layer by layer, freeform objects with a wide range of materials can be built by a specific AM process, such as stereolithography (SL),[61,73,74] selective laser sintering (SLS),[75] direct ink writing,[76] inkjet printing,[77] and fused deposition modeling (FDM).[78] Compared with traditional fabrication processes, AM technologies present significant advantages with respect to design freedom, assembly complexity, processable material diversity, as well as fabrication cost and efficiency.

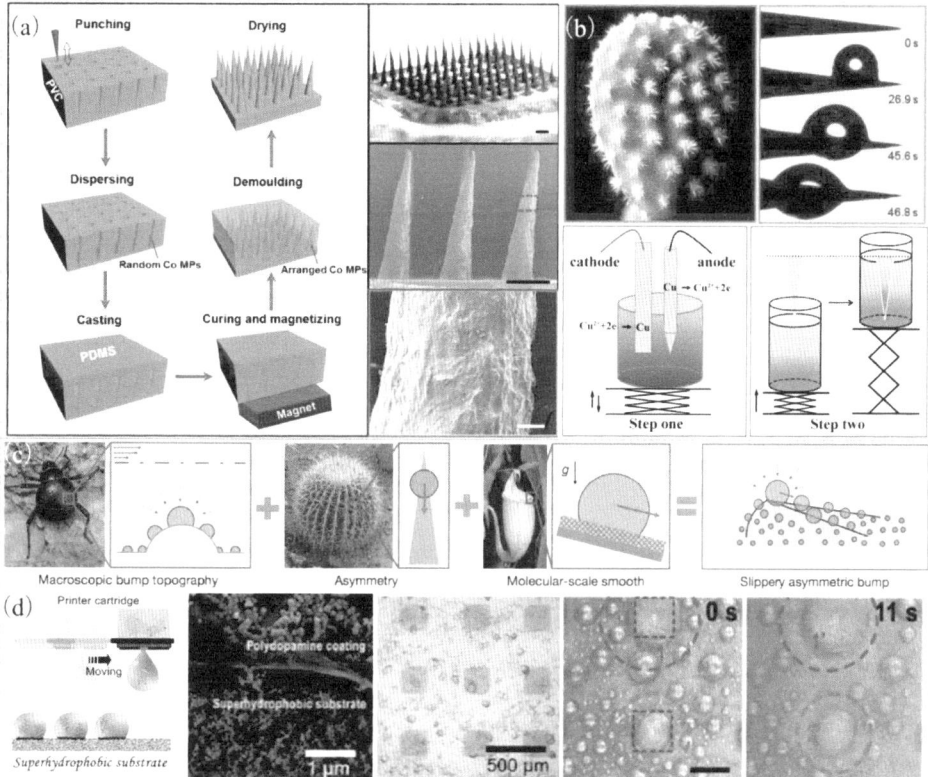

Fig. 6.9. Water harvesting via bio-inspired structures. (a) The fabrication of cactus-inspired magnetically flexible conical arrays and corresponding SEM images;[70] (b) appearance of natural cactus spines and the scheme of the fabrication process of the artificial cactus cone;[32] (c) the design ideal of convex millimetric bump inspired by desert beetles, cactus and pitcher;[69] and (d) superhydrophobic surface for biomimetic fog collection via Inkjet printing.[71]

AM technology has also been widely applied to the fabrication of biomimetic reinforced mechanics, shape-changing and hydrodynamic structures, as well as optical, electrical devices.[79] Hence AM technology creates new opportunities for manipulating and mimicking the complex structures in nature, which are intrinsically multi-scale, multi-material and multi-functional. For instance, inspired by the fog-harvesting behavior of desert beetle's back texture, superhydrophilic micro-patterns were fabricated on top of the superhydrophobic surface by the one-step inkjet printing technology (refer to Fig. 6.9(d)). Compared with a sole superhydrophilic or superhydrophobic surface, alternative patterned wettability shows enhanced water collection efficiency. Bioinspired water harvest structures can also be built by a hybrid process that combines traditional fabrication processes and 3D printing technologies.[69] As shown in Fig. 6.9(c), slippery asymmetric bumps, derived from desert beetles, cactus, and pitch plant, were designed and fabricated to achieve high condensation rate with max vapor diffusion flux. Such hybrid process provides more possibilities to fabricate multi-scale and multi-material bioinspired structures for water collection.

Recent advances in AM technology have shown progressive achievements in the production of new bioinspired water collection devices. Traditionally, conical cone arrays inspired by cactus were made with single micro-tip in several previously reported methods.[32,70] The replication of natural cactus, which possesses multi-cone arrays with controllable arrangement, remains a challenging task due to the fabrication restrictions. With the help of 3D printing technologies, the fabrication of such water collector with multiple artificial spines was enabled by using a nanocomposite 3D printing method. Taking such cactus inspired water collector as an example, the AM approach for biomimetic application is illustrated in the following section. Specifically, the design principle of artificial cactus is firstly introduced, with a focus on optimizing apex angle for water condensation rate and water droplet transportation velocity. The development of material and coating is briefed discussed afterwards. To increase the wettability of artificial cactus, multi-wall Carbon Nano-tube (CNT) is added to the hydrophilic photo-curable polymer as the material of artificial spines. The fabrication process, using a novel 3D-printing approach called Immersed Surface Accumulation (ISA), will also be described in the section. To further accelerate the water transportation rate of artificial cactus spines, the vapor-phase process is integrated to recoat a layer of hydrophobic monolayer on the outside surface of the 3D printed spines. Lastly, the performance of the artificial spines was measured under different flow directions to identify the relationship between the layout of artificial spines and the efficiency of water collection. Design optimization was conducted based on the theory of aerodynamics, and a special patterned array with artificial multi-directional spines was developed. Based on the developed AM process for biomimetic applications, the functional advantages that biomimetic design and fabrication bring will be discussed.

6.1.1.1 *Design of artificial multiple spines*

In nature, the condensed water droplet driven by the gradient Laplace pressure moves along cactus spines from the top tip to the bottom root no matter how spines are arranged. The Laplace pressure is determined by the surface tension of water and the radii of cactus spine at both sides of water drops.[32] The gravity of water drop can be ignored because it is insignificant compared with the

Laplace force. From the microscopic observation, the radius of condensed water droplet ranges from 0.25mm to 0.75mm (Fig. 6.10(c)). The apex angle of cactus spines is from 10 degrees to 15 degrees (refer to Fig. 6.10(a) and 6.10(c)). The angle of artificial cactus spines is studied in terms of water condensation rate and transportation velocity.[32] The apex angle of artificial spines is inversely proportional to the length of artificial spines. For optimal water collection efficiency, the apex angle of artificial spines (1.5mm) should be set at 15 degrees, and arrays of bionic spines bundle are designed with gap of 300 μm in order to generate well-ventilated air (See Fig. 6.10(e)).

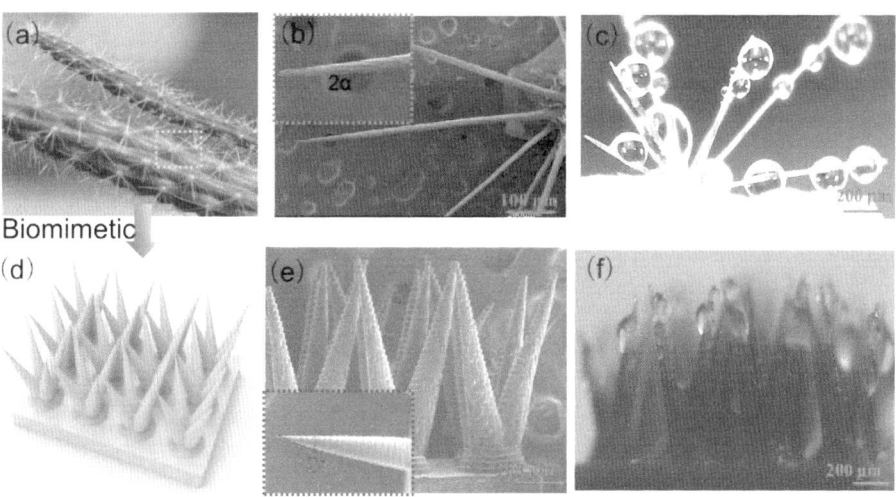

Fig. 6.10. An illustration of design principle of cactus inspired multiple spines. (a) The image of natural cactus; (b) the SEM image of cactus's spine; (c) the water collection of nature cactus spines under microscopic observation; (d) the CAD model of water collection device with multiple artificial spines; (e) the SEM image of multiple artificial spines; and (f) the water collection of artificial cactus spines under microscopic observation.

6.6.2 *Material and Methods*

Some relative studies have revealed that the hydrophobic appearance of artificial cactus spine can promote the growth of water droplets, and the hydrophilic appearance of artificial spine will accelerate transportation of water droplets.[32] Therefore, the wettability of artificial cactus spine should be controlled by managing water droplet growth and transportation. In order to achieve high efficiency of water collection, nano composite is prepared with the mixture of multi-wall carbon Nano-tube (MWCNT) and photopolymer. The pure photopolymer is a hydrophilic material, and the artificial cactus made by pure polymer show advantages in water condensation. The hydrophilic behavior will be enhanced by adding the multi-wall carbon nano-tube (MWCNT) due to the increase of surface roughness. Furthermore, superhydrophobic monolayer coated on the surface of artificial spines improves the water droplet growth rate, and the superhydrophobic thin film further reduces the water re-evaporation. With nano-coating, condensed water is prone to merge and form stream, then flows down along the artificial spine. Such speed of water collection on artificial spine is more than four times of the one without nano-coating.[108]

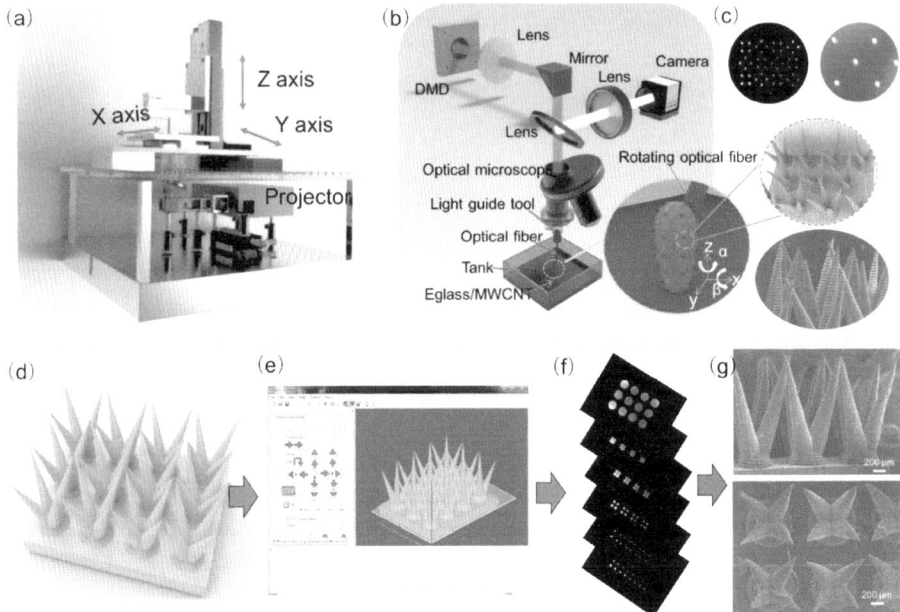

Fig. 6.11. An illustration of fabrication procedures of artificial spines via Immersed surface accumulation process. (a) A schematic diagram of Immersed Surface Accumulation printing system; (b) the layout of optical system in the immersed surface accumulation system; (c) the mask image and projection light beam for one layer fabrication of artificial spines; (d) CAD model of artificial spines; (e) the software used to print artificial spines; (f) sliced mask images; and (g) the SEM images of the 3D-printed artificial spines.

The surfaces with artificial cactus spines can be replicated by using a novel nanocomposite 3D printing process named immersed surface accumulation (*ISA*) (Fig. 6.11(a)). In order to print micro-scale sharp tip, the laser-based CNC accumulation is extended to the *ISA* based 3D printing process (Fig. 6.11(a)).[80–82] The light guide tool is immersed into the resin tank, and material is cured and accumulated layer by layer under the exposure of 2D patterned light beam transmitted by a light guide tool. With the movement of the light guide tool inside the photo-curable resin tank, the ISA process enables the fabrication of multi-scale structures on the surface of an inserted object (Fig. 6.11(b)).

A vision system containing a beam splitter, a convex lens and a CMOS camera can be used to observe the light beam during the printing process (Fig. 6.11(b)). To fabricate the artificial cactus spine, a designed model of artificial spines is processed by a model slicing software system to generate a series of 2D patterned mask images (Fig. 6.11 (d)-(f)). The dimension of 2D patterned light beam used in the ISA based 3D printing system is 3.67×2.75 mm; hence the resolution of the light beam is 2.5 µm/pixel. The curing performance of material needs to be studied, and the layer thickness should be set at a suitable value, so that the following cured layer can attach onto the previously built layers. Using the ISA based 3D printing process, the top tip of the 3D printed artificial cactus spines can achieve 8µm dimension (refer to Fig. 6.11(g)).

6.6.3 *Water Collection of Artificial Cactus Spines*

For artificial cactus spines with multiple tips, the arrangement of spines plays an important role for water collection, as such layout impacts the airflow through each artificial cactus spine. The water collection efficiency of artificial spines is determined by such airflow through each spine. Compared with artificial cactus spine array arranged in a square pattern, the artificial cactus array with hexagonal arrangement showed better water collection rate when they are placed in the same condition of ambient airflow.[108] Water collection occurs after the diameter of water drop, which is intercepted by obstacle, rise to 8–40 µm in an airflow. The obstacle will cause the fog flow to deviate from its original path, and the turbulence will occur when the velocity of air flow exceeds the threshold level. The velocity of fog flow reaches the maximum along the path without any obstacles, and the uniform fog flow stream around each spine is crucial on achieving high fog collection efficiency.[33] The improper arrangement of artificial spines bundles will induce non-uniformed velocity distribution of airflow. Natural cactus shows hexagonal arrangement of spines bundles, and each spine bundle allows a uniform contact area of airflow. Each bundle of artificial spine is consisted of four spines, and the airflow through artificial spines is much more complex compared with traditional arrays of single spine. As shown in Fig. 6.12, more water droplets grow on the artificial spines with hexagonal arrangement from 0 sec to 4 min. Water droplets can not only deposit on the spines of windward side but also on the spines of the leeward side in the simple orderly rank. However, the velocity distribution of both the leeward and windward sides of the artificial spine are the same when bundles are arranged in hexagonal arrangement. The simulation of airflow can be simulated using *Comsol Multiphysics* (refer to Fig. 6.12), and the velocity distribution of airflow varies with the arrangement of bundles of artificial spines.[108]

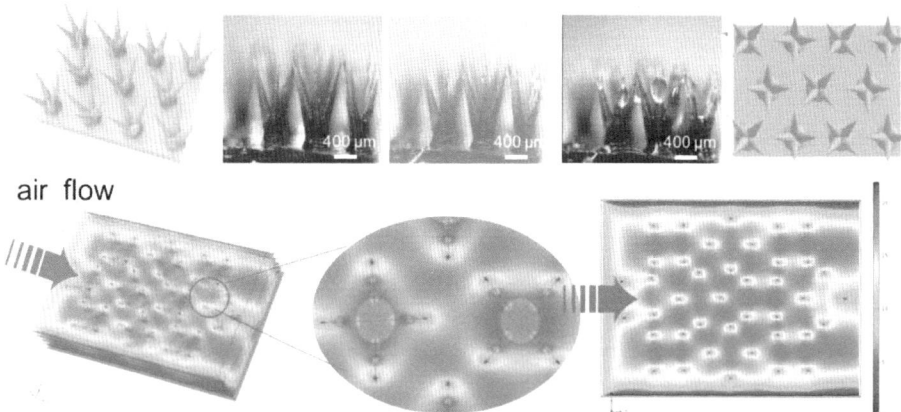

air flow

Fig. 6.12. Optical observation of water collection and the simulation of airflow velocity with hexagonal layout of artificial spines.

In this section, an AM process called ISA-based 3D printing is introduced for fabricating multiple spines inspired by cactus. The AM process development and its material and design optimization are discussed in details. The 3D printed multiple spines surface, with nano-coating and optimized arrangement, showed obvious advantages compared to other surfaces. The underlying mechanism has two major advantages: first, more turbulent and complicated flow field around the optimized spines increase the effective deposition area, facilitating more tiny water drops' deposition and ultimately the fog collection; second, the directional transport of water drops on the Nano coating surface benefits quick rebirth of the deposition sites, and further favors efficient fog collection. The investigation of this optimized bioinspired design and 3D nanocomposite printing method may provide a new avenue to collecting water efficiently and potentially be useful for relieving the water crisis in arid regions.

6.6.3.1 *3D printing of biomimetic superhydrophobic and superlipophilic structures*

In nature, many kinds of plants and animals possess hydrophobic or superhydrophobic surfaces that are attributed to complex micro- and nanostructures on the surfaces (refer to Fig. 6.13).[15,83,84] One of the interesting inspirations is the superhydrophobic surface, which has attracted great attention because of its potential in various applications.[85] One example is the superhydrophobic surface with lotus effect for self-cleaning. Water droplet forms almost perfect spheres and slides easily on the hierarchically structured lotus leaf consisting of nanoscale level hydrophobic wax crystals (refer to Fig. 6.13(a)).[110] Similar phenomena were discovered on the leaves and the petal of some plants, such as rose, rice, and silver ragwort (refer to Fig. 6.13 (b)-(d)). Looking deeper to the 'lotus effect', the water droplet has a large contact angle (CA~161 degrees) and a low sliding angle (2 degrees) on the surface.[86] In contrast, for the 'petal effect', the Rose petals show super-hydrophobicity (CA~152 degrees) that pins water droplets with strong adhesion.[87] Besides, micro- and nano-scale superhydrophobic structures can be observed on the feather, wing, skin, and feet of animals (Fig. 6.13 (e)-(h)). These superhydrophobic structures found in nature provide us inspirations and design principles for the construction of multifunctional systems. Inspired by such structures, artificial superhydrophobic surfaces have attracted much interest due to their potential applications, e.g. self-cleaning, oil/water separation, anti-reflection, and microdroplet transportation.[85] In these applications, controlling the liquid adhesion toward the target surface is particularly important, as it is the adhesive property that eventually determines the dynamic action of liquid on the surface.[88]

Fig. 6.13. The superhydrophobic micro-structures of animals and plants in nature. (a) Image of lotus leaf and its micro-structures; (b) image of rice leaf and its micro structures; (c) image of rose petal and its micro structures; (d) image of silver ragwort leaf and its micro-structures; (e) image of water strider leg and its micro-structures; (f) image of butterfly and the micro structures on its wing; (g) image of bird feather and its micro structures; and (h) image of dragonfly (*Odonata Hemicordulia tau*) and micro-structures on its skin.

However, the fabrication of such multi-scale biological hydrophobic surfaces is a big challenge for traditional manufacturing methods.[39] In most of these methods, the hydrophobic surfaces are fabricated with simple micro-structures which significantly restrict the functional performance.[112] 3D printing presents a new way to study the bioinspired surface by replicating the complex bionic structures, and many efforts have been performed to produce artificial superhydrophobic surfaces using 3D printing technology. As shown in Fig. 6.14(a), an oil-skimmer was developed by integrating 3D printed low surface energy mesh cap onto commercial vessel. The fabrication freedom of 3D printing enables the oil permeation rate study of artificial mesh with different pore size.[89] The 3D printed mesh showed high efficiency oil removal, and the oil captured can be restored inside the vessel even in ambitious stirring or reversion conditions, providing a solution of oil cleaning in harsh environment.

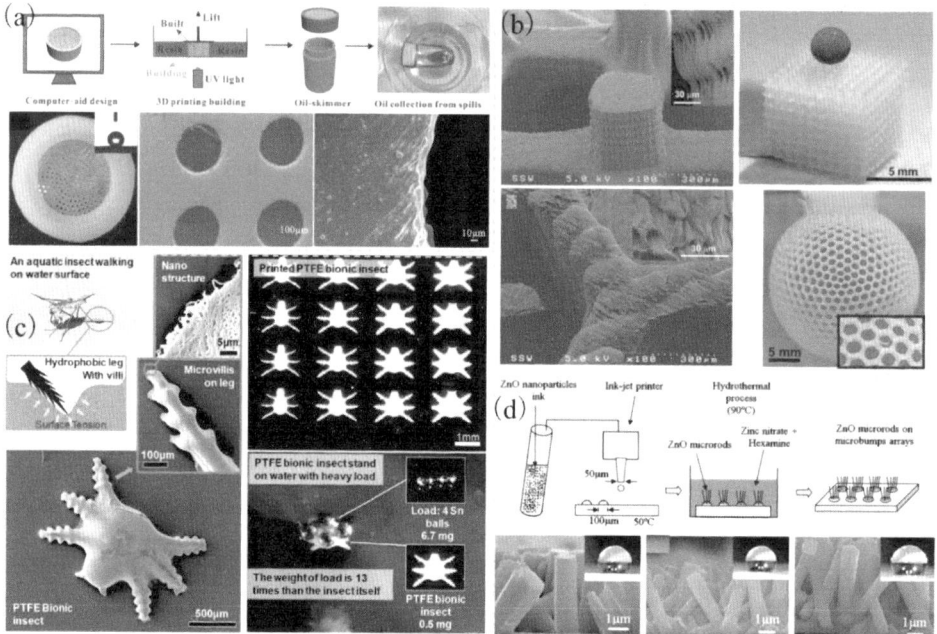

Fig. 6.14. Superhydrophobic structures via 3D printing. (a) 3D printed mesh cap on vessel for oil collection from spills;[89] (b) superhydrophobic iLattice and hollow iBall fabricated by photo-polymerization based 3D printing process;[90] (c) the printed PTFE biomimetic insect with super-hydrophobic nanopored PTFE microvilli legs;[91] and (d) superhydrophobic surface with ZnO microrods fabricated by hybrid process.[92]

3D printing is also a facile method, providing more possibility to fabricate functional materials. For example, the surface properties of printed material can be changed from hydrophobic to hydrophilic by integrating initiator. Taking advantage of photo-polymerization based stereolithography, poly(PEGMA)-grafted sphere shaped lattice and poly (PFMA)-grafted cube shaped micro-lattices were fabricated.[90] As shown in Fig. 6.14(b), a perfect water droplet forms on the top surface of poly(PFMA)-grafted cube shaped micro-lattices due to superhydrophobicity, and also the water droplet is able to be kept inside the poly(PEGMA)-grafted, sphere shaped lattice, due to high hydrophilicity. In addition, inspired by water strider, a bionic insect with six legs was designed and printed by the super-hydrophobic PTFE material (Fig. 6.14(c)), and it has capability of holding load that is more than ten times of its own weight.[91]

Superhydrophobic surface can also be printed by hybrid processes that combine both 3D printing and other traditional manufacturing methods.[92] As shown in Fig. 6.14(d), hydrophilic ZnO can be used to create hydrophobic surface, with micro structures fabricated and roughness enhanced by jointly using inkjet printer and the hydrothermal process.

A particular example regarding this bio-inspired structure topic is given here. In nature, the hierarchical architecture of the Salvinia Molesta is dominated by complex elastic eggbeater-shaped hairs coated with nanoscopic wax crystals and hydrophilic patches on the terminal cells, and forms a unique combination of hydrophilic path on superhydrophobic surfaces ('Salvinia effect').[93] Inspired by Salvinia Molesta leaves, 100 times smaller artificial superhydrophobic surface with eggbeater shaped structures were reproduced by 3D laser-based lithography.[94] Even the contact angle of such micro structure is near 122°, which is not super-hydrophobic, the bionic structure showed promising air retention behavior, and could be potentially used for drag force reduction. From there, an innovative way to produce the artificial surfaces with biomimetic superhydrophobic eggbeater structure is introduced for the purpose of fully replicating the eggbeater structure in Salvinia Molesta hairs. In the following section, the material development and fabrication procedures of artificial superhydrophobic eggbeater structure will be discussed, and several applications based on artificial superhydrophobic surface will be investigated.

6.6.4 *Design of Superhydrophobic and Superolophilic Structures*

Salvinia paradox shows significant superhydrophobic properties with special hydrophilic patches. Since the upper side of floating leaves is densely covered with complex multicellular hairs, and top hydrophilic patches pin the air-water interface to the tips of the eggbeater hairs, the surface of salvinia paradox can achieve big retention layer of air when it is immerged into water. Superhydrophobic structures of salvinia paradox have four hairs, which are converged with hydrophilic patches on the top tip, forming an eggbeater-shaped structure with a total height of 2 mm (refer to Fig. 6.15 (a)-(c)). Inspired by this natural superhydrophobic structure, biomimetic eggbeater structures are designed according to the real size of the eggbeater and pillars of salvinia paradox in nature, with their arms bent together at the terminal ends to create a minimization of the contact area (Fig. 6.15(d)). The height of the stalk of an artificial hair is 700 μm, and the radius of stalk at each side is 300 μm and 150 μm, respectively.

Different number of hairs with a diameter of 35 µm forms the artificial eggbeater head, whose height is 250 µm (Fig. 6.15(d)). The artificial eggbeater structures show petal effect. Its hydrophilic material creates zero velocity conditions at the surface of the wetted area, promoting the stabilization of the air-water interface at a predefined level at the top of the hairs (Fig. 6.15 (f)-(g)). The tunable effect of the adhesion is ascribed to different morphologies (numbers of eggbeater arms, and gap between eggbeater structures) on the surfaces[95]. The energy required for the water to penetrate the region between the hairs needs to be maximized to stabilize the air-water interface. The hairs are split into four arms to create as much surface per height difference as possible. From the theoretical point of view, air is trapped between the solid and liquid phases in the artificial eggbeater structure (refer to Fig. 6.15(e)).

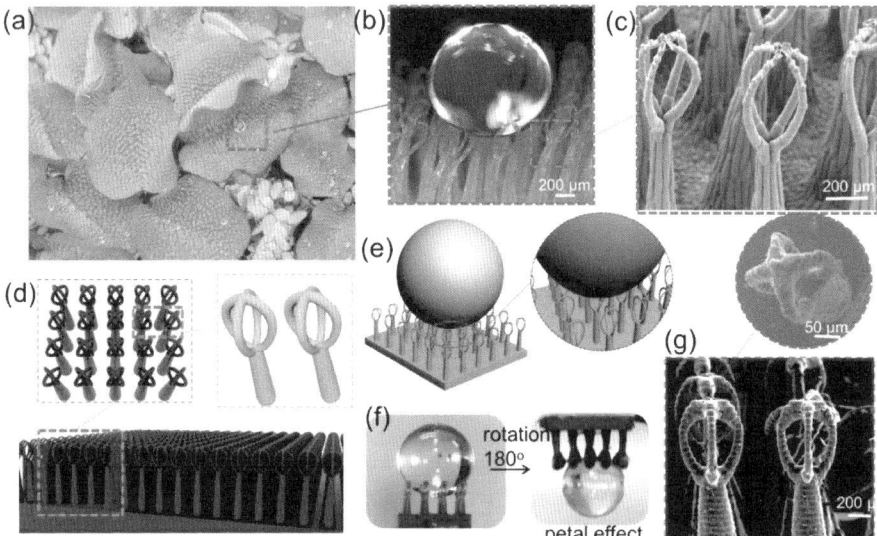

Fig. 6.15. Schematic illustration of biomimetic design of superhydrophobic and superlipophilic structure with eggbeater. (a) Morphology of Salvinia molesta floating leaf, upper side of the leaf surface densely covered with hairs; (b) a spherical water droplet on the surface; (c) SEM image of the eggbeater hair structure; (d) schematic diagram of surface covered with large scale of bionic eggbeater structures and each eggbeater structure is designed with four hairs; (e) the schematic diagram of the interface between water and eggbeater structures; (f) the petal effect of the biomimetic eggbeater shaped structure; and (g) SEM images of 3D printed arrays of eggbeater structures with four hairs, and there are MWCNT on the surface of the tip of eggbeater structures.

6.6.5 *Material and Methods*

The ISA-based nanocomposite 3D printing process is applied to the fabrication of biomimetic eggbeater structures. Light guide tool is immersed in the material tank filled with liquid resin, and high-resolution light beam is projected at the end surface of light guide tool. A series of mask images are obtained by slicing the design model of biomimetic eggbeater array, and the 2D patterned light beam generated by mask images is further calibrated to achieve uniformed light distribution (refer to Fig. 6.16 (a)-(c)). During the movement of the light guide tool, biomimetic eggbeater structures are gradually formed with continuous exposure of the controlled light beam.[95] Taking advantage of both dynamically controlled light projection and 5-axis tool movement, the *ISA-3D* printing system is capable of selectively constructing microscale biomimetic functional structures on the surface of freeform objects.[112] Therefore, using the ISA-based 3D printing, the interface performance of an inserted object can be modified by adding biomimetic micro-scale structures based on functionality requirements.[113] For instance, the surface wettability of the 3D printed lotus was changed from originally hydrophilic to superhydrophobic after the biomimetic structures were printed on its surface (refer to Fig. 6.16 (d)).[111]

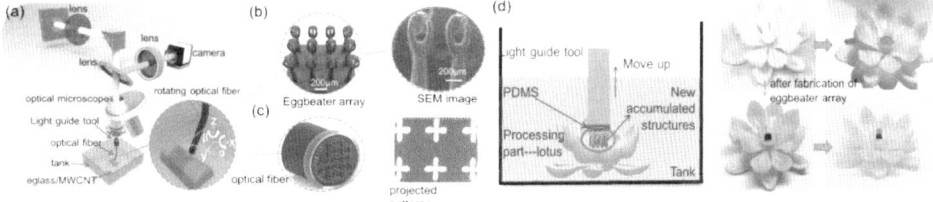

Fig. 6.16. Schematic diagram of the immersed surface accumulation (ISA) based 3D printing process. (a) The optical system in the *ISA-3D* printing system (insert shows the magnification of light guide tool and optical fiber in (c) with projected 2D micro patterns); (b) models and a SEM image of the 3D-printed eggbeater arrays; and (d) an illustration of how to add the micro-scale eggbeater structures on the freeform surface of a lotus, and the wettability of the lotus surface with and without adding the superhydrophobic eggbeater structures.

Recent research shows that even hydrophilic material can macroscopically behave as hydrophobic, if surface is decorated with special functional micro-structured features.[94] The wettability of the 3D-printed flat surface with and without micro-structures is shown in Fig. 6.17. The photocurable polymer exhibits hydrophilic and superolephilic characteristics, and the surface wettability can be enhanced by using composite resin, composed of multiwall carbon nanotube (MWCNT) and photocurable polymer.[95] After coating superhydrophobic nano particles, the water droplet can easily roll off the flat surface of polymer, and the contact angle of oil droplet is increased dramatically (refer to Fig. 6.17(c)). Plenty of studies also indicate that robust hydrophobic property can be achieved by adding micropillar array.[92] Because water and oil droplets on the artificial eggbeater structures are in the Wetzel state, there is no air layer on the top of the 3D printed micropillar array (Fig. 6.17(d)). 3D printed micro-scale eggbeater structures show superhydrophobic and olephobic property (refer to Fig. 6.17 (e)). Surface roughness and surface energy determines the water droplet performance on the micro-structured features. It is noticed that the surface roughness of 3D printed eggbeater structures can be increased with the addition of MWCNT, and the increased roughness will lead to the increment of water contact angles and lipophilicity.[95] Besides, the air-trapping capability of 3D printed eggbeater structures is enhanced by increasing the surface roughness. The water can be held by artificial hairs of 3D printed eggbeater structures, and this petal effect of printed eggbeater structures is attributed to high adhesion of its hydrophilic material. After spraying a thin layer of superhydrophobic nanoparticles on the surface of the 3D-printed eggbeater structures, this unique character of grabbing water will be diminished. The surface character of eggbeater structures is transformed from lipophilic to olephobic, making it possible to manipulate the droplet of oily substance (refer to Fig. 6.17(f)).

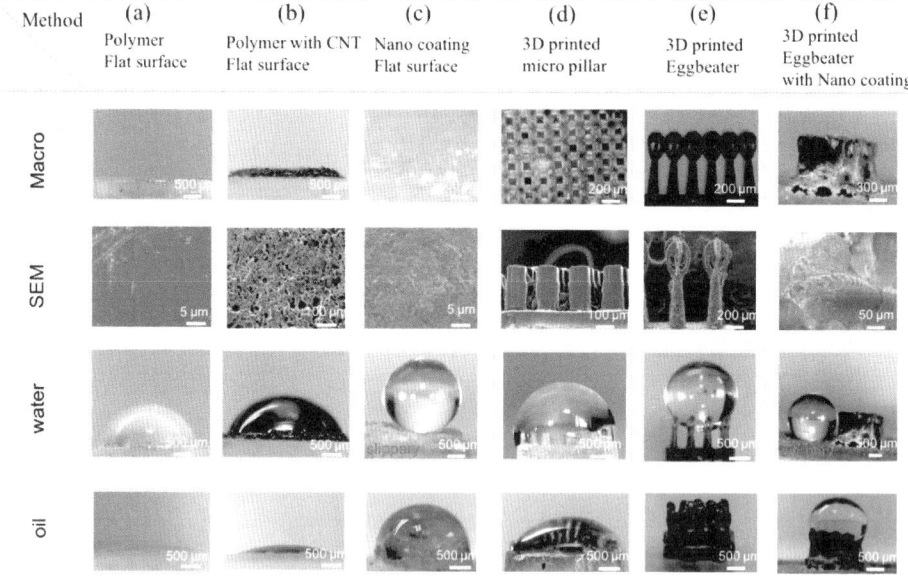

Fig. 6.17. Hydrophilic and lipophilic property of surface with micro-structured features. Optical microscopy images, SEM images, contact angles of oil and water droplets of (a) flat polymer surface; (b) flat surface printed with composite material; (c) flat surface printed with composite material after nano spray; (d) 3D printed micropillar array; (e) 3D printed eggbeater structure with composite material; and (f) 3D printed eggbeater structure with composite material after nano spray.

6.6.6 *Applications of Superhydrophobic and Superlipophilic Structures*

The superhydrophobic property of the 3D-printed eggbeater structure, also called Salvinia effect, is due to the surface tension and low surface fraction of the contact area between water droplet and the eggbeater tip. The study of Salvinia effect gives insight to the development of artificial superhydrophobic and superolephilic surfaces, which have a wide variety of applications in fields such as self-cleaning, droplet control, micro droplet transportation, drag force reduction, and oil/water separation. In this section, several applications of 3D-printed eggbeater structures will be discussed.

(1) *Oil cleaning*

The cleanup of oil from water is a challenging topic and therefore motivates many specific designs of devices. Conventional oil sorbents are not efficient because they absorb large amounts of water in addition to oil, and sometimes secondary pollution may be incurred by some cleanup methods. Biological functional surfaces present some hints for the construction of effective oil cleaning devices.[96,97] For example, Salvinia paradox leaf can be used to absorb oil due to its superhydrophobic property.[96] The selective wetting properties are ideal for natural oil absorbent applications and bioinspired oil sorbent materials. A reusable superhydrophobic and lipophilic surface like 3D-printed artificial eggbeater structures can therefore be applied to high efficiency oil cleaning. As shown in Fig. 6.18, the oil droplet forms a round shape on the surface of water liquid due to low density and surface tension. When oil droplet contacts the printed eggbeater arms, the oil quickly spreads and penetrates through the 3D-printed artificial eggbeater structures and attaches to the structure firmly, thus being able to be removed from water. Such oil removal capability utilizes the superlipophilicity property of the 3D-printed eggbeater arms for oil wetting. It can achieve an average absorption rate (weight gain per second) of 12 $g.g^{-1}.s^{-1}$, which is lower than the carbon aerogel ($68g.g^{-1}.s^{-1}$),[89] but is higher than Graphene sponge ($0.57g.g^{-1}.s^{-1}$),[98] Cu foams ($\sim 1g.g^{-1}.s^{-1}$),[97] and NPs/PTEE-coated PU sponge ($2g.g^{-1}.s^{-1}$).[99] As an environment friendly material with separation efficiency greater than 99.9%, 3D-printed biomimetic eggbeater structures can be applied to surfaces used for applications such as clean-up of oil spills, wastewater treatment, and oil–water separation under harsh conditions.

Recent Advances in Additive Manufacturing

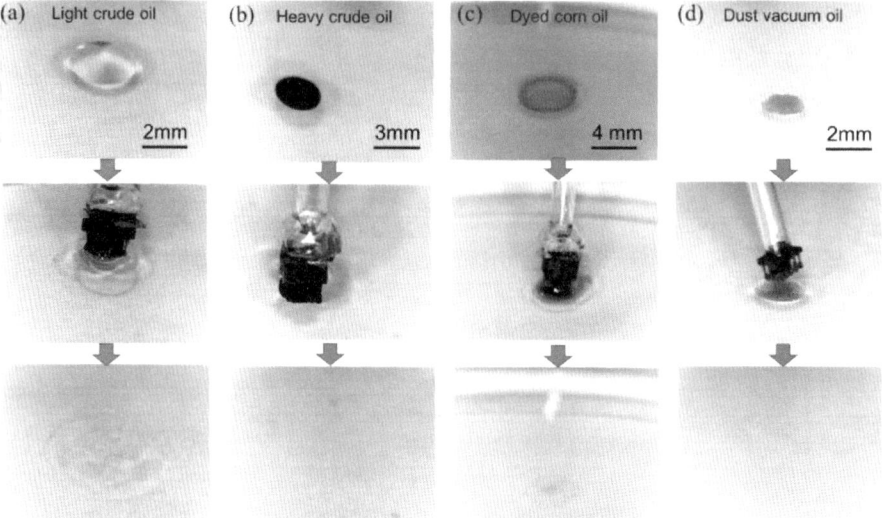

Fig. 6.18. The absorption process of different oils using the 3D printed eggbeater structure ($N = 4$, $d = 0.4$mm). (a) A light, paraffin base crude oil; (b) a heavy, asphalt base crude oil; (c) dyed corn oil with oil red dye; and (d) dust vacuum oil, respectively.

(2) *Droplet manipulations*

Liquid droplets are transported by a flow carrier filled with mutually immiscible liquid, and microdroplet manipulation is usually conducted with the help of certain auxiliary system,[100,101] where cumbersome and complex operation procedures are usually required. In contrast, 3D-printed eggbeater structures, featuring special high adhesion of liquid, can be used as an effective solution of manipulating droplets in a much easier way (see in Fig. 6.19). For instance, based on stable retention of micro-droplet, 3D printed eggbeater structures can be applied as microreactors to observe the chemical reaction or cell growth. With several outstanding characteristics such as high flexibility, significant material saving, and low cost, such microreactors show advances in protein crystallization, enzymatic kinetics, and other biochemical reactions.[102] Apart from microreactors, the 3D-printed eggbeater structures can be used to split water droplets by tuning its surface adhesion to different levels (see in Fig. 6.19(b)). The control of adhesion force is achieved through switching between different wetting states and changing the contact area between the eggbeater arms and water droplet. Specifically, the water adhesive force of eggbeater structure can be adjusted by changing the number of eggbeater arms (e.g. from 23 μN for $N = 2$ to 55 μN for $N = 8$). The non-loss water droplet transport can be achieved by using the eggbeater structures with significantly different adhesive forces (refer to

Fig. 6.19(c)). Such maneuverability allows the 3D-printed eggbeater structures to be used as a micro-gripper that can split micro droplets without loss or contamination, hence with potential applications in the area such as micro-sample analysis.[103] Lastly, such eggbeater structures can also be used to separate mixtures of water and olephilic materials, at a high absorption rate and only single unit is needed. The difference of capillary forces acting on the two phases between water and oil greatly helps the separation (see in Fig. 6.19(d)).

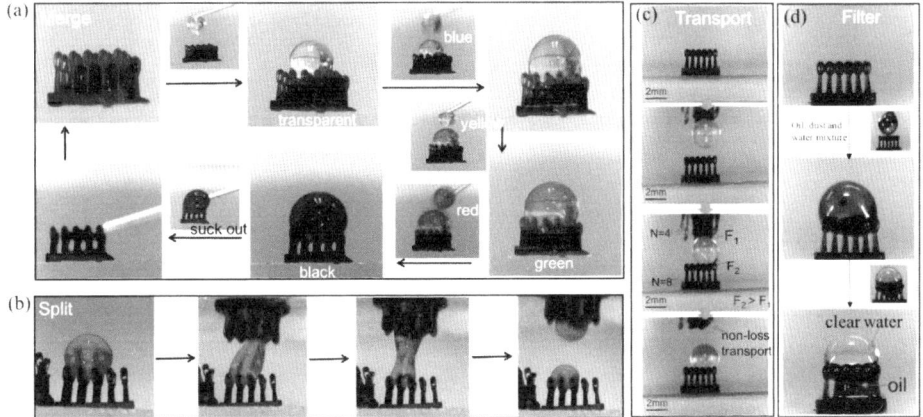

Fig. 6.19. The droplet manipulation application of superhydrophobic surface with the 3D-printed eggbeater structures. (a) The merge of several water droplets with different colors on the surface of artificial eggbeater structures; (b) water droplet is split by two eggbeater structures, both with $N = 4$; (c) the non-loss water transportation from an eggbeater structure ($N = 4$) to another with larger adhesive force ($N = 8$); and (d) oil/water filtering by using the 3D-printed eggbeater structure.

To replicate the morphology of the Salvinia Molesta leaves, the immersed surface accumulation 3D printing (ISA-3D) process has been developed and utilized for the fabrication of complex bioinspired eggbeater patterns. The process development covers the areas of material selection, design optimization, as well as detailed process preparation. The ISA-3D process development reveals the intriguing possibility of designing artificial surfaces on top of biomimetic structure to form a superhydrophobic surface. Specifically, the artificial eggbeater structure in the previous example shows remarkable superhydrophobic property and petal effect. The superhydrophobic property is associated with the number of eggbeater arms and the gap distance between each hair. The wetting characteristics of the 3D-printed hydrophobic surface are governed by both chemical composition and geometric structure.[104,105] The successful replication of natural structure gives the 3D-printed piece some outstanding properties. The water droplet control capability, for instance, indicates various potential applications in biomedical engineering and microfluidic devices. The petal effect of hydrophobic surface enables 3D cell culture to mimic avascular tumor naturally with inherent metabolic and proliferative gradients.[106] Such performance can potentially be used for building low cost toolbox for 3D culture in cell or tissue engineering. Meanwhile, because of different interfacial effects of oil and water, the 3D-printed superhydrophobic structures show an efficient separation of various kinds of oil/water mixtures and high absorption rate for oil spill cleanup. It facilitates an efficient and low-cost approach for oil spill cleanup and oil/water separation, offering a prospective solution for oil cleaning.[107]

6.7 Summary

Nature has developed high-performance materials and structures over millions of years' evolution, which present valuable sources of inspiration for engineering material systems. Biomimicry by learning from nature's concepts and design principles is driving a paradigm shift in modern engineering science and technology development. To address the fabrication challenges presented by biomimicry study, additive manufacturing (3D printing) could be a powerful tool for building hypothetical models in order to understand the biological function and its interaction with the environmental constraints in nature and in engineering systems.[109,110] The recent developments on bioinspired 3D-printing technology have demonstrated the potential of fabricating more sustainable materials and structures that were inspired by nature.[9,65,110,114,115] Here, we presented the introduction of biomimetic mechanically reinforced structures and hydrodynamic structures enabled by 3D printing. Understanding natural structures for various engineering applications and replicating them using 3D printing will lead us to drive the biomimicry field forward. At the same time, the fabrication challenges presented by biomimicry will lead to the development of more novel additive manufacturing processes. The future study on bioinspired 3D printing will fall in the category of multifunctional, multiscale, multimaterial, and multidimensional (e.g. 4D printing) fabrication. The development of biomimetic additive manufacturing technology will further lead to breakthroughs in constructing next-generation functional materials and structures for future engineering systems.

6.8 Outlook

Despite significant progresses in the fabrication of bioinspired materials and structures as well as the related applications, there are still many challenges that need to be addressed. The synthesis mechanisms of nature need to be further studied for the design of bioinspired structures that can be used in engineering systems. For the purpose, 3D printing together with other technologies needs to be developed to reveal the underlying mechanisms behind the exciting properties and phenomena that are observed in nature. Recently, the bioinspired 3D printing researches are focused on the duplication of single function using single material. The natural structure always exhibit multi-functionalities with multi-materials. The challenge lies in the solution of how to fabricate the multiple materials, especially the ceramics and polymers together to generate reinforced properties. For example, the excellent strong nacre structure is attributed to the aligned $CaCO_3$ platelet with the protein in between. In order to replicate the multimaterial structure, the 3D printing of ceramics needs sintering in high temperature that is usually well beyond the melting point of polymer. Hence 3D printing technology in the future needs to be further improved in order to fabricate complex biological structures. Another potential direction is to develop new multiscale 3D printing technology by integrating different 3D printing processes that were developed for different size scales to address the multiscale challenge required by the bioinspired structure fabrication. In addition, the hybrid process by integrating the 3D printing process with traditional fabrication technology is also a promising direction in future.

References

1. U. G. Wegst, H. Bai, E. Saiz, A. P. Tomsia, R. O. Ritchie, Bioinspired structural materials, *Nat. Mater.* **14**(1), 23 (2015).
2. N. Huebsch, D. J. Mooney, Inspiration and application in the evolution of biomaterials, *Nature* **462**(7272), 426 (2009).
3. Y. Mengüç, S. Y. Yang, S. Kim, J. A. Rogers, M. Sitti, Gecko-inspired controllable adhesive structures applied to micromanipulation, *Adv. Funct. Mater.* **22**(6), 1246–1254 (2012).
4. J. Sun, B. Bhushan, Hierarchical structure and mechanical properties of nacre: a review, *Rsc Adv.* **2**(20), 7617–7632 (2012).
5. Z. Liu, Z. Zhang, R. O. Ritchie, On the Materials Science of Nature's Arms Race, *Adv. Mater.* **30**(32), 1705220 (2018).
6. X. Li, Y. Chen, Micro-scale feature fabrication using immersed surface accumulation, *J. Manuf. Process.* **28**, 531–541 (2017).
7. E. A. Zimmermann, B. Gludovatz, E. Schaible, N. K. Dave, W. Yang, M. A. Meyers, R. O. Ritchie, Mechanical adaptability of the Bouligand-type structure in natural dermal armour, *Nat. Commun.* **4**, 2634 (2013).
8. S. E. Naleway, M. M. Porter, J. McKittrick, M. A. Meyers, Structural design elements in biological materials: application to bioinspiration, *Adv. Mater.* **27**(37), 5455–5476 (2015).
9. Y. Yang, X. Song, X. Li, Z. Chen, C. Zhou, Q. Zhou, Y. Chen, Recent progress in biomimetic additive manufacturing technology: From materials to functional structures, *Adv. Mater.*, **30**(36), 1706539 (2018).

10. L. Wen, J. C. Weaver, P. J. Thorn, Hydrodynamic function of biomimetic shark skin: Effect of denticle pattern and spacingycroft, *Bioinspir. Biomim.* **10**(6), 066010 (2015).

11. D. J. Babu, M. Mail, W. Barthlott, J. J. Schneider, Superhydrophobic vertically aligned carbon nanotubes for biomimetic air retention under water (Salvinia effect), *Adv. Mater. Interfaces* **4**(13), 1700273 (2017).

12. S. Zhang, J. Huang, Z. Chen, Y. Lai, Bioinspired special wettability surfaces: from fundamental research to water harvesting applications, *Small* **13**(3), 1602992(2017).

13. P. S. Brown, B. Bhushan, Bioinspired materials for water supply and management: water collection, water purification and separation of water from oil, *Phil. Trans. R. Soc. A* **374** (2073), 20160135 (2016).

14. Q. Wang, X. Yao, H. Liu, D. Quéré, L. Jiang, Self-removal of condensed water on the legs of water striders, *Proc. Natl. Acad. Sci.* **112**(30), 9247–9252 (2015).

15. C. Song, Y. Zheng, Wetting-controlled strategies: From theories to bio-inspiration, *J. Colloid. Interf. Sci.* **427**, 2–14 (2014).

16. A. R. Parker, C. R. Lawrence, Water capture by a desert beetle, *Nature* **414**, 33–34 (2001).

17. C. Dorrer, J. R. Rühe, Mimicking the Stenocara beetle dewetting of drops from a patterned superhydrophobic surface, *Langmuir* **24**(12), 6154–6158 (2008).

18. Z. Yu, F. F. Yun, Y. Wang, L. Yao, S. Dou, K. Liu, L. Jiang, X. Wang, Desert beetle‐inspired superwettable patterned surfaces for water harvesting, *Small* **13**(36), 1701403 (2017).

19. D.-J. Huang, T.-S. Leu, Fabrication of a wettability-gradient surface on copper by screen-printing techniques, *J. Micromech. Microeng* **25**(8), 085007 (2015).

20. M. Cao, J. Xiao, C. Yu, K. Li, L. Jiang, Hydrophobic/hydrophilic cooperative Janus system for enhancement of fog collection, *Small* **11**(34), 4379–4384, (2015).

21. L. Hong, T. Pan, Photopatternable superhydrophobic nanocomposites for microfabrication, *J. Microelectromech. S.* **19**(2), 246–253 (2010).

22. A. Nakajima, Y. Nakagawa, T. Furuta, M. Sakai, T. Isobe, S. Matsushita, Sliding of water droplets on smooth hydrophobic silane coatings with regular triangle hydrophilic regions, *Langmuir* **29**(29), 9269–9275 (2013).

23. H. Zhu, R. Duan, X. Wang, J. Yang, J. Wang, Y. Huang, F. Xia, Prewetting dichlormethane induced aqueous solution adhered on Cassie superhydrophobic substrates to fabricate efficient fog-harvesting materials inspired by Namib desert beetles and mussels, *Nanoscale,* **10**(27), 13045–13054 (2018).

24. X. Tian, Y. Chen, Y. Zheng, H. Bai, L. Jiang, Controlling water capture of bioinspired fibers with hump structures, *Adv. Mater.* **23**(46), 5486–5491 (2011).

25. H. Dong, N. Wang, L. Wang, H. Bai, J. Wu, Y. Zheng, Y. Zhao, L. Jiang, Bioinspired electrospun knotted microfibers for fog harvesting, *ChemPhysChem* **13**(5), 1153–1156 (2012).

26. M. Du, Y. Zhao, Y. Tian, K. Li, L. Jiang, Electrospun multiscale structured membrane for efficient water collection and directional transport, *Small* **12**(8), 1000–1005 (2016).

27. P. Comanns, P. C. Withers, F. J. Esser, W. Baumgartner, Cutaneous water collection by a moisture-harvesting lizard, the thorny devil (Moloch horridus), *J. Exp. Biol.* **219**(21), 3473–3479 (2016).

28. P. Comanns, K. Winands, K. Arntz, F. Klocke, W. Baumgartner, Laser-based biomimetic functionalization of surfaces: from moisture harvesting lizards to specific fluid transport systems, *Int. J. Des. Nat. Ecodyn.* **9**(3), 206–215 (2014).

29. Z. Pan, W. G. Pitt, Y. Zhang, N. Wu, Y. Tao, T. T. Truscott, The upside-down water collection system of Syntrichia caninervis, *Nat. Plants* **2**(7), 16076 (2016).

30. H. Andrews, E. Eccles, W. Schofield, J. Badyal, Three-dimensional hierarchical structures for fog harvesting, *Langmuir,* **27**(7), 3798–3802 (2011).

31. J. Ju, H. Bai, Y. Zheng, T. Zhao, R. Fang, L. Jiang, A multi-structural and multi-functional integrated fog collection system in cactus, *Nat. Commun.* **3**, 1247 (2012).

32. J. Ju, K. Xiao, X. Yao, H. Bai, L. Jiang, Bioinspired conical copper wire with gradient wettability for continuous and efficient fog collection, *Adv. Mater.* **25**(41), 5937–5942 (2013).

33. J. Ju, X. Yao, S. Yang, L. Wang, R. Sun, Y. He, L. Jiang, Cactus stem inspired cone‐arrayed surfaces for efficient fog collection, *Adv. Funct. Mater.* **24**(44), 6933–6938 (2014).

34. Y. Huang, B. B. Stogin, N. Sun, J. Wang, S. Yang, T. S. Wong, A switchable cross‐species liquid repellent surface, *Adv. Mater.* **29**(8), 1604641 (2017).

35. S. V. Murphy, A. Atala,3D bioprinting of tissues and organs, *Nat. Biotechnol.* **32**(8), 773 (2014).

36. Y. M. Song, Y. Xie, V. Malyarchuk, J. Xiao, I. Jung, K.-J. Choi, Z. Liu, H. Park, C. Lu, R.-H. Kim, Digital cameras with designs inspired by the arthropod eye, *Nature* **497**(7447), 95 (2013).

37. B. Su, Y. Tian, L. Jiang, Bioinspired interfaces with superwettability: from materials to chemistry, *J. Am. Chem. Soc.* **138**(6), 1727–1748 (2016).

38. L. Pu, R. Saraf, V. Maheshwari, Bio-inspired interlocking random 3-D structures for tactile and thermal sensing, *Sci. Rep.* **7**(1), 5834 (2017).

39. S. H. Huang, P. Liu, A. Mokasdar, L. Hou, Additive manufacturing and its societal impact: a literature review, *Int. J. Adv. Manuf. Tech.* **67**(5-8), 1191–1203 (2013).

40. W. Gao, Y. Zhang, D. Ramanujan, K. Ramani, Y. Chen, C. B. Williams, C. C. Wang, Y. C. Shin, S. Zhang, P. D. Zavattieri, The status, challenges, and future of additive manufacturing in engineering, *CAD,* **69**, 65–89 (2015).

41. S. M. Thompson, L. Bian, N. Shamsaei, A. Yadollahi, An overview of Direct Laser Deposition for additive manufacturing; Part I: Transport phenomena, modeling and diagnostics, *Addit. Manuf.* **8**, 36–62 (2015).

42. S. Mellor, L. Hao, D. Zhang, Additive manufacturing: A framework for implementation, *Int. J. Prod. Econ.* **149**, 194–201, (2014).

43. M. Vaezi, H. Seitz, S. Yang, A review on 3D micro-additive manufacturing technologies, *Int. J. Adv. Manuf. Tech.* **67**(5–8), 1721–1754 (2013).

44. B. G. Compton, J. A. Lewis, 3D-printing of lightweight cellular composites, *Adv. Mater.* **26**(34), 5930–5935 (2014).

45. J. Taboas, R. Maddox, P. Krebsbach, S. Hollister, Indirect solid free form fabrication of local and global porous, biomimetic and composite 3D polymer-ceramic scaffolds, *Biomaterials* **24**(1), 181–194 (2003).

46. Z. Quan, A. Wu, M. Keefe, X. Qin, J. Yu, J. Suhr, J.-H. Byun, B.-S. Kim, T.-W. Chou, Additive manufacturing of multi-directional preforms for composites: opportunities and challenges, *Mater. Today* **18**(9), 503–512 (2015).

47. J. P. Lewicki, J. N. Rodriguez, C. Zhu, M. A. Worsley, A. S. Wu, Y. Kanarska, J. D. Horn, E. B. Duoss, J. M. Ortega, W. Elmer, 3D-Printing of meso-structurally ordered carbon fiber/polymer composites with unprecedented orthotropic physical properties, *Sci. Rep.* **7**, 43401 (2017).

48. A. S. Gladman, E. A. Matsumoto, R. G. Nuzzo, L. Mahadevan, J. A. Lewis, Biomimetic 4D printing, *Nat. Mater.* **15**(4), 413 (2016).

49. D. E. Yunus, W. Shi, S. Sohrabi, Y. Liu, Shear induced alignment of short nanofibers in 3D printed polymer composites, *Nanotechnology* **27**(49), 495302 (2016).

50. A. E. Jakus, E. B. Secor, A. L. Rutz, S. W. Jordan, M. C. Hersam, R. N. Shah, Three-dimensional printing of high-content graphene scaffolds for electronic and biomedical applications, *ACS Nano,* **9**(4), 4636–4648 (2015).

51. M. M. Porter, M. Yeh, J. Strawson, T. Goehring, S. Lujan, P. Siripasopsotorn, M. A. Meyers, J. McKittrick, Magnetic freeze casting inspired by nature, *Mater. Scie. Eng.: A* **556**, 741–750 (2012).

52. M. M. Porter, L. Meraz, A. Calderon, H. Choi, A. Chouhan, L. Wang, M. A. Meyers, J. McKittrick, Torsional properties of helix-reinforced composites fabricated by magnetic freeze casting, *Compos. Struct.* **119**,174–184 (2015).

53. J. J. Martin, B. E. Fiore, R. M. Erb, Designing bioinspired composite reinforcement architectures via 3D magnetic printing, *Nat. Commun.* **6**, 8641 (2015).

54. R. M. Erb, R. Libanori, N. Rothfuchs, A. R. Studart, Composites reinforced in three dimensions by using low magnetic fields, *Science* **335**(6065), 199–204 (2012).

55. T. P. Niebel, F. J. Heiligtag, J. Kind, M. Zanini, A. Lauria, M. Niederberger, A. R. Studart, Multifunctional microparticles with uniform magnetic coatings and tunable surface chemistry, *Rsc Adv.* **4**(107), 62483–62491 (2014).

56. D. Kokkinis, M. Schaffner, A. R. Studart, Multimaterial magnetically assisted 3D printing of composite materials, *Nat. Commun.* **2015**, *6*, 8643.

57. D. Van den Ende, S. Van Kempen, X. Wu, W. Groen, C. Randall, S. Van der Zwaag, Dielectrophoretically structured piezoelectric composites with high aspect ratio piezoelectric particles inclusions, *J. Appl. Phys.* **111**(12), 124107 (2012).

58. S. Armon, E. Efrati, R. Kupferman, E. Sharon, Geometry and mechanics in the opening of chiral seed pods, *Science,* **333**(6050), 1726–1730 (2011).

59. H. O. Fabritius, C. Sachs, P. R. Triguero, D. Raabe, Influence of structural principles on the mechanics of a biological fiber-based composite material with hierarchical organization: the exoskeleton of the lobster Homarus americanus, *Adv. Mater.* **21**(4), 391–400 (2009).

60. Z. Chen, X. Song, L. Lei, X. Chen, C. Fei, C. T. Chiu, X. Qian, T. Ma, Y. Yang, K. Shung, 3D printing of piezoelectric element for energy focusing and ultrasonic sensing, *Nano Energy* **27**, 78–86 (2016).

61. Y. Pan, X. Zhao, C. Zhou, Y. Chen, Smooth surface fabrication in mask projection based stereolithography, *J. Manuf. Process.* **14**(4), 460–470 (2012).

62. X. Song, Y. Chen, T. W. Lee, S. Wu, L. Cheng, Ceramic fabrication using mask-image-projection-based stereolithography integrated with tape-casting, *J. Manuf. Process.* **20**, 456–464 (2015).

63. C. Park, J. Wilkinson, S. Banda, Z. Ounaies, K. E. Wise, G. Sauti, P. T. Lillehei, J. S. Harrison, Aligned single-wall carbon nanotube polymer composites using an electric field, *J. Polym. Sci. Polym. Phys.* **44**(12), 1751–1762 (2006).

64. T. Takahashi, T. Murayama, A. Higuchi, H. Awano, K. Yonetake, Aligning vapor-grown carbon fibers in polydimethylsiloxane using dc electric or magnetic field, *Carbon,* **44**(7), 1180–1188 (2006).

65. Y. Yang, Z. Chen, X. Song, Z. Zhang, J. Zhang, K. K. Shung, Q. Zhou, Y. Chen, Biomimetic anisotropic reinforcement architectures by electrically assisted nanocomposite 3D printing, *Adv. Mater.* **29**(11), 1605750 (2017).

66. M. A. Meyers, J. McKittrick, P.-Y. Chen, Structural biological materials: critical mechanics-materials connections, *Science,* **339**(6121), 773–779 (2013).

67. S. Suresh, Graded materials for resistance to contact deformation and damage, *Science,* **292**(5526), 2447–2451 (2001).

68. B. J. Bruet, J. Song, M. C. Boyce, C. Ortiz, Materials design principles of ancient fish armour, *Nat. Mater.* **7**(9), 748 (2008).

69. K.-C. Park, P. Kim, A. Grinthal, N. He, D. Fox, J. C. Weaver, J. Aizenberg, Condensation on slippery asymmetric bumps, *Nature* **531**(7592), 78 (2016).

70. Y. Peng, Y. He, S. Yang, S. Ben, M. Cao, K. Li, K. Liu, L. Jiang, Magnetically induced fog harvesting via flexible conical arrays, *Adv. Funct. Mater.* **25**(37), 5967–5971 (2015).

71. L. Zhang, J. Wu, M. N. Hedhili, X. Yang, P. Wang, Inkjet printing for direct micropatterning of a superhydrophobic surface: toward biomimetic fog harvesting surfaces, *J. Mater. Chem. A* **3**, 2844–2852 (2015).

72. X. Li, B. Xie, J. Jin, Y. Chai, Y. Chen, 3D Printing Temporary Crown and Bridge by Temperature Controlled Mask Image Projection Stereolithography, *Procedia. Manuf.* **26**, 1023–1033 (2018).

73. Y. Pan, C. Zhou, Y. Chen, A fast mask projection stereolithography process for fabricating digital models in minutes, *J. Manuf. Sci. Eng.* **134**(5), 051011 (2012).

74. C. Zhou, Y. Chen, Z. Yang, B. Khoshnevis, Digital material fabrication using mask-image-projection-based stereolithography, *Rapid Prototyp. J.* **19**(3), 153–165 (2013).

75. J.-P. Kruth, P. Mercelis, J. Van Vaerenbergh, L. Froyen, M. Rombouts, Binding mechanisms in selective laser sintering and selective laser melting, *Rapid Prototyp. J.* **11**(1), 26–36 (2005).

76. J. A. Lewis, Direct ink writing of 3D functional materials, *Adv. Funct. Mater.* **16**(17), 2193–2204 (2006).

77. H. Sirringhaus, T. Kawase, R. Friend, T. Shimoda, M. Inbasekaran, W. Wu, E. Woo, High-resolution inkjet printing of all-polymer transistor circuits, *Science* **290**(5499), 2123–2126 (2000).

78. S.-H. Ahn, M. Montero, D. Odell, S. Roundy, P. K. Wright, Anisotropic material properties of fused deposition modeling ABS, *Rapid Prototyp. J.* **8**(4), 248–257 (2002).

79. D. Lin, Q. Nian, B. Deng, S. Jin, Y. Hu, W. Wang, G. J. Cheng, Three-dimensional printing of complex structures: man made or toward nature?, *ACS Nano* **8**(10), 9710–9715 (2014).

80. Y. Chen, C. Zhou, J. Lao, A layerless additive manufacturing process based on CNC accumulation, *Rapid Prototyp. J.* **17**(3), 218–227 (2011).

81. X. Zhao, Y. Pan, C. Zhou, Y. Chen, C. C. Wang, An integrated CNC accumulation system for automatic building-around-inserts, *J. Manuf. Process.* **15**(4), 432–443 (2013).

82. Y. Pan, C. Zhou, Y. Chen, J. Partanen, Multi-tool and multi-axis cnc accumulation for fabricating conformal features on curved surfaces, *ASME J. Manuf. Sci. Eng.* **136**(3), 031007 (2014).

83. K. Liu, Y. Tian, L. Jiang, Bio-inspired superoleophobic and smart materials: design, fabrication, and application, *Prog. Mater. Sci.* **58**(4), 503–564 (2013).

84. C. Valero, H. Amaveda, M. Mora, J. M. García-Aznar, Combined experimental and computational characterization of crosslinked collagen-based hydrogels, *PloS One* **13**, e0195820 (2018).

85. G. X. Gu, I. Su, S. Sharma, J. L. Voros, Z. Qin, M. J. Buehler, Three-dimensional-printing of bio-inspired composites, *J. Biomech. Eng.* **138**(2), 021006 (2016).

86. T. J. Hinton, Q. Jallerat, R. N. Palchesko, J. H. Park, M. S. Grodzicki, H.-J. Shue, M. H. Ramadan, A. R. Hudson, A. W. Feinberg, Three-dimensional printing of complex biological structures by freeform reversible embedding of suspended hydrogels, *Sci. Adv.* **1**(9), e1500758 (2015).

87. T. Bhattacharjee, S. M. Zehnder, K. G. Rowe, S. Jain, R. M. Nixon, W. G. Sawyer, T. E. Angelini, Writing in the granular gel medium, *Sci. Adv.* **1**(8), e1500655 (2015).

88. Z. Qin, B. G. Compton, J. A. Lewis, M. J. Buehler, Structural optimization of 3D-printed synthetic spider webs for high strength, *Nat. Commun.* **6**, 7038 (2015).

89. C. Yan, Z. Ji, S. Ma, X. Wang, F. Zhou, 3D Printing as Feasible Platform for On-Site Building Oil-Skimmer for Oil Collection from Spills, *Adv. Mater. Interfaces* **3**(13), 1600015 (2016).

90. X. Wang, X. Cai, Q. Guo, T. Zhang, B. Kobe, J. Yang, i3DP, a robust 3D printing approach enabling genetic post-printing surface modification, *Chem. Commun.* **49**(86), 10064–10066 (2013).

91. Y. Zhang, M. Yin, O. Xia, A. P. Zhang, H.-Y. Tam. Optical 3D µ-printing of polytetrafluoroethylene (PTFE) microstructures. In *IEEE Micro Electro Mechanical Systems (MEMS)*, Belfast, UK (Jan 2018).

92. M. T. Z. Myint, R. Kitsomboonloha, S. Baruah, J. Dutta, Superhydrophobic surfaces using selected zinc oxide microrod growth on ink-jetted patterns, *J. Colloid Interf. Sci.* **354**(2), 810–815 (2011).

93. W. Barthlott, T. Schimmel, S. Wiersch, K. Koch, M. Brede, M. Barczewski, S. Walheim, A. Weis, A. Kaltenmaier, A. Leder, The Salvinia paradox: superhydrophobic surfaces with hydrophilic pins for air retention under water, *Adv. Mater.* **22**(21), 2325–2328 (2010).

94. O. Tricinci, T. Terencio, B. Mazzolai, N. M. Pugno, F. Greco, V. Mattoli, 3D micropatterned surface inspired by salvinia molesta via direct laser lithography, *ACS Appl. Mater. Inter.* **7**(46), 25560–25567 (2015).

95. Y. Yang, X. Li, X. Zheng, Z. Chen, Q. Zhou, Y. Chen, 3D-Printed Biomimetic Super-Hydrophobic Structure for Microdroplet Manipulation and Oil/Water Separation, *Adv. Mater.* **30**(9), 1704912 (2018).

96. C. Zeiger, I. C. R. da Silva, M. Mail, M. N. Kavalenka, W. Barthlott, H. Hölscher, Microstructures of superhydrophobic plant leaves-inspiration for efficient oil spill cleanup materials, *Bioinspir. Biomim.* **11**(5), 056003 (2016).

97. J. Song, Y. Lu, J. Luo, S. Huang, L. Wang, W. Xu, I. P. Parkin, Barrel-Shaped Oil Skimmer Designed for Collection of Oil from Spills, *Adv. Mater. Interfaces* **2**(15), 1500350 (2015).

98. H. Bi, X. Xie, K. Yin, Y. Zhou, S. Wan, L. He, F. Xu, F. Banhart, L. Sun, R. S. Ruoff, Spongy graphene as a highly efficient and recyclable sorbent for oils and organic solvents, *Adv. Funct. Mater.* **22**(21), 4421–4425 (2012).

99. P. Calcagnile, D. Fragouli, I. S. Bayer, G. C. Anyfantis, L. Martiradonna, P. D. Cozzoli, R. Cingolani, A. Athanassiou, Magnetically driven floating foams for the removal of oil contaminants from water, *ACS Nano* **6**(6), 5413–5419 (2012).

100. A. J. Demello, Control and detection of chemical reactions in microfluidic systems, *Nature,* **442**(7101), 394–402 (2006).

101. B. Su, S. Wang, Y. Song, L. Jiang, A heatable and evaporation-free miniature reactor upon superhydrophobic pedestals, *Soft Matter* **8**(3), 631–635 (2012).

102. H. Song, D. L. Chen, R. F. Ismagilov, Reactions in droplets in microfluidic channels, *Angew. Chem. Int. Ed.* **45**(44), 7336–7356 (2006).

103. Y. Lai, F. Pan, C. Xu, H. Fuchs, L. Chi, In situ surface-modification-induced superhydrophobic patterns with reversible wettability and adhesion, *Adv. Mater.* **25**(12), 1682–1686 (2013).

104. A. Lafuma, D. Quéré, Superhydrophobic states, *Nat. Mater.* **2**(7), 457 (2003).

105. R. N. Wenzel, Resistance of solid surfaces to wetting by water, *Ind. Eng. Chem.* **28**(8), 988–994 (1936).

106. J. Friedrich, C. Seidel, R. Ebner, L. A. Kunz-Schughart, pheroid-based drug screen: considerations and practical approach, *Nat. Protoc.* **4**(3), 309 (2009).

107. S. Lee, B. Kim, S. H. Kim, E. Kim, J. H. Jang, Superhydrophobic, Reversibly Elastic, Moldable, and Electrospun (SupREME) Fibers with Multimodal Functions: From Oil Absorbents to Local Drug Delivery Adjuvants, *Adv. Funct. Mater.* **27**(37), 1702310 (2017).

108. Li, X., Yang. Y, Liu, L., Chen, Y., Chu, M., Sun, H., Chen, Y. 3D-Printed Cactus-Inspired Spine Structures for Highly Efficient Water Collection, *Adv. Mater. Interfaces* **7**(3), 1901752 (2020).

109. Lipson, H., & Kurman, M., *Fabricated: The New World of 3D Printing*. John Wiley & Sons (2013).

110. Chua, C. K., & Leong, K. F., *3D Printing and Additive Manufacturing: Principles and Applications (with Companion Media Pack) of Rapid Prototyping, 4th Edition* World Scientific (2014).

111. Li, X., Chen, Y., Multi-scale 3D printing of bioinspired structures for functional surfaces, *Proc. of International Symposium on Flexible Automation* (ISFA), pp.13–20, Kanazawa, Japan (July, 2018).

112. Li, X., Yang. Y, Chen, Y., Bio-inspired Micro-Scale Texture Fabrication based on immersed surface accumulation process, *Proc. of the World Congress on Micro and Nano Manufacturing* (WCMNM) pp. 33–36, Kaohsiung, Taiwan (March 2017).

113. Li, X., Baldacchini T, Song, X., Chen, Y., Multi-scale additive manufacturing: an investigation on building objects with macro-, micro- and nano-scales features. *Proc. of the 11th International Conference on Micro Manufacturing (ICOMM)*, Orange County, California, USA (March 2016).

114. Leung, Y.S., Kwork, T.H., Li, X., Yang, Y., Wang, C.C., Chen, Y., Challenges and status on design and computation for emerging additive manufacturing technologies, *Journal of Computing and Information Science in Engineering* **19**(2) (2019).

115. Yang Y, Li X, Chu M, Sun H, Jin J, Yu K, Wang Q, Zhou Q, Chen Y, Electrically assisted 3D printing of nacre-inspired structures with self-sensing capability, *Science Advances* **5**(4), eaau 9490 (2019).

© 2020 World Scientific Publishing Company
https://doi.org/10.1142/9789811222825_0007

Chapter 7

4D Printing Based on Multi-Material Design

Devin J. Roach, Xiao Kuang, Craig M. Hamel,
Martin L. Dunn, and H. Jerry Qi

Recent advances in multi-material and smart material additive manufacturing have enabled a new generation of structures which can achieve complex shape-shifting. This additive manufacturing process is known as 4D printing, which allows 3D printed structures to change their configuration or function with time in response to external stimuli such as heat or light. The term 4D printing first appeared in 2013 and the field has seen significant progress in the following years. This chapter provides a review of the recent progress in this emerging field with a focus on the material selection and structural design.

7.1 Introduction

Additive Manufacturing (AM), or 3D printing, is defined by a range of technologies that are capable of translating virtual solid model data into physical models in a rapid process.[1] Due to its advantage of allowing structurally complex objects to be rapidly manufactured for immediate use, 3D printing has become one of the most notable inventions of the 21st century, finding applications ranging from visual prototypes[2] to tissue engineering, biomedicine,[3-5] electronic devices,[6,7] high-performance materials/metamaterials,[8-10] and space sciences.[11] According to the American Society for Testing and Materials (ISO/ASTM 52900:2015) there are over 50 different AM technologies that can be classified into seven different categories: binder jetting, material jetting, material extrusion, vat photo-polymerization, powder bed fusion, energy deposition, and sheet lamination[12]. Each technique shares a common feature: building a 3D structure through material addition in a layer-by-layer fashion. Another way to classify 3D printing technologies is based on how the raw material (or ink) is deposited, including fused filament fabrication (FFF), direct ink writing (DIW), stereolithography (SLA), selective laser sintering (SLS), binder jetting, and PolyJet printing, etc. In 2017, 320,000 AM-related US patents were filed,[13] representing a wide range of AM related research and application activities. The rapid commercialization of 3D printing has demanded extensive research to develop new, printable materials that have desirable attributes, such as strength, toughness, or color, depending on the end-use applications.[14]

In 2013, at a TED talk, Tibbits of MIT first developed the term 4D printing,[15] which combines 3D printing and smart (or active) materials to form objects whose shape could change after printing.[16-21] In this initial definition, 4D printing was "3D printing + time", with the fourth dimension being time.[16,22-24] Currently, one popular definition of 4D printing is that the shape, property, and functionality of a 3D printed structure could evolve with time when it is exposed to a predetermined stimulus.[24-30]

4D printing relies on the fast growth and interdisciplinary research of 3D printing, smart materials and design.[20,31] Compared to the static objects created by 3D printing, 4D printing allows a 3D printed structure to change its configuration or function with time in response to external free energy or stimulus such as temperature (Fig. 7.1A). The basic mechanism of shape-shifting is either by using a smart material directly, or by creating localized eigenstrains (e.g. mismatch strain) within a printed object, usually a composite, during or after printing.[20,32,33] A schematic example of 4D printing is shown in Fig. 7.1B, which demonstrates the combination of PolyJet 3D printing and smart materials to yield a 4D printed structure. Here, a rubber/plastics bilayer composite was printed by PolyJet 3D printing where the eigenstrain was directly embedded in the composite during printing. After heating above the glass transition of the plastics layer, the release of eigenstrain led to a deformation mismatch between the two layers and resulted in a controlled shape transformation.

4D printing adds the additional dimension of time to 3D printing and thus has a few notable advantages: (1) allowing the fabrication of intelligent devices which can detect environmental signals by changing its shape or function; (2) saving printing time and materials, especially in the case of printing thin or lattice structures;[35,36] (3) saving space for storage and transportation of the printed parts. It should be noted that pursuing materials, structures, or devices that can change their shape, properties, or functionalities has been a focus by the smart material research community for several decades.[37-40] Recent advances in 3D printing have greatly assisted the development of new smart structures and devices by significantly simplifying the fabrication process, which was a primary barrier for smart materials research in the past. The emergence of 4D printing has drawn

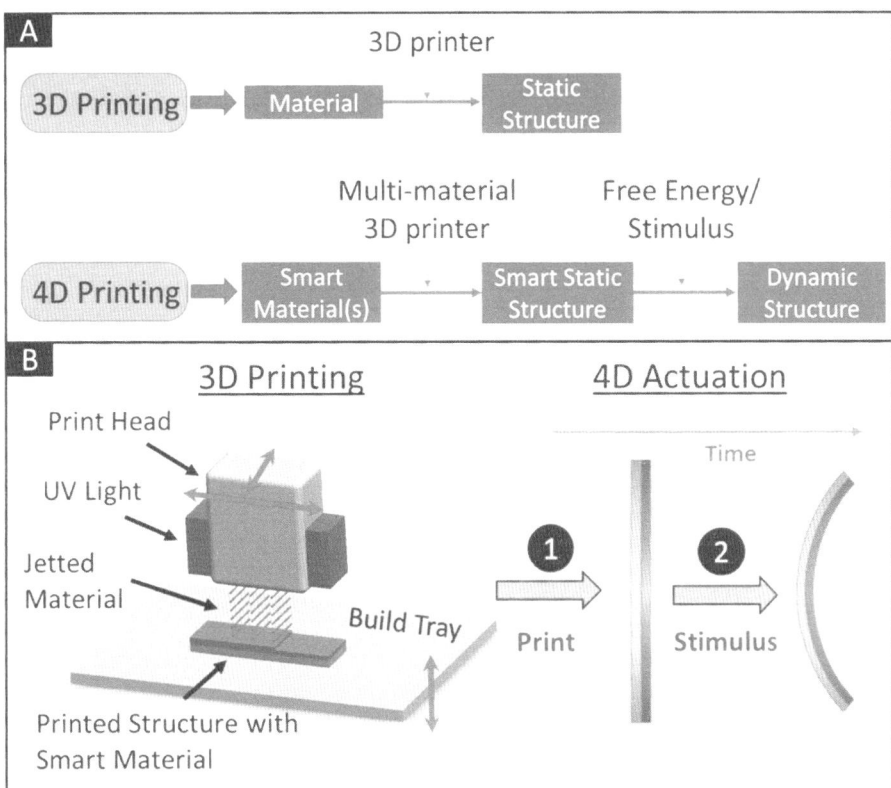

Fig. 7.1. (A) Flowchart demonstrating the difference between 3D and 4D printing. (B) Schematic of 4D actuation of an object upon stimulus directly after a standard PolyJet 3D printing process (reproduced from Ding et al.[34] with permission from HDIAC).

significant attention from the smart materials community as well as the advanced manufacturing community. This led to the intensive 4D printing research with the first academic paper in this field appearing in 2013,[22] which demonstrated how static objects could transform their shape after printing. The 4D printing of metals,[54] and ceramics[41,42] demonstrates the ever increasing scope of potential applications offered by 4D printing. Compared to smart metals and ceramics, however, smart polymers have the advantage of diverse stimulus-responsive groups, large deformability, and abundant material choices,[43] and thus have been widely used in 4D printing. Although single 3D printed smart materials, such as shape memory polymers (SMPs), have been used for shape changing,[44-47] the majority of 4D printing techniques are achieved by 3D printing composites using multi-material design. For this reason, the primary focus of this chapter will be on polymeric multi-material 3D printing. We will begin by outlining the basic elements of 4D printing and extend this to the major polymeric material systems and mechanisms for 4D printing. We will end with comments on the future perspectives of 4D printing.

7.2 The Basic Elements of 4D Printing

7.2.1. *Multi-material 3D Printing*

4D printing requires at least one active (or smart) material or a composite of smart materials. The energy input or external stimulus used to induce or release eigenstrain (or strain mismatch) within a printed object leads to shape shifting. Multi-material 3D printing techniques, where different materials can be precisely placed in 3D space, are highly desirable to achieve 4D printing. The most common and successful multi-material 3D printing method is the PolyJet method,[48] where multiple inkjet printheads are used to deposit different material liquid resins onto the printing platform, followed by a rapid polymerization, or solidification, of the jetted resins using ultraviolet (UV) light. To date, the 3D Systems ProJet MultiJet 3D Printer (Rock Hill, SC, USA) and the Stratasys PolyJet Connex series (Eden Prairie, MN, USA) are the most commonly used commercial multi-material 3D printing platforms. However, both rely on inkjet technology and only work with the company's proprietary inks, all of which are acrylate-based photopolymers. Due to the thermoset nature of the inkjet printed polymers, the shape memory effects can be observed and utilized as 3D printed shape memory polymers (SMPs). Polymer extrusion-based 3D printing techniques such as fused filament fabrication (FFF) and direct ink-write (DIW) are also suitable for multi-material 3D printing by using multiple deposition nozzles.[49-53] Digital light processing (DLP) 3D printing has emerged in recent years as a rapid AM approach. Some methods were developed where multiple polymer resin vats[46] or grayscale light [54,55] are used to create a single, multi-material object. Most recently, researchers have developed multi-material printers by integrating multiple AM techniques into a single printing platform to realize novel structures with a wide array of material properties and functionalities.[56,57]

7.2.2 *Stimulus-responsive Smart Materials*

To-date, two primary smart materials exploited in 4D printing are hydrogels and SMPs which can be activated by water/moisture,[24,58] temperature,[25,59-61] light,[62,63] or a combination of these.[27,29,64] Hydrogels that swell in water or solutions have been extensively used to produce shape change. When hydrogels are used along with non-swelling polymers, the resulting swelling ratio mismatch can lead to a biased deformation field and shape shifting.[27,65] Water is the most commonly used stimulus for hydrogels. In addition, chemical,[66] pH,[30] and temperature responsive hydrogels[67,68] have been used. A primary advantage of using hydrogels is that the swelling and de-swelling behavior can result in reversible shape changes, although typically occurring over a very long timescale.[24,69]

The second type of smart materials commonly used in 4D printing is SMPs. SMPs are a class of smart polymers that can hold a temporary shape and recover its initial shape in the presence of an external stimulus.[40,70-74] Compared to hydrogels, SMPs have the advantages of excellent mechanical properties and facile programmability. SMPs have been studied extensively for their use in stimulus-responsive shape change finding applications in artificial muscles, soft robotics, implantable medical devices, and consumer products.[75-83] To achieve shape-shifting, an SMP requires a programming step and a recovery step. For the most widely studied thermally activated SMPs, the programming steps are as follows: (1) the SMP is first deformed at a temperature above a transition temperature (T_t) (normally the polymer glass transition temperature), (2) then the SMP is cooled down to below the T_t, and (3) finally unloaded. This process fixes the deformed, or programmed, shape. To achieve the shape change, the SMP can be heated above the T_t, and the initial shape is recovered due to its entropic elasticity[84-87]. This shape recovery process is known as the shape memory effect (SME) and is the baseline physical phenomenon occurring in all 4D printed structures relying on SMPs. It is noted that thermally triggered SMPs can be achieved not only by direct heating, but also by indirect heating, such as Joule heating,[88] the photothermal effect,[89-93] and the hysteresis effect,[94-97] offering a wide variety of choices for actuation. When using SMPs in multi-material 4D printing, through specific distribution of SMPs in the composite matrix, many interesting shape changing properties can be achieved.[59,98] SMPs have been used as the basic active component in printed active fiber composites (PACs), bilayers, and laminates for 4D printing,[22,59,60,75,98-100] which will be discussed later.

In addition to SMP and hydrogels, liquid crystal elastomers (LCE) have recently began to generate intensive interest in 4D printing. LCEs have the advantage of rapid and reversible actuation: two issues currently facing SMP and hydrogel-based systems. Previously, LCE had to be mold-fabricated and complex programming steps were involved in order to achieve reversible actuation.[101,102] However, in recent years, direct ink-write (DIW) 3D printing has been utilized to 3D print LCEs able to be activated directly after printing.[103,104] As such, multi-material LCE-based structures have been fabricated utilizing novel 3D printing platforms. These structures can achieve novel activation sequences hardly attainable using SMP and hydrogel-based composites.

Some other shape shifting smart materials and composites can rely on light or magnetic fields to achieve remote actuation.[45,89,91,93,105-107] These materials, therefore, can be useful when specific knowledge about the system is required, resulting in a type of smart structure encryption preventing unauthorized structure change.

7.3 Composites Hydrogel-Based 4D Printing

7.3.1 *Water-responsive Hydrogel Composites*

One of the first investigations into 4D printing was by the Tibbits' group at MIT. Using the Connex 500 3D printer (Stratasys, Eden Prairie, Minnesota, USA), 4D printing was obtained by combining a rigid polymer base and a hydrogel to create a structure that could change its shape upon immersing in water.[9,21,22] They used a hydrophilic polymer, or hydrogel, that absorbed water and increased in size with a volume expansion up to 150%. Through the precise placement of the hydrogel in the composite structure, strain gradients between the two materials could be generated to cause a large shape change (Fig. 7.2A). Figure 7.2B shows the linear expansion of the hydrogel/rigid polymer composites. Different shapes with

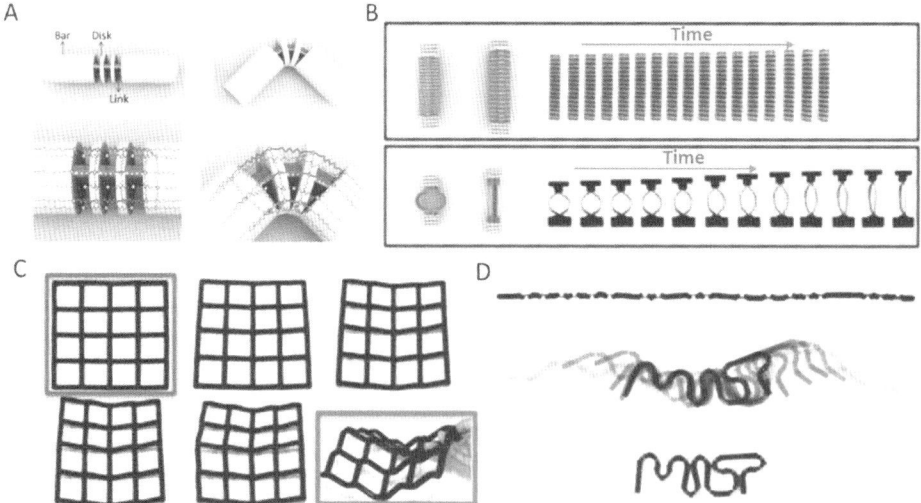

Fig. 7.2. (A) Renderings of an initial joint and its folding with their corresponding spring-mass systems. (B) Rendered illustration of the linear stretching primitive and ring stretching primitive. (C) Deformation of a grid into a hyperbolic surface.[24] (D) Sequence showing the self-folding of a 4D-printed multi-material single strand into a three-dimensional cube.[23]

complex geometries could be fabricated with a self-evolving shape including 1D folding and 2D folding as well as 2D folding and stretching. In Fig. 7.2C, a grid could self-evolve into a hyperbolic surface in water. They also demonstrated a 1D strand that could transform into a three-dimensional cube as seen in Fig. 7.2D.

Gladman et al.[27] of the Lewis group at Harvard University developed a biomimetic hydrogel that could be 4D printed by using the DIW method. There,

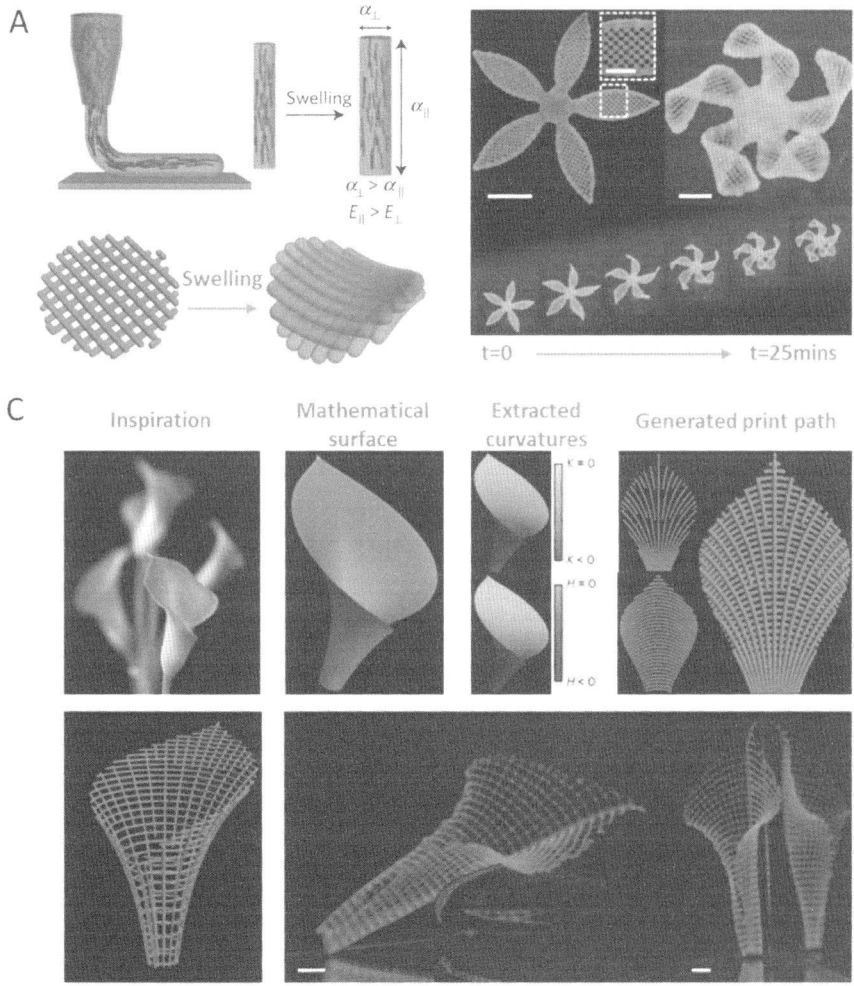

Fig. 7.3. (A) Schematic of the shear-induced alignment of cellulose fibrils in a hydrogel during DIW 3D printing and subsequent effect on anisotropic stiffness and swelling for controlled bending. (B) 3D printed bilayers oriented with respect to the long axis of each petal, with time-lapse sequences of the flowers during the swelling process (scale bars = 15 mm). (C) 3D printed hydrogel fibrils used to imitate the shape and surface topography of a lily flower after actuation.[27]

the ink consisted of hydrogel resin as well as nanofibrillated cellulose (NFC). As shown in Fig. 7.3A, NFC underwent shear-induced alignment as the hydrogel ink was extruded through the deposition nozzle. By doing so, the printed hydrogel architectures were encoded with localized, anisotropic swelling behavior which was controlled by NFC alignment along the prescribed printing pathways. This anisotropic swelling could be utilized to allow precise control over the curvature in bilayer hydrogel composite structures.[27] For example, Fig. 7.3B shows a 4D printed flower comprised of a −45°/45° bilayer lattice yielding a twisted configuration of the petals. Bio-4DP was further demonstrated by mimicking the complexity of the orchid Dendrobium helix by encoding multiple shape-changing domains. Figure 7.3C shows the resulting 3D morphology following swelling in water to resemble the orchid that could exhibit four distinct shapes (three different petal types and one flower center), based on the design of 3D printing pathways.

7.3.2 Temperature Responsive Composite Hydrogels

Thermally responsive hydrogels are attractive in many biomedical applications as the swelling/shrinking of the hydrogel can be controlled by aqueous temperature. In addition, reversible shape change can be achieved by cycling the temperature. These features have attracted research of thermally responsive hydrogel 4D printing to achieve reversible shape change. Naficy et al.[69] achieved bi-directional bending using a thermoresponsive and non-thermoresponsive bilayer hydrogel structure. Different hydrogel inks, i.e., a thermoresponsive poly(N-isopropylacrylamide) (PNIPAm) layer and a non-responsive hydrogel layer were printed using the DIW method. After printing and swelling below 32°C, the different swelling ratios of the two hydrogel layers shifted the sheet from a flat shape to a bent one. As the temperature was increased to 60°C, passing PNIPAm's critical temperature, it started to shrink, which led to a new equilibrium shape. As shown in Fig. 7.4A, a printed hinge exhibited a reversible bending behavior triggered by the temperature change. At low temperatures, the structure bent in one direction while at high temperatures, the swelling ratios of the two hydrogels changed, and the structure bent in the opposite direction. Due to the fabrication flexibility offered by 3D printing, more complex shapes can be easily achieved. Figure 7.4B shows a thin sheet that could fold into a box after swelling at 20°C, then open up, and finally transform back into a flat structure after heating to 60°C.

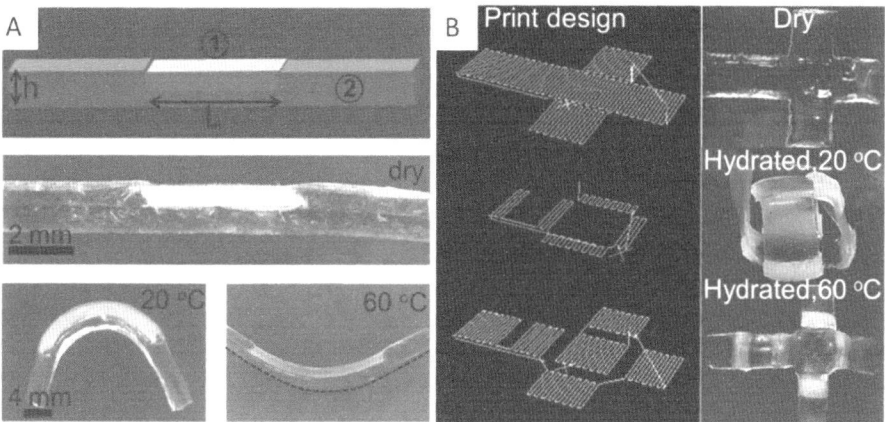

Fig. 7.4. (A) The design of a thermally responsive hydrogel bilayer hinge and images of the printed composite in the as printed state, at 20°C in water, and 60°C in water demonstrating reversible shape changing. (B) The 3D printing pattern used to fabricate a folding box and the resulting print at various temperatures.[69]

In 2016, Mao et al.[108] developed a new design that combined a hydrogel, an SMP, and an elastomer to create a composite that demonstrated two reversible, stable, and stiff shapes with a rapid transition speed. In this work, they utilized the hydrogel swelling as the driving force and the temperature dependent modulus change of the SMP as a switch. Figure 7.5A shows the basic concept of their design where the hydrogel was sandwiched between an SMP layer and an elastomer layer. The structure was first immersed in cold water (~0°C) for ~12hours (S1 in Fig. 7.5B) so that the hydrogel could absorb water. The high modulus of the SMP at this low temperature prevented the stripe from deforming. After the hydrogel absorbed enough water, the stripe was placed in hot water. The modulus of the SMP dropped quickly by ~three orders of magnitude, which led to a large and rapid shape change (S2). Then, the stripe was moved out of the hot water bath (S3) and it maintained the bent shape. After leaving the stripe in a low temperature environment and allowing the hydrogel to dry (S4) the stripe became very stiff and could carry a load of up to 25g (S4). The stripe was then heated and returned to its initial flat configuration (S5). Due to the vast design freedom offered by multi-material 4D printing, this basic concept could be easily applied to more complicated designs. For example, a flower was printed as shown in Fig. 7.5C. The flower closed its petals after it was immersed in cold water. After following the steps shown in Fig. 7.5B, the flower could also carry a load of 25g at low temperatures (Fig. 7.5D) and return to its initial configuration. Using this robust multi-material design, rapid, reversible, and robust shape changing structures could be 3D printed.

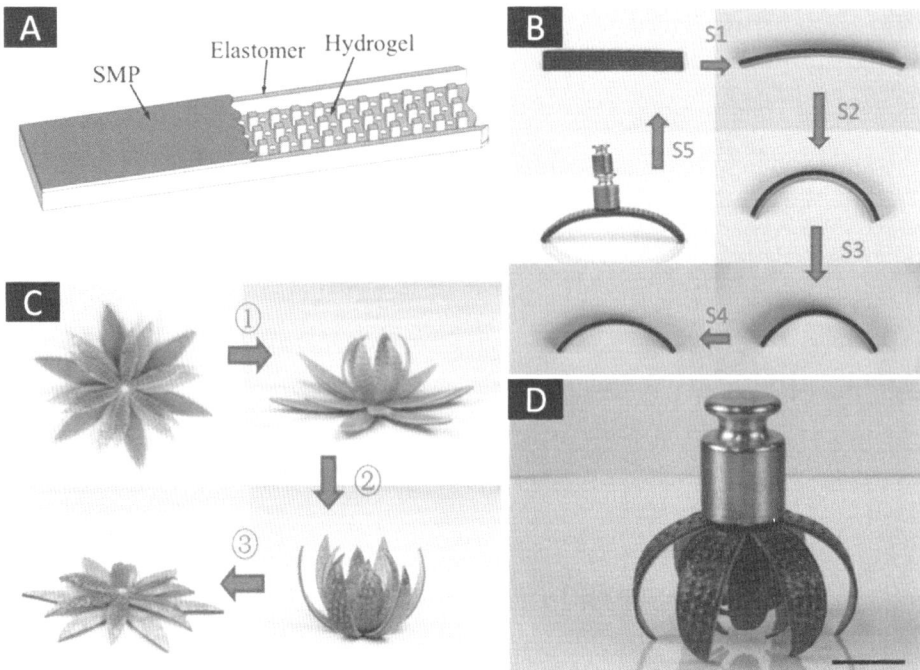

Fig. 7.5. (A) The design concept of the reversible actuator which combines an SMP, a hydrogel, and an elastomer. (B) The sequence of shape changes as the stripe went through different environmental conditions. (C) The sequence of reversible actuation of a printed flower. (D) The dried configuration of the flower was stiff and could carry a load of 25g (scale bar =12.5 mm).[108]

In addition, hydrogels that respond to other stimuli, such as pH[109] and light,[110] have been used for 4D printing. For example, Nadgorny et al. 4D-printed a recyclable pH-responsive hydrogel using an FFF 3D printer.[30] The resulting structures could swell linearly at a specific pH level and then shrink at another pH level. 4D printing of hydrogels has opened the door for potential applications for rapidly fabricated and environment-sensitive smart structures where water or solvent are permitted. For example, intelligent objects can be fabricated and deployed in deep-sea exploration, or in dangerous, or potentially poisonous environments. In addition, bio-compatible 4D structures can find applications as medical stents,[111] biomimetic tissue,[112] and targeted drug delivery.[21]

7.4 SMP Composites-Based 4D Printing

SMPs are another group of smart polymers that have been widely studied. Compared to hydrogels, SMPs have the advantages of high modulus, large deformability, and no requirement to use water. Due to these advantages, SMPs are also widely studied in 4D printing. In the following section we will delve further into the mechanisms of using SMPs in multi-material 4D printing.

7.4.1 *Printed Active Composites*

One of the first examples in 4D printing was the printed active composite (PAC) demonstrated by Ge et al. in 2013 by printing SMP fibers within an elastomer matrix as shown in Fig. 7.6A.[22] This bilayer PAC consisted of a pure elastomeric matrix layer and a layer with SMP fibers. By stretching the as-printed PAC at a high temperature, then cooling to a low temperature, and releasing the mechanical load, bending was observed. Upon heating, the PAC could recover its initial flat shape. They also demonstrated the effect of the underlying fiber architecture on the resulting PAC shape changes, seen in Fig. 7.6B. In a follow-up work by the same group, the PAC design was further used to design more complicated origami folding to form model airplanes[113,114] (Fig. 7.6C). The 4D printed active origami structures show the potential of using 4D printing to solve engineering problems related to the packing of large structures into small volumes, for the sake of storage and transportation.

Wu et al. moved this method a step further by incorporating two different SMP fibers.[48] The two fibers utilized in this system had different activation temperatures of ~ 37°C and ~57°C. After printing, the PAC was programmed with the same simple thermal-mechanical programming step as described earlier. After being programmed, the PAC could memorize two temporary shapes. Many interesting applications for active structures could be realized. Figure 7.6D shows an active hook activated in 30°C water to lift a small box.

Subsequently, upon immersion in 70°C water, the hook returned to a straight shape and the box was released. The capability of shape changing under mild temperatures then returning to the initial shape under a higher temperature offers a significant advantage. For example, such a design could be used as suture to tighten wounds or as a fixture to attach devices using moderate temperatures in the field and once the patient or the device returns to the hospital or the service center, a higher temperature can be applied to quickly release the fixture. In addition, such a design can be used for soft robots, where the temporary intermediate shape can be used to pass certain obstacles, such as a barricade, or a small opening, then recover to its original shape.

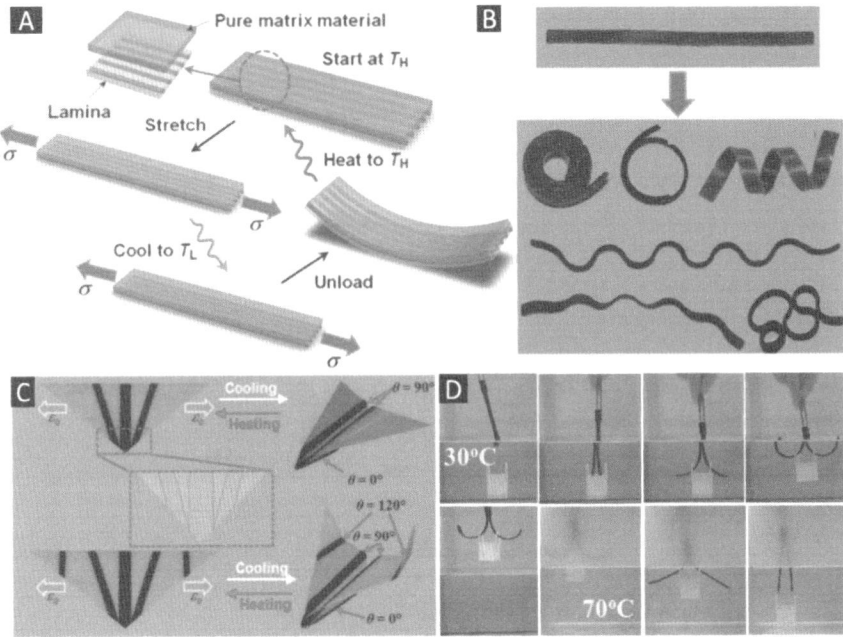

Fig. 7.6. (A) A schematic demonstration of a PAC laminate architecture involving printed SMP fibers during a programming and recovery cycle.[22] (B) The wide array of resulting bending deformations, depending on the placement of the SMP fibers in the composite.[22] (C) Active origami airplanes could be 4D printed with precise placement of SMP fibers.[115] (D) The shape changing behavior of a 4D printed PAC hinge which could be actuated at different temperatures to create a smart hook.[48]

7.4.2 *3D Printing of Multi-material SMPs*

Multi-material 3D printing can create digital materials with tunable glass transition temperatures (T_gs) (or digital SMPs), which can be exploited for multiple shape-shifting. When heated to the same recovery temperature, hinges using SMPs with higher T_gs needed a longer time to recover, leading to sequential shape shifting. Figure 7.7A shows the sequential folding of these hinges to create a multi-material self-locking SMP device.[116] Ge et al.[46] exploited high-resolution projection micro-stereolithography (PµSL) to print tailorable SMPs with T_g from ~−50°C to ~180°C. They demonstrated sequential shape recovery by printing a multi-material flower whose inner and outer petals had different T_gs. After programming the petals at 70°C and cooling down to 20°C, the external force was removed, and the flower was fixed at a temporary budding state, where the petals stayed closed. The sequential recovery of outer and inner petals was triggered by raising the temperature, first to 50°C, and then to 70°C, leading to the full shape recovery of the flower to its original blooming state (Fig. 7.7B).

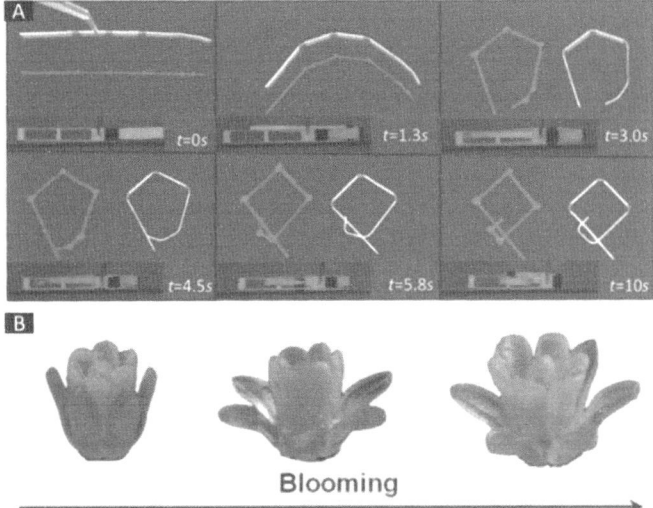

Fig. 7.7. 4D printing by using multi-material SMPs. (A) The comparison of experiments and the simulations of interlocking SMP multi-material device by 3D printing.[116] (B) The sequential blooming of a 3D printed multi-material flower.[117]

7.4.3 *Bilayer SMPs Composites for Direct 4D Printing*

The PACs discussed above were successful in creating active structures; however, they must be programmed using several steps, including heating, deforming, cooling, and removing the load, before they can be activated. Recently, Ding et al.[25] developed a new method allowing the printed structure to be activated directly after printing and maintain a new shape through a simple heating process.

To achieve this, a bilayer laminate strip was printed by using Stratasys Connex3 Objet500 with TangoBlack as the elastomer and VeroClear as the SMP. After heating, the strip bent and the bending curvature could be controlled by various printing parameters. This simple bilayer design and the resulted bending could be used for more sophisticated designs for more complex shape changes. Figure 7.8A shows a printed lattice of bilayer laminates with a collapsed configuration that could be deployed into an open configuration after printing and heating above the T_g of the SMP. In Fig. 7.8B, a printed flower was shown to bloom into a configuration where petals have different curvatures upon heating due to their various T_g. Similarly, Ding et al.[35] printed a bilayer mesh which could transform into a buckyball shape as shown in Fig. 7.8C. The mechanism of this method was discussed in Yuan et al.[118] and relies on the embedded residual stresses developed during the layer-by-layer fabrication process. Based on this direct 4D printing method, Sundaram et al.[119] developed 3D-printed self-folding electronics.

The bilayer composite, together with the conductive circuit, was printed using multi-material 3D-printing on a home-made inkjet printer. After being removed from the printing stage, the composite structure gradually self-folded to achieve a new configuration (Fig. 7.8D).

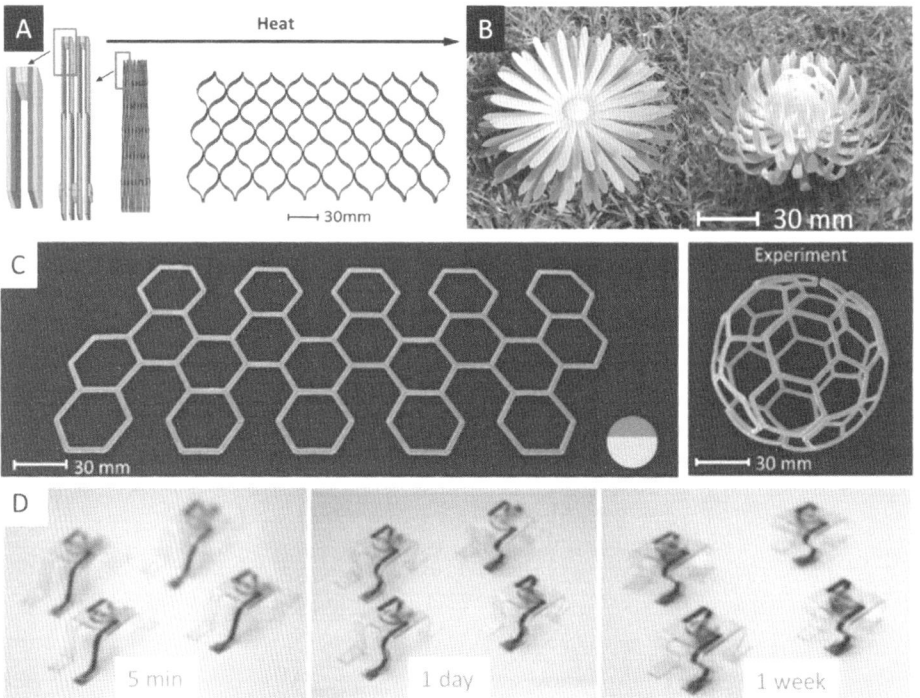

Fig. 7.8. (A) Bilayer SMP lattice structure printed in a collapsed configuration that deploys into an open configuration directly after printing and heating;[120] (B) Printed flower consisting of multiple petals at multiple layers that blooms into a configuration with petals showing different curvatures upon heating.[120] (C) A Bucky ball directly transformed from a planar mesh upon heating.[35] (D) Electrochromic elements integrated in a fully printed composite that can change its folding angle over time due to the direct 4D printing approach.[120]

7.4.4 *LCE-Based 4D Printing*

Currently, 4D printed objects capable of producing reversible, or two-way, shape changes are a primary research focus. This is because SMPs typically require a programming step and thus the actuation is generally one-time and non-reversible. Hydrogels, on the other hand, can yield large and reversible volume changes, however, their response speed is relatively slow.[24,27,69] Liquid crystal elastomers (LCEs) are a class of soft-active materials which can achieve rapid and reversible shape changes,[91,121-126] making them an attractive candidate for 4D printing. LCEs achieve their actuation through a transition of the molecular chains between an ordered liquid crystal (nematic) state and a random (isotropic) state in response to light,[127-132] heat,[101,133] electrical[134] or magnetic fields.[135] In 2017, Ware et al. developed a method to 3D print LCE that could be rapidly and reversibly activated directly after printing by heating.[125] The basic printing scheme is shown in Fig. 7.9A. There, the macroscopic LCE oligomer chain alignment was achieved due to the shearing generated by high temperature DIW 3D printing. Due to the alignment and subsequent shrinkage strain of the LCE in the printing direction, as shown in Fig. 7.9B, some interesting shape changes could be programmed into structures based on the 3D print path. Figure 7.9C shows a bilayer LCE sample

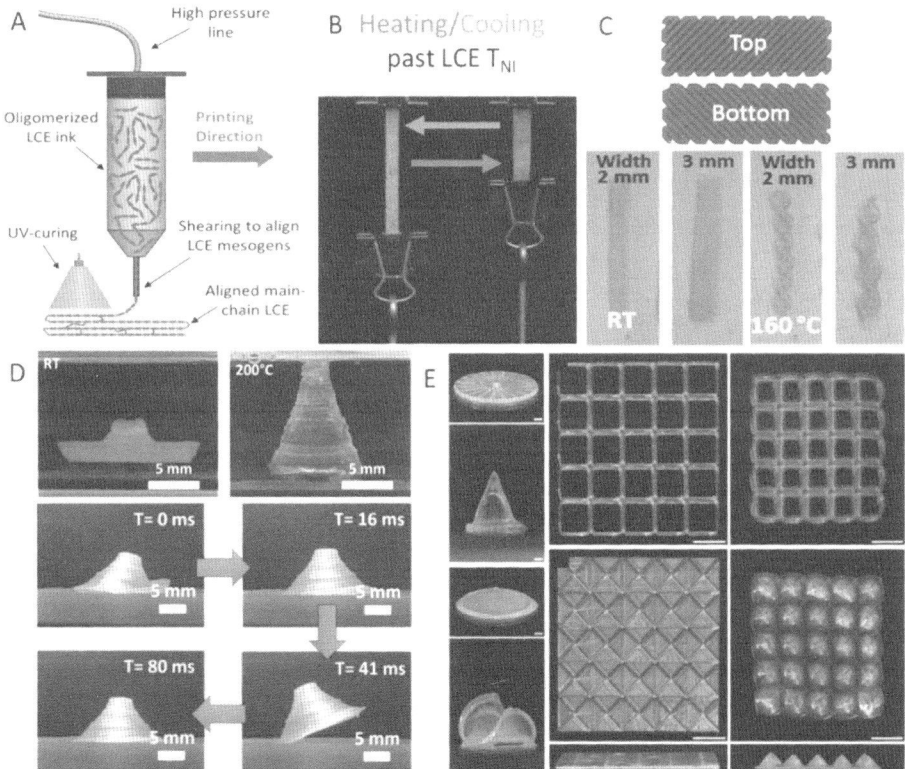

Fig. 7.9. (A) Schematic of the direct-write 3D printing of liquid crystal elastomer (LCE) which aligns the LCE chains and activation direction in the printing direction. (B) The reversible actuation of the LCE upon heating/cooling past the nematic-isotropic transition temperature (TNI). (C) Bilayer LCE sample that can twist due to varying printing path between the top and bottom layers. (D) Snapping actuation of a 3D printed LCE. (E) An array of LCE structures that can transform after printing depending on their printing path.[137,138]

with varying print paths on the top and bottom layers and its subsequent twisting activation upon heating. Following Ware et al. Vs work, other researchers adjusted the chemistry and overall LCE ink formulation to achieve lower activation and printing temperatures.[26,136] Figure 9D-E show printed LCE structures capable of obtaining unique deformation depending on the printing path of the DIW 3D printer including a structure which could achieve a snap-through transition on the order of 80ms.

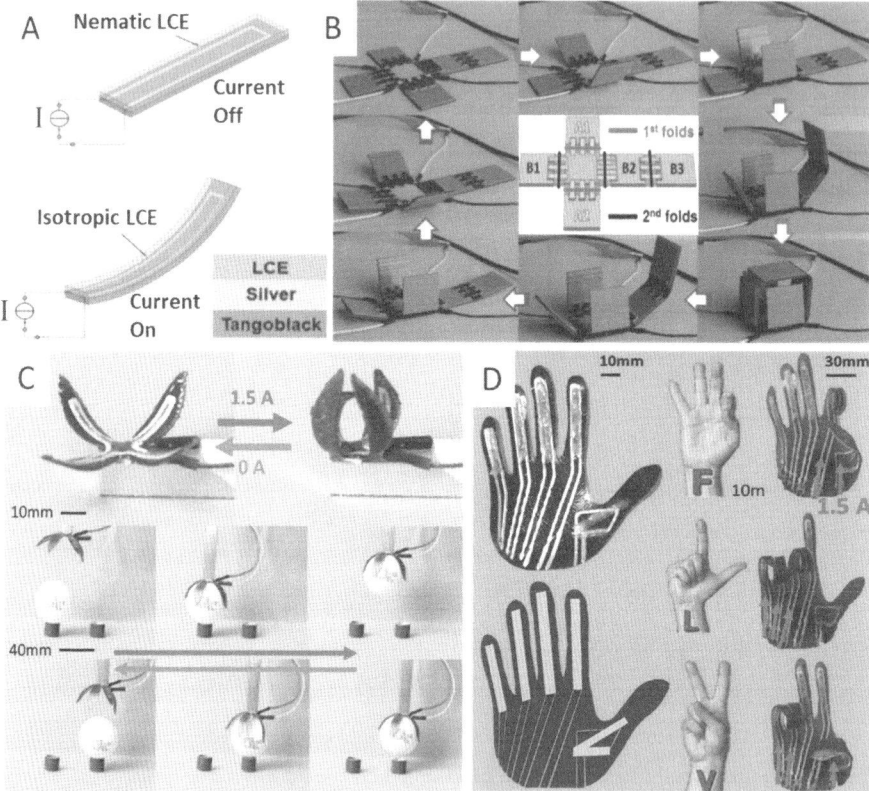

Fig. 7.10. (A) Schematic of the multi-material LCE-based actuator hinge relying on Joule heating for LCE activation.[77] (B) Multi-material box which has two separate printed conductive lines to achieve sequential activation of two separate folds.[77] (C) Soft-robotic gripper pick-and-placing a ping pong ball.[140] (D) Fully 3D printed hand where each finger can be activated separately to form American sign language letters.[140]

Yuan et al. first introduced the concept of combining the advantages of multi-material 3D printing and LCEs to create reversible actuation of complicated structures.[139] Here, prefabricated LCE was placed in a multi-material 3D printed structure containing an elastomer and silver conductive wires. By 3D printing conductive wires, electrical current could be applied to the structure and Joule heating could be used to heat the LCE past its T_{NI}. Figure 7.10A shows the basic multi-material hinge design. Figure 7.10B shows a box that could be sequentially activated due to separately 3D printed conductive traces. Roach et al. further demonstrated room temperature LCE 3D printing and utilized this method to fabricate fully 3D printed LCE-based multi-material soft robotics[140]. The basic hinge design shown in Fig. 7.10A was entirely 3D printed and utilized to create structures that could achieve complex folding and activation sequences as seen in Fig. 7.10C-D. Here, a soft robotic gripper was printed and used to pick up and place a ping pong ball where the red arrows represented Joule heating above the LCE T_{NI} and the blue arrows represented cooling back to room temperature. In addition, a hand was 3D printed such that each finger could be activated separately to produce letters from the American sign language alphabet.[140]

7.5 Multi-Material 4D Printing Based on Other Methods

7.5.1 *4D Printing by Using Shrinkage*

When a polymer is cured, volume shrinkage can be observed due to the conversion of molecular interaction from Van der Waals interaction to covalent bonds. Therefore, volume shrinkage is common in many 3D printing processes. Although volume shrinkage generally should be avoided in 3D printing, if it can be precisely controlled, it can be advantageous for creating 4D printed shape changing structures.[61,141-143]

Fused filament fabrication is one of the most common 3D printing methods for polymers. The rapid heating and cooling of the filaments, along with constraints imposed by the print platform or previous layers, can cause a large internal stress. Zhang et al. observed that, upon heating above the T_g, the internal stresses of aligned polymer chains could be released, causing large shrinkages, which can be used for shape shifting.[61] As seen in Fig. 7.11A, the printed lattices contracted upon heating, and the circular rings could transform into hexagons. By printing PLA fibers on material insensitive to temperature change (such as paper), and heating the composite above the T_g after printing, large bending deformation was achieved and could be used to create interesting shape changes. For example, a PLA/paper bi-layer composite could deform into a flower, a twisting stripe, or a sinusoidal architecture as seen in Fig. 7.11B-C. Manen et al.[144] exploited a single-step printing process with different in-plane alignment of the 3D printed PLA to create shrinkage. The printing parameters, such as filament plying angle, thickness, as well as the activation temperatures could be used to control the shrinkage of multi-ply panels to achieve sequential folding (Fig. 7.11D).

Fig. 7.11. (A) Shrinkage-induced shape change of an FFF 3D printed pattern upon heating above the material Tg. (B) Bi-layer composite of FFF thermosetting PLA and paper to create a flower shape after heating. (C) Interesting shape changes such as twisting and sinusoidal shapes utilizing the same method described in (B). (D) 3D printed PLA filaments simultaneously decrease in length and increase in thickness once heated above their glass transition temperature to achieve two-step folding of the initially flat petals to create a tulip.[145]

Zhao et al. used the volume shrinkage stress induced during photopolymerization to create shape changes.[62] A schematic of this approach is shown in Fig. 7.12A. Here, photoabsorbers were added into the polymer resin to attenuate the light, creating a light intensity gradient across the thickness direction inside the liquid resin. The material directly exposed to the light was cured faster than the material further from the light, resulting in a nonuniform volume shrinkage and a stress gradient in the cured polymer. Once the printed part was removed from the substrate, the cured polymer film bent towards the less cured portion (top surface in Fig. 7.12A). Further studies revealed that the bending curvature depended on the illumination time and the light intensity, or the grayscale in the CAD drawing used to build the structure (Fig. 7.12B). By combining the design of the spatial distribution of the grayscale pattern, complex 3D origami structures could be easily created. For example, a flat polymer sheet after photopolymerization using the pattern shown in the inset (Fig. 7.12C) bent into the shape shown in Fig. 7.12D. A complex polyhedron structure (Fig. 7.12E) could also be fabricated by using the pattern shown in Fig. 7.12F. To extend the application of this method, a two-sided illumination method was used as shown in the schematic in Fig. 7.12G. After the first illumination process, the material was flipped and a secondary grayscale pattern was projected. Using this method, origami structures which required bending deformation in different directions could be realized, such as the Miura-Ori structure seen in Fig. 7.12H.

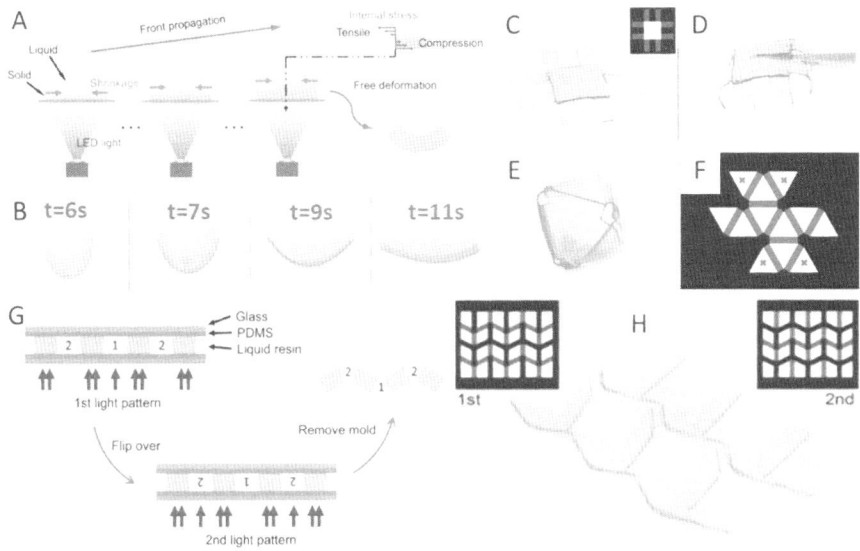

Fig. 7.12. (A) The schematic graph showing the volume shrinkage–induced bending by frontal photopolymerization. (B) Samples cured in a petri dish with different irradiation time showing different bending curvatures. (C) Polymer sheet right after photopolymerization and (D) the corresponding free bending of spatial, differently cured sheet. (E) An octahedron structure after illumination with the grayscale pattern in (F). (G) Two-sided illumination process to create multi-directional bending deformation forming more complex origami structures, such as Miura-Ori structure (H).[62]

7.5.2 *Dissolvation-induced Multi-material 4D Printing*

The volume shrinkage induced by dissolution of residual unreacted polymeric components has also been exploited to achieve shape changing after 3D printing and subsequent curing. Zhao et al.[141] used the DLP 3D printing method to create complex, free-standing, and reversible origami structures via dissolution-induced self-folding. The projected light grayscale pattern and the photoabsorbers added to the polymer are combined to control the reaction degree in the x-y plane and the z-axis, respectively. As shown in Fig. 7.13A, the as-cured flat film using different grayscale light patterns could change to 3D origami structures after dissolution in water and would return to its initial flat shape in acetone by swelling again. Wu et al.[146] extended the grayscale curing to printing 3D structures by DLP printing. Fig. 7.13B shows the digital design of a self-expanding/shrinking structure with location-specific grayscale values during printing. After dissolvation in a water/acetone (v/v=15:1) solution, the printed object could expand to two times of its original shape in the vertical direction, then recover to a compact shape after drying. The reversible shape-shifting could occur with the absorption and evaporation of acetone. In addition, Huang et al.[142] controlled the light exposure time at different locations on a 2D flat sheet to enable location-specific curing. Figure 7.13C shows that the polymer network with spatially variable monomer reaction conversion and cross-linking density led to different dissolvation and swelling ratios in water. As a result, a cured flat sheet could change into a unique 3D architecture, such as a cartoon face, after dissolvation in water (Fig. 7.13D).

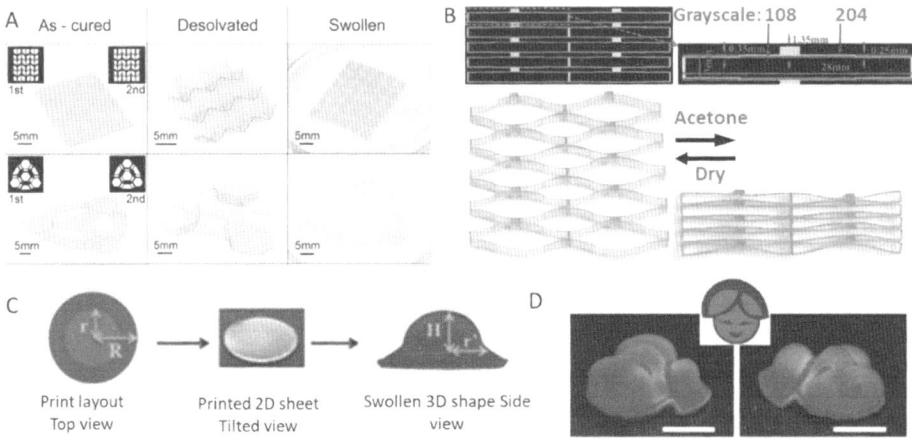

Fig. 7.13. (A) The as-cured flat pattern, dissolved shape, and swollen flat shape of different 3D origami structures.[146] (B) The design pattern with different light grayscale in each part for printing and the deformed structure after dissolvation.[146] (C) A schematic graph showing how a planar sheet with patterned concentric circle could swell into a cap-shape 3D structure;[141] (D) 3D cartoon face viewed from two different angles right after photopolymerization and the corresponding free bending of a spatial, differently cured sheet.[141]

7.5.3   *Light-induced 4D Printing*

Currently, light is an attractive stimulus for 4D printing as it is abundantly available and can be applied remotely. Light-driven activation has already demonstrated a vast potential for use within smart materials realizing applications in self-assembly, soft robotics, UV filters, and smart surfaces. By incorporating photo-responsive smart materials into 4D printed objects, rapid activation and tunable deformation by varying the wavelength or location of the applied light,[147-149] Liu et al. reported the programmed self-folding of two-dimensional (2D) polymer sheets into 3D objects in a sequential manner using external light.[147,148] Figure 7.14A shows that the out-of-plane and in-plane structures could be sequentially actuated by blue light and red light. Yang et al. demonstrated light-responsive 4D printed structures by FFF printing an SMP with carbon black fillers that could generate heat upon exposure to light.[150] Figure 7.14B shows the resulting shape change of a printed cube under direct sunlight. Similarly, Hua et al.[151] reported the 3D printing of photo-responsive shape changing composites based on polylactic acid (PLA) and multi-walled carbon nanotubes (MWCNTs) on paper substrates with FFF printing for the construction of flexible photothermal-responsive shape changing actuators. In Fig. 7.14C, the 3D printed flower shifted from the closed to open state like the blooming of flowers. During illumination, the flower turned from bud to bloom and then recovered to its shape in the form of closed petals upon turning off the light.

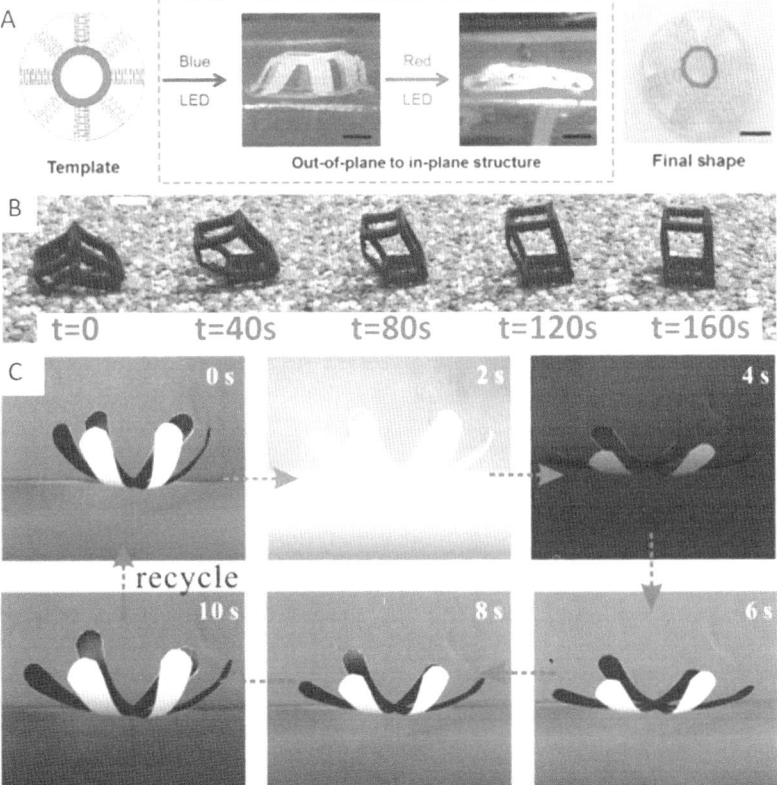

Fig. 7.14. (A) The out-of-plane and in-plane structures are actuated sequentially by shrinking the outer ring first to achieve out-of-plane motion and then shrinking the inner ring to flatten the sample again.[93] (B) Sunlight-activated recovery of an SMP/carbon black composite. (C) Photo-activated shape changing behavior of a 3D printed flower from the close state to the open state.[151]

7.6 The Future of 4D Printing

Since its inception in 2013, 4D printing has realized unprecedented advances. Nevertheless, for 4D printing to realize its full potential in real-world applications, much research is still needed in the field, including 3D printing techniques, materials science, modeling and design tools. First, high-resolution, high-speed 3D printing technologies should be developed to fabricate materials and devices with multiscale complex geometries. More sophisticated multi-material 3D printing technologies are especially desirable in this domain. Advances such as large-scale printing[152] or multi-method printing will inevitably assist in advancing 4D printing towards further practical applications. Second, the development of novel inks or stimulus-responsive materials that can enable multi-functional material printing will have a profound impact on the future of 4D printing. To date, SMP and hydrogel composites are the most widely utilized materials systems for 4D printing. LCE is a promising smart material enabling reversible shape shifting. The merging of LCE and multi-material printing will open a new avenue for multi-functional 4D printing finding broad applications in robotics, optics, and biomedicine.[136,153] In addition, each unique application presents different requirements that the material system must satisfy for proper 3D printing and deployment. For instance, biomedical devices must be fabricated from a bio-compatible polymer so that they may be introduced into the body without harm.[111,154] Another such example exists in the world of wearable electronics where a major roadblock in their wide implementation is a proper energy storage system that can withstand constant mechanical deformation as the wearer moves.[155] It should be noted that most of the existing 3D printing approaches may not be applicable to some smart or functional materials. As such, the development of new materials for 4D printing also requires the modification of existing printing techniques. Finally, more sophisticated and robust design methodologies and advanced theoretical models can further advance 4D printing. A model-based design tool could be utilized to accurately predict the shape transformation of the printed shape under external stimulus. Topology optimization has recently been presented as a promising technique for the design of smart structures based on the target architecture.[156-163] Artificial intelligence (AI) based design approaches have also very recently been developed and shown great promise for exploring a large design space.[164]

3D printing of smart and functional materials will enable innovative applications that can hardly be achieved with conventional manufacturing processes. The potential applications of 4D printing are well beyond the feature of simple shape-shifting. For example, taking advantage of basic shape shifting, 4D printing can aid in the fabrication of active truss structures capable of gigantic shape change that can be used as free-standing, self-erecting tents or soft micro-robots that can be fabricated for therapeutic drug delivery at desired locations within the body.[80,165] In addition, printed functional objects such as electronic devices can be combined with shape-changing to achieve 4D printing of functional devices. Moreover, other stimulus-responsive properties, such as self-healing and self-sensing, are beginning to be incorporated into 3D/4D printed structures.[44,166-168] Further applications can be realized if this technology is implemented with other advances such as color-shifting textiles that can help individuals hide in certain environments[169,170] or computing for active health-monitoring.[171]

7.7 Conclusions

In this chapter we have introduced the basic elements of the emerging field of multi-material 4D printing which has achieved rapid development in recent years. In addition, we have introduced several 4D printing approaches based on their shape changing mechanisms. 4D printing enables the targeted evolution of 3D printed structures' shape, property, and functionality as a function of time. Like many other emerging technologies, there are still many challenges facing 4D printing. To better implement and maximize the potential applications of 4D printing, further research needs to be done, including the development of new 4D printing concepts and techniques, printable smart materials, design methods, and theoretical models. However, with ever-increasing multidisciplinary research, many exciting applications of 4D printing will be witnessed by exploiting this technology. Multi-material 4D printing advancements will inevitably transform industries such as robotics, biomedicine, electronics, defense and security.

Acknowledgement

The support of an AFOSR grant (FA9550-16-1-0169; Dr. B.-L. "Les" Lee, Program Manager). We also acknowledge the support of the NSF award (CMMI-1462894, 1462895). Gift funds from HP, Inc and Northrop Grumman Corporation are also greatly appreciated.

References

1. I. Gibson, D. Rosen and B. Stucker, *Additive Manufacturing Technologies, 2nd Edition.* Springer (2015).
2. J.P. Kruth, M.C. Leu and T. Nakagawa, Progress in Additive Manufacturing and Rapid Prototyping. *CIRP Annals - Manufacturing Technology* **47**(2), 525–540 (1998).

3. C. Colosi, S.R. Shin, V. Manoharam, S. Massa, M. Costantini, A. Barbetta, M.R. Dokmeci, M. Dentini and A. Khademhosseini, Microfludic bioprinting of heterogeneous 3D tissue constructs using low-viscosity bioink. *Adv Mater.* **28**(4), 677–684 (2016).

4. D.B. Kolesky, R.L. Truby, A.S. Gladman, T.A. Busbee, K.A. Homan, J.A. Lewis, 3D bioprinting of vascularized, heterogeneous cell - laden tissue constructs, *Adv Mater.* **26**(19), 3124–3130 (2014).

5. B. Derby, Printing and prototyping of tissues and scaffolds, *Science* **338**(6109), 921–926 (2012).

6. J.H. Cho, J. Lee, Y. Xia, B. Kim, Y. He, M.J. Renn, T.P. Lodge and C.D. Frisbie, Printable ion-gel gate dielectrics for low-voltage polymer thin-film transistors on plastic, *Nat Mater.* **7**(11), 900–906 (2008).

7. T.S. Wei, B.Y. Ahn, J. Grotto and J.A. Lewis, 3D printing of customized Li-ion batteries with thick electrodes, *Adv Mater.* **30**(16), 1703027 (2008).

8. F. Kotz, K. Arnold, W. Bauer, D. Schild, N. Keller, K. Sachsenheimer, T.M. Nargang C. Richter, D. Helmer and B.E. Rapp, Three-dimensional printing of transparent fused silica glass, Nature **544**, 337–339 (2017).

9. Z.C. Eekel, C. Zhou, J.H. Martin, A.J. Jacobsen, W.B. Carter, T.A. Schaedler, Additive manufactguring of polymer-derived ceramics, *Science* **351**(6268) 58–62 (2016).

10. X. Zheng, H. Lee, T.H. Weisgraber, M. Shusteff, J. DeOtte, E.B. Duoss, J.D. Kuntz, M.M. Biener, Q. Ge, J.A. Jackson, S.O. Kucheyev, N.X. Fang and C.M. Spadaccini, Ultralight, ultrastiff mechanical metamaterials, *Science* **344**(6190), 1373–1377 (2014).

11. T.J. Prater, Q.A. Bean, R.D. Beshears, T.D. Rolin, N.J. Werkheiser, Summary report on Phase I Results from the 3D printing in zero-G technology demonstrataion mission, *NASA Technical Reports Server* **1**, 156 (2016).

12. J-Y. Lee, W.S. Tan, J. An, C.K. Chua, C.Y. Tang, A.G. Fane and T.H. Chong, The potential to enhance membrane module design with 3D printing technology, *Journal of Membrane Science* **499**, 480–490 (2016).

13. M. Petch, New study of 3D printing patents reveals second fastest growing technology of 2017, *3D Printing Industry* (January 10, 2018).

14. 3D printing and additive manufacturing state of the industry, Annual Worldwide Progress report. Wohlers Report 2017.

15. S. Tibbits, The emergence of 4D printing, TED conference (2013).

16. P. Eujin, 4D printing: Dawn of an emerging technology cycle, *Assembly Automation* **34**(4), 310–314 (2014).

17. J. Choi, O.C. Kwon, W. Jo, H.J. Lee and M-W. Moon, 4D printing technology: A review, 3D Printing and Additive Manufacturing **2**(4), 159–167 (2015).

18. D-G. Shin, T-H. Kim and D-E. Kim, Review of 4D printing materials and their properties, *International Journal of Precision Engineering and Manufacturing-Green Technology* **4**(3), 349–357 (2017).

19. F. Momeni, S.M. Mehdi Hassani, X. Liu and J. Ni, A review of 4D printing, *Materials & Design* **122**, 42–79 (2017).

20. B. Gao Q. Yang, X. Zhao, G. Jin, Y. Ma and F. Xu, 4D bioprinting for biomedical applications, *Trends in Biotechnology* **34**(9), 746–756 (2016)

21. Q. Ge, H.J. Qi and M.L. Dunn, Active materials by four-dimension printing, *Applied Physics Letters* **103**(13), 131901 (2013).

22. S. Tibbits, 4D printing: multi-material shape change, *Architectural Design* **84**(1), 116–121 (2014).

23. D. Raviv, W. Zhao, C. McKnelly, A. Papadopoulou, A. Kadambi, B. Shi, S. Hirsh, D. Dikovsky, M. Zyracki, C. Olguin, R. Raskar and S. Tibbits, Active printed materials for complex self-evolving deformations, *Sci Rep.* **4**, 7422 (2015)

24. Z. Ding, C. Yuan, X. Peng, T. Wang, H.J. Qi and M.L. Dunn, Direct 4D printing via active composite materials, *Sci Adv.* **3**(4) (2017).

25. A. Kotikian, R.L. Truby, J.W. Boley, T.J. White and J.A. Lewis, 3D printing of liquid crystal elastomeric actuators with spatially programed nematic order, *Adv Mater.* **30**(10) (2018).

26. A.S. Gladman, E.A. Matsumoto, R.G. Nuzzo, L. Mahadevan and J.A. Lewis, Biomimetic 4D printing, *Nat Mater.* **15**(4), 413–418 (2016).

27. G. Yang, H.S. Patanwala, B. Bognet and A.K.W. Ma, Inkjet and inkjet based 3D printing: Connecting fluid properties and printing performance, *Rapid Prototyping Journal* **23**(3), 562–576 (2017).

28. O. Kukenok and A.C. Balazs, Stimuli-responsive behavior of composites integrating thermo-responsive gels with photo responsive fibers, *Materials Horizon* **3**(1), 53-62 (2016).

29. M. Nadgorny, Z. Xiao, C. Chen and L.A. Connal, Three dimensional printing of pH-responsive and functional polymers on an affordable desktop printer, *ACS Appl Mater Interfaces* **8**(42), 28946–28954 (2016).

30. X. Kuang, D.J. Roach, J. Wu, C.M. Hamel, Z. Ding, T. Wang, M.L. Dunn and H.J. Qi, Advances in 4D printing: Materials and applications, *Adv Funct Mater.* **29**(2), 1804290 (2019).

31. Z.X. Khoo, J.E.M. Teoh, Y. Liu, C.K. Chua, S. Yang, J. An, K.F. Leong and W.Y. Yeong, 3D printing of smart materials: A review on recent progresses in 4D printing, *Virtual and Physical Prototyping* **10**(3), 103–122 (2015).

32. J. Lee, H.C. Kim, J.W. Choi, I.H. Lee, A review on 3D printed smart devices for 4D printing, *International Journal of Precision Engineering and Manufacturing Green Technology*, **4**(3) 373–383 (2017).

33. D.J. Roach, C. Hamel, J. Wu, X. Kuang, M.L. Dunn and H.J. Qi, 4-D printing: Potential applications of 3-D printed active composite materials, *Journal of the Homeland Defense & Security Information Analysis Center* **4**(4) (2017).

34. Z. Ding, O. Weeger, H.J. Qi and M.L. Dunn, 4D rods: 3D structures via programmable 1D composite rods, *Mater Des.* **137**, 256–265 (2018)

35. B. An, Y. Tao, J. Gu, T. Cheng, X.A. Chen, X. Zhang W. Zhao, Y. Do, S. Takahashi, H-Y. Wu, T. Zhang and L. Yao, Thermorph: Democratizing 4D printing of self-folding materials and interfaces, *Proceedings of the 2018 CHI Conference on Human Factors in Computing Systems*, pp. 1–12 (April 2018).

36. R. Bogue, Smart materials: A review of capabilities and applications, *Assembly Automation* **34**(1), 16–22 (2014).

37. D.J. Leo, *Engineering Analysis of Smart Material Systems*. John Wiley & Sons (2007).

38. A. Lendlein and V.P. Shastri, Stimuli sensitive polymers, *Adv Mater.* **22**(31) 3344–3347 (2010).

39. D. Roy, J.N. Cambre and B.S. Sumerlin, Future perspectives and recent advances in stimuli-responsive materials, *Prog Polm Sci* **35**(1), 278–301 (2010).

40. F.L. Bargardi, H. Le Ferrand, R. Libanori and A.R. Studard, Bio-inspired self-shaping ceramics, *Nat Commun.* **7**, 13912 (2016).

41. G. Liu, Y. Zhao, G. Wu and J. Lu, Origami and 4D printing of elastomer-derived ceramic structures, *Sci Adv* **4**(8) (2018).

42. J. Ma, B. Franco, G. Tapia, K. Karayagiz, L. Johnson, J. Liu, R. Arroyave, I. Karaman and A. Elwany, Spatial control of functional response in 4D-printed active metallic structures, *Sci Rep* **7**, 46707 (2017).

43. X. Kuang, K. Chen, C.K. Dunn, J. Wu, V.C. Li and H.J. Qi, 3D printing of highly stretchable, shape-memory and self-healing elastomer toward novel 4D printing, *ACS Appl Mater Interfaces* **10**(8), 7381–7388 (2018).

44. H. Wei, Q. Zhang, Y. Yao, L. Liu, Y. Liu and J. Leng, Direct-write fabrication of 4D active shape-changing structures based on a shape memory polymer and its nanocomposite, *ACS Appl Mater Interfaces* **9**(1), 876–883 (2017).

45. Q. Ge, A.H. Sakhaei, H. Lee, C.K. Dunn, N.X. Fang and M.L. Dunn, Multimaterial 4D printing with tailorable shape memory polymers, *Sci Rep* **6**, 31110 (2016).

46. K. Chen, X. Kuang, V. Li, G. Kang and H.J. Qi, Fabrication of tough epoxy with shape memory effects by UV-assisted direct-ink write printing, *Soft Matter* **14**(10), 1879–1886 (2018).

47. J. Wu, C. Yuan, Z. Ding, M. Isakov, Y. Mao, T. Wang, M.L. Dunn and H.J. Qi, Multi-shape actie composites by 3D printing of digital shape memory polymers, *Sci Rep.* **6**, 24224 (2016).

48. D. Espalin, A.J. Ramirez, F. Medina, R. Wicker, Multi-material, multi-technology FDM: Exploring build process variations, *Rapid Prototyping Journal* **20**(3), 236–244 (2014).

49. D. Kokkinis, F. Bouville, A.R. Studard, 3D printing of materials with tunable failure via bioinspired mechanical gradients, *Adv Mater.* (2018).

50. J. Mueller, J.R. Raney, K. Shea and J.A. Lewis, Architected lattices with high stiffness and toughness via multicore-shell 3D printing, *Adv Mater.* **30**(12), 1705001 (2018).

51. N. Zhou, C. Liu, J.A. Lewis and D. Ham, Gigahertz electromagnetic structures via direct ink writing for radio-frequency oscilltor and transmitter applications, *Adv Mater.* **29**(15), 1605198 (2017).

52. X. Liu, H. Yuk, S. Lin, G.A. Parada T-C. Tang, E. Tham, C. Fuente-Nunez, T.K. Lu and X. Zhao, 3D printing of living responsive materials and devices, *Adv Mater.* **30**(4) 1704821 (2018).

53. G.I. Peterson, J.J. Schwartz, D. Zhang, B.M. Weiss, M.A. Ganter, D.W. Storti and A.J. Boydston, Production of materials with spatially-controlled cross-link density via vat photopolymerization, *ACS Appl Mater Interfaces* **8**(42), 29037–29043 (2016).

54. K. Qiu, Z. Zhao, G. Haghiashtiani, S.Z. Guo, M. He, R. Su, Z. Zhu, D.B. Bhuiyan, P. Murugan, F. Meng, S.H. Park, C-C. Chu, B. M. Ogle, D.A. Saltzman, D. R. Konety, R.M. Sweet and M.C. McAlpine, 3D printed organ models with physical properties of tissue and integrated sensors, *Adv Mater Technol.* **3**(3) 1700235 (2018).

55. D.J. Roach, C.M. Hamel, C.K. Dunn, M.V. Johnson, X. Kuang, H.J. Qi, The m4 3D printer: A multi-material multi-method additive manufacturing platform for future 3D printed structures, *Additive Manufacturing* **29**, 100819 (2019).

56. VC-F. Li, X. Kuang, C.M. Hamel, D. Roach, Y. Deng and H.J. Qi, Cellulose nanocrystals support material for 3D printing complexly shaped structures via multi-materials-multi-methods printing, *Additive Manufacturing* **28**, 14–22 (2019).

57. M. Jamal, S.S. Kadam, R. Xiao, F. Jivan, T-M. Onn, R. Fernandes, T.D. Nguyen and D.H. Gracias, Bio-origami hydrogel scaffolds composed of photocrosslinked PEG bilayers, *Advanced Healthcare Materials* **2**(8), 1142–1150 (2013).

58. Y. Mao, K. Yu, M.S. Isakov. J. Wu, M.L. Dunn and J.H. Qi, Sequential self-folding structures by 3D printed digital shape memory polymers, *Scientific Reports* **5**, 13616 (2015).

59. K. Yu, M.L. Dunn and H.J. Qi, Digital manufacture of shape changing components, *Extreme Mechanics Letters* **4**, 9–17 (2015).

60. Q. Zhang, K. Zhang and G. Hu, Smart three dimensional lightweight structure triggered from a thin composite sheet via 3D printing technique, *Scientific Reports* **6**, 22431 (2016).

61. Z. Zhao, J. Wu, X. Mu, H. Chen, H.J. Qi and D. Fang, Origami by frontal photopolymerization, *Sci Adv.* **3**(4) (2017).

62. K.E. Laflin, C.J. Morris, T. Muqeem and D.H. Gracias, Laser triggered sequential folding of microstructures, *Applied Physics Letters* **101**(13), 131901 (2012).

63. S.E. Bakarich, R. Gorkin, M. Panhuis and G.M. Spinks, 4D printing with mechanically robust, thermally actuating hydrogels, *Macromol Rapid Commun.* **36**(12), 1211–1217 (2015).

64. H. Therien-Aubin, Z.L. Wu, Z. Nie, E. Kumacheva, Multiple shape transformations of composite hydrogel sheets, *J Am Chem Soc* **135**(12), 4834–4839 (2013).

65. Z.L. Wu, M. Moshe, J. Greener, H. Therien-Aubin, Z. Nie, E. Sharon and E. Kumacheva, *Nat Commun* **4**, 1586 (2013).

66. S. Naficy, R. Gately, R. Gorkin, H. Xin and G.M. Spinks, 4D printing of reversible shape morphing hydrogel structures, *Macromol Mater Eng.* **302**(1) (2017).

67. D. Han, Z. Lu, S.A. Chester, H. Lee, Micro 3D printing of a temperature-responsive hydrogel using projection micro-stereolithography, *Sci Rep.* **8**(1), 1963 (2018).

68. S. Naficy, R. Gately, R. Gorkin, H. Xin and G.M. Spinks, 4D printing of reversible shape morphing hydrogel structures, Macromolecular materials and engineering **302**(1), 1600212 (2017).

69. A. Lendlein and S. Kelch, Shape-memory polymers, *Angew Chem Int Edit.* **41**(12), 2034–2057 (2002).

70. J. Hu, Y. Zhu, H. Huang and J. Lu, Recent advances in shape memory polymers: Structure, mechanism, functionality, modeling and applications, *Prog Polym Sci.* **37**(12), 1720–1763 (2012).

71. Q. Zhao, H.J. Qi and T. Xie, Recent progress in shape memory polymer: New behavior, enabling materials, and mechanistic understanding, *Prog Polym Sci.* **49-50**, 79–120 (2015).

72. X. Kuang, G. Liu, X. Dong and D. Wang, Triple-shape memory epoxy based on Diels-Alder adduct molecular switch, *Polymer* **84**, 1–9 (2016).

73. M. Behl and A. Lendlein, Shape-memory polymers, *Mater Today*, **10**(4), 20–28 (2007).

74. A. Miriyev, K. Stack and H. Lipson, Soft material for soft actuators, *Nat. Commun.* **8**(1), 596 (2017).

75. R.F. Shepherd, F. Ilievski, W. Choi, S.A. Morin, A.A. Stokes, A.D. Mazzeo, X. Chen, M. Wang and G.M. Whitesides, Multigait soft robot, *Proceedings of the National Academy of Sciences* **108**(51) 20400–20403 (2011).

76. C. Yuan, D.J. Roach, C.K. Dunn, Q. Mu, X. Kuang, C.M. Yakacki, T.J. Wang, K. Yu and H.J. Qi, 3D printed reversible shape changing soft actuators assisted by liquid crystal elastomers, *Soft Matter* **13**(33), 5558–5568 (2017).

77. T. Umedachi, V. Vikas and B.A. Trimmer, Highly deformable 3-D printed soft robot generating inching and crawling locomotions with variable friction legs, Proceedings of the IEEE/RSJ International Conference on Intelligent Robots and Systems (Nov 2013).

78. S. Kim, M. Spenko, S. Trujillo, B. Heyneman, D. Santos and M.R. Cutkosky, Smooth vertical surface climbing with directional adhesion, *IEEE Transactions on Robotics* **24**(1), 65–74 (2008).

79. H.W. Huang, A.J. Petruska, M.S. Sakar, M. Skoura, F. Ullrich, Q. Zhang, S. Pane and B.J. Nelson, Self-folding hydrogel bilayer for enhanced drug loading, encapsulation, and transport, *Proceedings of the 2016 38th Annual International Conference of the IEEE Engineering in Medicine and Biology Society (EMBC)* (Aug 2016).

80. M. Zarek, N. Mansour, S. Shapira and D. Cohn, 4D printing of shape memory-based personalized endoluminal medical devices, *Macromol Rapid Commun.* **38**(2), 1600628 (2017).

81. A. Lendlein and R. Langer, Biodegradable, elastic shape-memory polymers for potential biomedical applications, **296**(5573), 1673–1676 (2002).

82. K. Liu, J. Wu, G.H. Paulino and H.J. Qi, Programmable deployment of tensegrity structures by stimulus-responsive polymers, *Sci Rep.* **7**(1), 3511 (2017).

83. K. Yu, Q. Ge, H.J. Qi, Reduced time as a unified parameter determining fixity and free recovery of shape memory polymers, *Nat. Commun.* **5**, 3066 (2014).

84. Q. Ge, X. Luo, C.B. Iversen, P.T. Mather, M.L. Dunn and H.J. Qi, Mechanisms of triple-shape polymeric composites due to dual thermal transitions, *Soft Matter.* **9**(7), 2212–2223 (2013).

85. K. Yu, H. Li, A.J.W. McCLung, G.P. Tandon, J.W. Baur and H.J. Qi, Cyclic behaviors of amorphous shape memory polymers, *Soft Matter.* **12**(13), 3234–3245 (2016).

86. K. Yu, T. Xie, J. Leng, Y. Ding and H.J. Qi, Mechanisms of multi-shape memory effects and associated energy release in shape memory polymers, *Soft Matter.* **8**(20), 5687–5695 (2012).

87. L. Jinsong, L. Xin, L. Yanju and D. Shanyi, Electroactive thermoset shape memory polymer nanocomposite filled with nanocarbon powders, *Smart Mater Struct.* **18**(7), 074003 (2009).

88. Q. Guo, C.J. Bishop, R.A. Meyer, D.R. Wilson, L. Olasov, D.E. Schlesinger, P.T. Mather, J.B. Spicer, J.H. Elisseeff and J.J. Green, Entanglement-based thermoplastic shape memory polymeric particles with photothermal actuation for biomedical applications, *ACS Appl Mater Interfaces* **10**(16), 13333–13341 (2018).

89. H. Zhang and Y. Zhao, Polymers with dual light-triggered functions of shape memory and healing using gold nanoparticles, *ACS Appl Mater Interfaces* **5**(24), 13069–13075 (2013).

90. X. Liu, R. Wei, P.T. Hoang, X. Wang, T. Liu and P. Keller, Reversible and rapid laser actuation of liquid crystalline elastomer micropillars with inclusion of gold nanoparticls, *Adv. Funct Mater.* **25**(20), 3022–3032 (2015).

91. D-D. Han, Y-L. Zhang, J-N. Ma, Y-Q. Liu, B. Han and H-B. Sun, Light-mediated manufacture and manipulation of actuators, *Adv Mater.* **28**(38), 8328–8343 (2016).

92. Y. Liu, B. Shaw, M.D. Dickey and J. Genzer, Sequential self-folding of polymer sheets, *Sci Adv.* **3**(3), e1602417 (2017).

93. T. Gong, W. Li, H. Chen, L. Wang, S. Shao and S. Zhou, Remotely actuated shape memory effect of electrospun composite nanofibers, *Acta biomaterialia* **8**(3), 1248–1259. (2012).

94. U.N. Kumar, K. Kratz, M. Heuchel, M. Behl and A. Lendlein, Shape-memory nanocomposites with magnetically adjustable apparent switching temperatures, *Adv Mater.* **23**(36), 4157–4162 (2011).

95. X. Zheng, S. Zhou, Y. Xiao, X. Yu, X. Li and P. Wu, Shape memory effect of poly(d,l-lactide)/Fe3O4 nanocomposites by inductive heating of magnetite particles, *Colliods and Surfaces B: Biointerfaces* **71**(1), 67–72 (2009).

96. S. Dadbakhsh, M. Speirs, J-P. Kruth, J. Schrooten, J. Luyten and J.V. Humbeeck, Effect of SLM parameters on transformation temperatures of shape memory nickel titanium parts, *Advanced Engineering Materials* **16**(9), 1140–1146 (2014).

97. K. Yu, A. Ritchie, Y. Mao, M.L. Dunn and H.J. Qi, Controlled sequential shape changing components by 3D printing of shape memory polymer multimaterials, *Procedia IUTAM* **12**, 193–203 (2015).

98. K. Gall, M.L. Dunn, Y. Liu, G. Stefanie and D. Balzar, Internal stress storage in shape memory polymer nanocomposites, *Applied Physics Letters* **85**(2), 290–292 (2004).

99. T. Zhao, R. Yu, X. Li, B. Cheng, Y. Zhang, X. Yang, X. Zhao, Y. Zhao and W. Huang, 4D printing of shape memory polyurethane via stereolithography, *European Polymer Journal*, **101**, 120–126 (2018).

100. M.O. Saed, A.H. Torbati, D.P. Nair, C.M. Yakacki, Synthesis of programmable main-chain liquid-crystalline elastomers using a two-stage thiol-acrylate reaction, *J. Visualized Exp.* (107), 53546 (2016).

101. J. Kupfer and H. Finkelmann, Nematic liquid single crystal elastomers, *Die Makromolekulare Chemis, Rapid Communications,* **12**(12), 717–726 (1991).

102. D.J. Roach, C. Yuan, X. Kuang, VC-F. Li, P. Blake, M.L. Romero, I. Hammel, K. Yu and H.J. Qi, Long liquid crystal elastomer fibers with large reversible actuation strains for smart textiles and artificial muscles, *ACS Appl. Mater. Interfaces* **11**(21), 19514–19521 (2019).

103. C.P. Ambulo, J.J. Burroughs, J.M. Boothby, H. Kim, M.R. Shankar and T.H. Ware, Four-dimensional printing of liquid crystal elastomers, *ACS Appl Mater Interfaces.* **9**(42), 37332–37339 (2017).

104. X. Yu, J. Pan, J. Deng, J. Zhou, X. Sun and H. Peng, A novel photoelectric conversion yarn by integrating photomechanical actuation and the electrostatic effect, *Adv Mater.* **28**(48) 10744-10749 (2016).

105. Y. Yang, Z. Pei, X. Zhang, L. Tao, Y. Wei and Y. Ji, Carbon nanotube-vitrimer composite for facile and efficient photo-welding of epoxy, *Chem Sci.* **5**(9), 3486–3492 (2014).

106. A. Lendlein, H. Jiang, O. Junger, R. Langer, Light-induced shape-memory polymers, *Nature* **434**(7035), 879–882 (2005).

107. Y. Mao, Z. Ding, C. Yuan, S. Ai, M. Isakov, J. Wu, T. Wang, M.L. Dunn and H.J. Qi, 3D printed reversible shape changing components with stimuli responsive materials, *Sci Rep.* **6**, 24761 (2016).

108. S. Naficy, G.M. Spinks, G.G. Wallace, Thin, tough, pH-sensitive hydrogel films with rapid load recovery, *ACS Appl Mater Interfaces* **6**(6), 4109–4114 (2014).

109. I. Tomatsu, K. Peng and A. Kros, Photoresponsive hydrogels for biomedical applications, *Advanced Drug Delivery Reviews* **63**(14), 1257–1266 (2011).

110. S, Irvine and S. Venkatraman, Bioprinting and differentiation of stem cells, *Molecules* **21**(9), 1188 (2016).

111. G. Villar, A.D. Graham and H. Bayley, A tissue-like printed material, *Science* **340**(6128), 48–52 (2013).

112. Q. Ge, K.D. Conner, H.J. Qi and L.D. Martin, Active origami by 4D printing, *Smart Mater Struct.* **23**(9), 094007 (2014).

113. C. Yuan, T. Wang, M.L. Dunn and H.J. Qi, 3D printed active origami with complicated folding patters, *International Journal of Precision Engineering and Manufacturing-Green Technology* **4**(3), 281–289 (2017).

114. G. Qi, K.D. Conner, H.J. Qi and L.D. Martin, Active origami by 4D printing, *Smart Mater Struct.* **23**(9), 094007 (2014).

115. Y. Mao, K. Yu, M.S. Isakov, J. Wu, M.L. Dunn and H.J. Qi, Sequential self-folding structures by 3D printed digital shape memory polymers, *Scientific Reports* **5**, 13616 (2015).

116. C. Yuan, Z. Ding, T. Wang, M.L. Dunn and H.J. Qi, Shape forming by thermal expansion mismatch and shape memory locking in polymer/elastomer laminates, *Smart Mater Struct.* **26**(10), 105027 (2017).

117. S. Sundaram, D.S. Kim, M.A. Baldo, R.C. Hayward, W. Matusik, 3D-printed self-folding electronics, *ACS Appl Mater Interfaces* **9**(37), 32290–32298 (2017).

118. Z. Ding, C. Yuan, X. Peng, T. Wang, H.J. Qi, M.L. Dunn, Direct 4D printing via active composite materials, *Sci Adv.* **3**(4), e1602890 (2017).

119. D. Kwak, H.H. CHoi, B. Kang, D.H. Kim, W.H. Lee and K. Cho, Tailoring morphology and structure of inkjet-printed liquid-crystalline semiconductor/insulating polymer blends for high-stability organic transistors, *Adv Funct Mater.* **26**(18) (2016).

120. Z. Pei, Y. Yang, Q. Chen, E.M. Terentjev, Y. Wei and Y. Ji, Mouldable liquid crystalline elastomer actuators with exchangeable covalent bonds, *Nat Mater.* **13**(1), 36–41 (2014).

121. J-a. Lv, Y. Liu, J. Wei, E. Chen, L. Qin and Y. Yu, Photocontrol of fluid slugs in liquid crystal polymer microactuators, *Nature* **537**(7619), 179–184 (2016).

122. Y. Yu, M. Nakano and T. Ikeda, Photomechanics: Directed bending of a polymer film by light, *Nature* **425**(6954),145 (2003).

123. T.H. Ware, Z.P. Perry, C.M. Middleton, S.T. Iacono and T.J. White, Programmable liquid crystal elastomers prepared by Thiol-ene photopolymerization, *ACS Macro Letters* **4**(9), 942–946 (2015).

124. S.W. Ula, N.A. Traugutt, R.H. Volpe, R.R. Patel, K. Yu and C.M. Yakacki, Liquid crystal elastomers: An introduction and review of emerging technologies, *Liquid Crystals Reviews,* **6**(1), 78–107 (2018).

125. X. Lu, H. Zhang, G. Fei, B. Yu, X. Tong, H. Xia and Y. Zhao, Liquid-crystalline dynamic networks dopes with gold nanorods showing enhanced photocontrol of actuation, *Adv Mater.* **30**(14), 1706597 (2018).

126. Y. Yu, M. Nakano and T. Ikeda, Directed bending of a polymer film by light, *Nature* **425**, 145 (2003).

127. C.L. van Oosten, C.W.M. Bastiaansen, D.J. Broer, Printed artificial cilia from liquid-crystal network actuators modularly driven by light, *Nature Materials* **8**, 677 (2009).

128. M. Camacho-Lopez, H. Finkelmann, P. Palffy-Muhoray and M. Shelley, Fast liquid-crystal elastomer swims into the dark, *Nature Materials* **3**, 307 (2004).

129. J-a. Lv, Y. Liu, J. Wei, E. Chen, L. Qin and Y. Yu, Photocontrol of fluid slugs in liquid crystal polymer microactuators, *Nature* **537**, 179 (2016).

130. H. Zeng, P. Wasylezyk, D.S. Wiersma and A. Priimagi, Light robots: Bridging the gap between microrobotics and photomechanics in soft materials, *Adv Mater.* **30**(24), 1703554 (2018).

131. L. Yang, K. Setyowati, A. Li, S. Gong and J. Chen, Reversible infrared actuation of carbon nanotube-liquid crystalline elastomer nanocomposites, *Adv Mater.* **20**(12), 2271–2275 (2008).

132. M. Brehmer, R. Zentel, G. Wagenblast and K. Siemensmeyer, Ferroelectric liquid-crystalline elastomers, *Macromolecular Chemistry and Physics* **195**(6), 1891–1904 (1994).

133. A. Kaiser, M. Winkler, S. Krause, H. Finkelmann and A.M. Schmidt, Magnetoactive liquid crystal elastomer nanocomposites, *Journal of Materials Chemistry* **19**(4), 538–543 (2009).

134. M. Lopez-Valdeolivas, D. Liu, D.J. Broer and C. Sanchez-Somolinos, 4D printed actuators with soft-robotic functions, *Macromol Rapid Commun.* **39**(5) (2018).

135. C.P. Ambulo, J.J. Burroughs, J.M. Boothby, H. Kim, M.R. Shankar and T.H. Ware, Four dimensional printing of liquid crystal elastomers, *ACS Appl Mater Interfaces* **9**(42), 37332–37339 (2017).

136. A. Koikian, R.L. Truby, J.W. Boley, T.J. White and J.A. Lewis, 3D printing of liquid crystal elastomeric actuators with spatially programed nematic order, *Adv Mater.* **30**(10) (2018).

137. Y. Chao, D. Zhen, T.J. Martin, H.J. Qi, Shape forming by thermal expansion mismatch and shape memory locking in polymer/elastomer laminates, *Smart Materials and Structures* **26**(10), 105027 (2017).

138. D.J. Roach, X. Kuang, C. Yuan, K. Chen and H.J. Qi, Novel ink for ambient condition printing of liquid crystal elastomers for 4D printing, *Smart Mater Struct.* **27**(12), 125011 (2018).

139. Z. Zhao, J. Wu, X. Mu, H. Chen, H.J. Qi and D. Fang, Desolvation induced origami of photocurable polymers by digit light processing, *Macromol Rapid Commun.* **38**(13), 1600625 (2017).

140. L. Huang, R. Jiang, J. Wu, J. Song, H. Bai, B. Li, Q. Zhao and T. Xie, Ultrafast digital printing toward 4D shape changing materials, *Adv Mater.* **29**(7) 1605390 (2017).

141. J. Hiller and H. Lipson, Design and analysis of digital materials for physical 3D voxel printing, *Rapid Prototyping Journal* **15**(2), 137–149 (2009).

142. T-v. Manen, S. Janbaz and AA. Zadpoor, Programming 2D/3D shape-shifting with hobbyist 3D printers, *Mater Horiz.* **4**(6), 1064–1069 (2017).

143. T-v. Manen, S. Janbaz and A.A. Zadpoor, Programming the shape-shifting of flat soft matter, *Mater Today.* **21**(2), 144–163 (2018).

144. J. Wu, Z. Zhao, X. Kuang, C.M. Hamel, D. Fang and H.J. Qi, Reversible shape change structures by grayscale pattern 4D printing, *Multifunctional Materials* **1**(1), 015002 (2018).

145. Y. Lee, H. Lee, T. Hwang, J-G. Lee and M. Cho, Sequential folding using light-activated polystyrene sheet, *Sci Rep.* **5**, 16544 (2015).

146. K.M. Lee, N.V. Tabiryan, T.J. Bunning and T.J. White, Photomechanical mechanism and structure-property considerations in the generation of photomechanical work in glass, azobenzene liquid crystal polymer networks, *Journal of Materials Chemistry* **22**(2), 691–698 (2012).

147. X. Zhang, Q. Zhou, H. Liu and H. Liu, UV light induced plasticization and light activated shape memory of spiropyran doped ethylene-vinyl acetate copolymers, *Soft Matter.* **10**(21), 3748–3754 (2014).

148. H. Yang, W.R. Leow, T. Wang, J. Wang. J. Yu, K. He, D. Qi, C. Wan and X. Chen, 3D printed photoresponsive devices based on shape memory composites, *Adv Mater.* **29**(33), 1701627 (2017).

149. D. Hua, X. Zhang, Z. Ji, C. Yan, B. Yu, Y. Li, X. Wang and F. Zhou, 3D printing of shape changing composites for constructing flexible paper-based photothermal bilayer actuators, *Journal of Materials Chemistry C* **6**(8), 2123–2131 (2018).

150. S. Curran, P. Chambon, R. Lind, L. Love, R. Wagner, S. Whitted, D. Smith, B. Post, R. Graves and C.A. Blue, Big area additive manufacturing and hardware-in-the-loop for rapid vehicle powertrain prototyping: A case study on the development of a 3-D printed shelby cobra, *SAE Technical Papers* (2016).

151. Y. Yang, Z. Pei, Z. Li, Y. Wei and Y. Ji, Making and remaking dynamic 3D structures by shining light on flat liquid crystalline vitrimer films without a mold, *J Am Chem Soc.* **138**(7), 2118–2121 (2016).

152. C.M. Yakacki, R. Shandas, D. Safranski, A.M. Ortega, K. Sassaman and G.K. Strong, Tailored biocompatible shape-memory polymer networks, *Adv Funct Mater.* **18**(16), 2428–2435 (2008).

153. M. Stoppa and A. Chiolerio, Wearable electronics and smart textiles, A critical review, *Sensors* **14**(7), 11957 (2014).

154. M. Langelaar, Topology optimization of 3D self-supporting structures for additive manufacturing, *Addit Manuf.* **12**, 60–70 (2016).

155. A. Clausen, F. Wang, J.S. Jensen, O. Sigmund and J.A. Lewis, Topology optimized architectures with programmable Poisson's ratio over large deformations, *Adv Mater.* **27**(37), 5523–5527 (2015).

156. J. Liu and A.C. To, Deposition path planning-integrated structural topology optimization for 3D additive manufacturing subject to self-support constraint, *Computer-Aided Design* **91**, 27–45 (2017).

157. A.M. Mirzendehdel and K. Suresh, Support structure constrained topology optimization for additive manufacturing, *Computer-Aided Design* **81**, 1–13 (2016).

158. A. Clausen, N. Aage and O. Sigmund, Exploting additive manufacturing infill in topology optimization for improved buckling load, *Engineering* **2**(2), 250–257 (2016).

159. K. Maute, A. Tkachuk, J. Wu, H.J. Qi, Z. Ding and M.L. Dunn, Level set topology optimization of printed active composites, *Journal of Mechanical Design* **137**(11), 111402 (2015).

160. J. Liu, A.T. Gaynor, S. Chen, Z. Kang, K. Suresh, A. Takezawa, L. Li, J. Kato, J. Tang, C.C.L. wang, L. Cheng, X. Liang and A.C. To, Current and future trends in topology optimization for additive manufacturing, *Structural and Multidisciplinary Optimization* **57**, 2457–2483 (2018).

161. D.W. Rosen, A review of synthesis methods for additive manufacturing, *Virtual and Physical Prototyping* **11**(4), 305–317 (2016).

162. C.M. Hamel, D.J. Roach, K.N. Long, F. Demoly, M.L. Dunn and H.J. Qi, Machine-learning based design of active composite structures for 4D printing, *Smart Mater Struct.* **28**(6), 065005 (2019).

163. V.I. Butvilovskaya, S.B. Popletaeva, V.R. Chechetkin, Z.I. Zubtsova, M.V. Tsybulskaya, L.O. Samokhina, L.I. Vinnitskii, A.A. Ragimov, E.I. Pozharitskaya, G.A. Grigor'eva, N.Y. Meshalkina, S..V. Golysheva, N.V. Shilova, N.V. Bovin, A.S. Zadedatelev and A.Y. Rubina, Multiplex determination of serological signatures in the sera of colorectal cancer patients using hydrogel biochips, *Cancer Med* **5**(7), 1361–1372 (2016).

164. D.L. Taylor and M-ih. Panhuis, Self-healing hydrogels, *Adv Mater.* **28**(41) 9060–9093 (2016).

165. G. Deng, F. Li, H. Yu, F. Liu, C. Liu, W. Sun, H. Jiang and Y. Chen, Dynamic hydrogels with an environmental adaptive self-healing ability and dual responsibr sol-gel transitions, *ACS Macro Lett.* **1**(2), 275–279 (2012).

166. T.A. Gisby, B.M. O'Brien and I.A. Anderson, Self sensing feedback for dielectric elastomer actuators, *Applied Physics Letters* **102**(19), 193703 (2013).

167. M. Qin, M. Sun, R. Bai, Y. Mao, X. Qian, D. Sikka, Y. Zhao, H.J. Qi, Z. Suo and X. He, Bioinspired hydrogel interferometer for adaptive coloration and chemical sensing, *Adv Mater.* **30**(21), 1800468 (2018).

168. B. Gauvreau, N. Guo, K. Schicker, K. Stoeffler, F. Boismenu, A. Ajji, R. Wingfield, C. Dubois and M. Skorobogatiy, Color-changing and color-tunable photonic bandgap fiber textiles, *Opt Express* **16**(20), 15677–15693 (2008).

169. S. Park, K. Mackenzie and S. Jayaraman, The wearable motherboard, A framework for personalized mobile information processing (PMIP), *Proceedings of the 39th Annual Design Automation Conference*, pp. 170–174, New Orleans, Louisiana, USA (2002).

170. B. Gauvreau, N. Guo, K. Schicker, K. Stoeffler, F. Boismenu, A. Ajji, R. Wingfield, C. Dubois, and M. Skorobogatiy, Color-changing and color-tunable photonic bandgap fiber textiles. *Opt Express.* **16**(20), 15677–15693 (2008).

171. S. Park, K. Mackenzie, S. Jayaraman, The wearable motherboard: a framework for personalized mobile information processing (PMIP). In *Proceedings of the 39th Annual Design Automation Conference*, New Orleans, Louisiana, USA. pp. 170–174 (2002).

Chapter 8

Functionalized Materials for Additive Manufacturing and 3D Printing

Tarek I. Zohdi

8.1 Introduction

Additive Manufacturing (AM) is usually defined as the process of joining materials to make objects from 3D model data, typically layer upon layer, as opposed to subtractive manufacturing methodologies, which remove material (American Society for Testing and Materials, ASTM). We refer the reader to the recent overview article by Huang et al.[43] on the wide array of activities in the manufacturing community in this area. One subclass of AM, so-called 3D Printing (3DP), has received a great deal of attention over the last few years. Typically such a process takes CAD drawings and slices them into layers, printing layer by layer. 3DP was pioneered by Hull[48] of the 3D Systems Corporation in 1984. 3DP was a 2.2 billion dollar industry in 2014, with applications ranging from motor vehicles, consumer products, medical devices, military hardware and the arts.

Key ingredients of these processes are the specialized materials and the precise design of their properties, enabled by the use of fine-scale "functionalizing" particles. The rapid rise in the use of particle-based materials has been made possible by the large-scale production of consistent, high-quality, particles, which are produced in a variety of ways, such as: (a) sublimation from a raw solid to a gas, which condenses into particles that are recaptured (harvested), (b) atomization of liquid streams into droplets by breaking jets of metal, (c) reduction of metal oxides and (d) comminution/pulverizing of bulk material. As mentioned in the preface, particle-functionalized materials play a central role in this field, in three main ways:

(1) To enhance overall filament-based material properties, by embedding particles within a binder, which is then passed through a heating element and the deposited onto a surface,

(2) To "functionalize" inks by adding particles to freely flowing solvents forming a mixture, which is then deposited onto a surface, and

Fig. 8.1. Typical printing ingredients: Top Left: Finely ground metallic powder (iron). Top Right: Extruded PLA. Bottom Left: ABS pellets and Bottom Right: Coarsely ground steel flakes (Zohdi[143]).

(3) To directly deposit particles, as dry powders, onto surfaces and then to heat them with a laser, e-beam or other external source, in order to fuse them into place.

In more detail, we have (see Fig. 8.1):

(1) **Heated filament based materials** (historically for prototyping) are comprised of thermoplastics. To extend the materials to applications beyond prototyping, second-phase particles are added to the heated mixture which solidify (cure) to form the overall material properties comprised of particles in a binding matrix when deposited on a substrate. The particles are used to "tune" the binding matrix properties to the desired overall state. Specifically, much of the commercial additive manufacturing processes are polymer-based, with second-phase particles added to enhance the properties of the binder, which is typically either (1) Polylactic acid or polylactide (PLA), which is a biodegradable thermoplastic aliphatic polyester or (2) Acrylonitrile butadiene styrene (ABS) which is a common thermoplastic polymer. In 2015, PLA had the second highest consumption volume

of any bioplastic of the world. PLA is derived from renewable resources, such as plants (corn starch, sugarcane, etc.). ABS is a terpolymer that is significantly stronger than PLA. It is made by polymerizing styrene and acrylonitrile in the presence of polybutadiene. The styrene gives the plastic a reflective surface while the rubbery polybutadiene endows toughness. The overall properties are created by rubber toughening, where fine particles of elastomer are distributed throughout the rigid matrix. Typically, metal and ceramic particles are also added to endow specific mechanical, thermal, electrical and magnetic effective overall properties.

(2) **Functionalized ink materials** (primarily for printed electronics) are comprised of particles in a solvent/lubricant which cure when deposited. Oftentimes, these inks are used to lay down electric circuit lines or to have some other specific electromagnetic function on a surface. One application where such functionalized inks are important is printed electronics on flexible foundational substrates, such as flexible solar cells and smart electronics. One important technological obstacle is to develop inexpensive, durable electronic material-units that reside on flexible platforms or substrates which can be easily deployed onto large surface areas. Ink-based printing methods involving particles are, in theory, ideal for large-scale electronic applications, and provide a framework for assembling electronic circuits by mounting printed electronic devices on flexible plastic substrates, such as polyimide and "PEEK" (Poly-Ether-Ether-Ketone, a flexible thermoplastic polymer) film. There are many variants of this type of technology, which is sometimes referred to as flexible electronics or flex circuits. Flex circuits can be, for example, screen printed silver circuits on polyester. For an early history of the printed electronics field, see Gamota.[26] In order to develop flexible micro/nanoelectronics for large area deployment, traditional methods of fabrication using silicon-based approaches have become limited for applications that involve large-area coverage, due to high cost of materials and equipment (which frequently need a vacuum environment). For flexibility and lower cost, the ability to develop these electronics on plastics is necessary. To accomplish this task, print-based technologies are starting to become popular for these applications. In many cases, this requires the development of nanoparticle-functionalized "inks". These nanoparticles include germanium (which has higher mobility and better tailorable absorption spectrum for ambient light than silicon) and silver (which is being studied due to the possibility to sinter the particles without the need of directly applied intense heating). Other semiconductor nanoparticles, including zinc and cadmium based compounds and

metals, such as gold and copper, can be considered. Precise patterning of (nanoparticle-functionalized) prints is critical for a number of different applications. For example, some recent applications include optical coatings and photonics (Nakanishi et al.[68]), MEMS applications (Fuller et al.,[25] Samarasinghe et al.[71] and Gamota et al.[26]) and biomedical devices (Ahmad et al.[2]). In terms of processing techniques, we refer the reader to Sirringhaus et al.,[96] Wang et al.,[101] Huang et al.,[44] Choi et al.[14−17] and Demko et al.[20,21] for details.[a] We further mention that electromagnetically sensitive fluids are typically constructed ("functionalized") by embedding charged or electromagnetically sensitive particles in a neutral fluid. Such fluids date back, at least, to Winslow[102,103] in 1947. While the most widely used class of such fluids are electrorheological fluids, which are comprised of extremely fine suspensions of charged particles (on the order of 50 microns) in an electrically neutral fluid, there has been a renewed interest in this class of materials because of so-called e-inks (electrically-functionalized inks) driven by printed electronics. Inkjet printing is attractive due to its simplicity, high throughput, and low material loss. However, patterning with inkjet printing is limited to a resolution of around 20-50 μm with current printers (Ridley et al.[88]) with higher resolution possible by adding complexity to the substrate prior to printing (Wang et al.[101]). Electrohydrodynamic printing has also been proposed to increase the resolution beyond the limits of inkjet printing, achieving a line resolution as small as 700 nm (Park et al.[81]).

(3) **Dry powder based materials** (primarily for sintered load-bearing structures) are deposited onto a surface and then heated by a laser, e-beam or other external source, in order to fuse them into place. These types of applications and associated technology are closely related to those in the area of spray coatings, and we refer the reader to the extensive works of Sevostianov and Kachanov[92−94] Nakamura and coworkers: Dwivedi et al.,[23] Liu et al.[57,58] Nakamura and Liu,[66] Nakamura et al.[67] and Qian et al.[84] and to Martin[61,62] for the state of the art in deposition technologies. In powder-based processes, after deposition, laser processing is applied to heat particles in a powder to desired temperatures either to subsequently soften, sinter, melt or ablate them. Selective laser sintering, was pioneered by Householder[47] in 1979 and Deckard and Beaman[19] in the mid-1980's.[b] Laser-

---

[a]For reviews of optical coatings and photonics, see Nakanishi et al.[68] and Maier and Atwater,[60] for biosensors see Alivisatos,[4] for catalysts, see Haruta[35] and for MEMS applications, see Fuller et al.[25] and Ho et al.[41]

[b]A closely related method, Electron Beam Melting, fully melts the material and produces dense solids that are void free.

based heating is quite attractive because of the degree of targeted precision that it affords.[c] Because of the monochromatic and collimated nature of lasers, they are a highly controllable way to process powdered materials, in particular with pulsing, via continuous beam chopping or modulation of the voltage. Carbon Dioxide (CO_2) and Yttrium Aluminum Garnett (YAG) lasers are commonly used. The range of power of a typical industrial laser is relatively wide, ranging from approximately 100-10,000 Watts. Typically, the initial beam produced is in the form of collimated (parallel) rays, which are then focused with a lens onto a small focal point of no more than about 0.00001 m in diameter. *However,* a chief concern of manufacturers are residual stresses and the microstructural defects generated in additively manufactured products, created by imprecisely controlled heat affected zones, brought on by miscalibration of the laser power needed for a specific goal. In particular, because many substrates can become thermally-damaged, for example from thermal stresses, ascertaining the appropriate amount of laser input is critical.

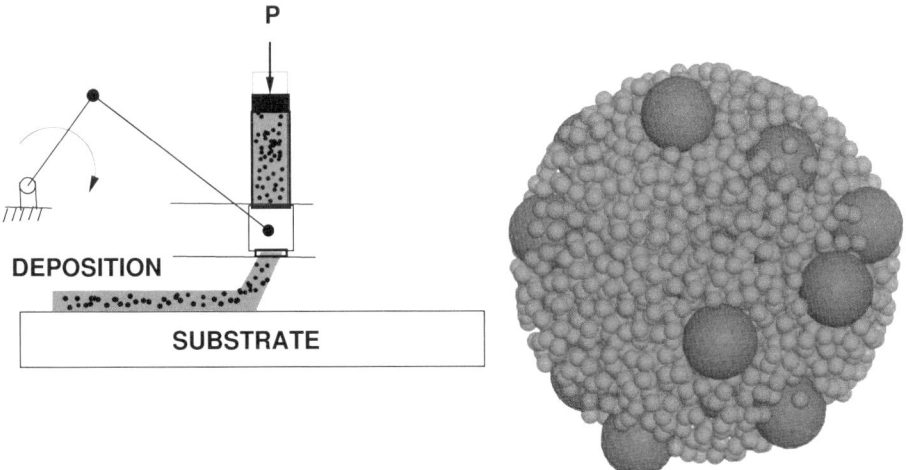

Fig. 8.2. Left: A linkage schematic of a 3D printer. Right: A multiphase droplet representation using the Discrete Element Method (Zohdi[143]).

---

[c]There are a variety of other techniques that may be involved in an overall additive manufacturing processes, such as: (a) electron beam melting, which is a process by where powder is bonded together layer per layer with an electron beam in a high vacuum, (b) aerosol jetting, which consists of utilizes directing streams of atomized particles at high velocities towards a substrate and (c) inkjet printing, which works by projecting small droplets of ink towards a substrate through a small orifice by pressure, heat, and vibration. The deposited material is then heated by UV light or other means to rapidly dry.

8.2   Objectives

In order for *emerging additive manufacturing approaches to succeed, such as the ones mentioned, one must draw upon rigorous, yet practical, methods to guide and simultaneously develop design rules for the proper selection of particle, binder and solvent combinations for upscaling to industrial manufacturing levels (Fig. 8.2). This motivates the content of this chapter.* During the development of new particulate-functionalized materials, experiments to determine the appropriate combinations of particulate and matrix phases are time-consuming and expensive. Therefore, "microstructure-macroproperty" methods have been generated over the last century in order to analyze and guide new material development. The overall properties of such materials are the aggregate response of the collection of inter-acting components (Fig. 8.3). The macroscopic properties can be tailored to the specific application, for example in structural engineering applications, by choos-ing a harder particulate phase that serves as a stiffening agent for a ductile, easy to form, base matrix material. "Microstructure-macroproperty" (micro-macro) meth-ods are referred to by many different terms, such as "homogenization", "regular-ization", "mean field theory", "upscaling", etc. in various scientific communities to compute effective properties of heterogeneous materials. We will use these terms interchangeably in this chapter. The usual approach is to compute a constitutive "relation between averages", relating volume averaged field variables, resulting in effective properties. Thereafter, the effective properties can be used in a macro-scopic analysis. The volume averaging takes place over a statistically representative sample of material, referred to in the literature as a representative volume element (RVE). The internal fields, which are to be volumetrically averaged, must be com-puted by solving a series of boundary value problems with test loadings. There is a vast literature on methods, dating back to Maxwell[63,64] and Lord Rayleigh,[86] for estimating the overall macroscopic properties of heterogeneous materials. For an authoritative review of the general theory of random heterogeneous media, see Torquato;[99] for more mathematical homogenization aspects, see Jikov et al.;[49] for solid-mechanics inclined accounts of the subject, see Hashin,[38] Mura,[65] Nemat-Nasser and Hori,[69] Huet;[45,46] for analyses of cracked media, see Sevostianov et al.[95] and for computational aspects, see Zohdi and Wriggers,[105–134] Ghosh[29] and Ghosh and Dimiduk.[30]

   This chapter, which is an abrievated version of the book: Zohdi,[143] *Model-ing and simulation of functionalized materials for additive manufacturing and 3D printing: continuous and discrete media*, focuses on some very basic concepts in this area, initially illustrated by a linear elasticity framework, where the mechan-ical properties of microheterogeneous materials are characterized by a spatially variable elasticity tensor $I\!E$. In order to characterize the effective (homogenized)

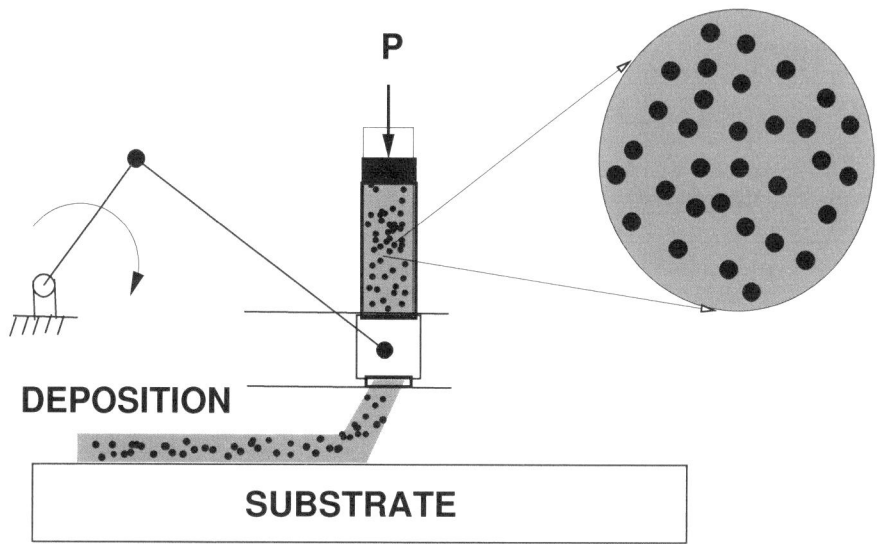

Fig. 8.3. A matrix binder and particulate additives (Zohdi[143]).

macroscopic response of such materials, a relation between averages,

$$\langle \boldsymbol{\sigma} \rangle_\Omega = \boldsymbol{I\!E}^* : \langle \boldsymbol{\varepsilon} \rangle_\Omega, \tag{8.1}$$

is sought, where

$$\langle \cdot \rangle_\Omega \overset{\text{def}}{=} \frac{1}{|\Omega|} \int_\Omega \cdot \, d\Omega, \tag{8.2}$$

and where $\boldsymbol{\sigma}$ and $\boldsymbol{\varepsilon}$ are the stress and strain tensor fields within a statistically representative volume element (RVE) of volume $|\Omega|$. The quantity $\boldsymbol{I\!E}^*$ is known as the effective property. It is the elasticity tensor used in usual structural analyses. Similarly, one can describe other effective quantities such as conductivity or diffusivity, in virtually the same manner, relating other volumetrically averaged field variables. However, for the sake of brevity, we restrict ourselves to linear elastostatics problems.

8.3   Notation

Throughout this work, boldface symbols denote vectors or tensors. Furthermore, we exclusively employ a Cartesian basis. For the inner product of two vectors (first order tensors), $\boldsymbol{u}$ and $\boldsymbol{v}$, we have $\boldsymbol{u} \cdot \boldsymbol{v} = u_i v_i = u_1 v_1 + u_2 v_2 + u_3 v_3$ in three dimensions, where Cartesian basis and Einstein index summation notation are used. In this introduction, for clarity of presentation, *we will ignore the difference between second order tensors and matrices.* Accordingly, if we consider the second order tensor $A = A_{ik} e_i \otimes e_k$, then a first order contraction (inner product) of two second order tensors $A \cdot B$ is defined by the matrix product $[A][B]$, with components of $A_{ij}B_{jk} = C_{ik}$. It is clear that the range of the inner index j must be the same for $[A]$ and $[B]$. For three dimensions, we have $i, j = 1, 2, 3$. The inner product of a tensor (matrix) with a vector is defined as $A \cdot v = A_{ij}v_j$. The second order inner (scalar) product of two tensors (matrices) is defined as $A : B = A_{ij}B_{ij} = tr([A]^T[B])$.

8.4 Basic Micro-macro Concepts

For a relation between averages to be useful, it must be computed over a sample containing a statistically representative amount of material. This is a requirement that can be formulated in a concise mathematical form. A commonly accepted macro/micro criterion used in effective property calculations is the so-called Hill's condition, $\langle \sigma : \epsilon \rangle_\Omega = \langle \sigma \rangle_\Omega : \langle \epsilon \rangle_\Omega$. Hill's condition (Hill,[40] 1952) dictates the size requirements on the RVE. The classical argument is as follows. For any perfectly bonded heterogeneous body, in the absence of body forces, two physically important loading states satisfy Hill's condition: (1) linear displacements of the form $u|_{\partial\Omega} = \mathcal{E} \cdot x \Rightarrow \langle \epsilon \rangle_\Omega = \mathcal{E}$ and (2) pure tractions in the form $t|_{\partial\Omega} = \mathcal{L} \cdot n \Rightarrow \langle \sigma \rangle_\Omega = \mathcal{L}$; where $\mathcal{E}$ and $\mathcal{L}$ are constant strain and stress tensors, respectively. Applying (1)- or (2)-type boundary conditions to a large sample is a way of reproducing approximately what may be occurring in a statistically representative microscopic sample of material in a macroscopic body. *The requirement is that the sample must be large enough to have relatively small boundary field fluctuations relative to its size and small enough relative to the macroscopic engineering structure. These restrictions force us to choose boundary conditions that are uniform.*

8.4.1 *Testing Procedures*

To determine $I\!\!E^*$, one specifies six linearly independent loadings of the form,

(1) $u|_{\partial\Omega} = \mathcal{E}^{(1\to6)} \cdot x$ or
(2) $t|_{\partial\Omega} = \mathcal{L}^{(1\to6)} \cdot n$,

where $\mathcal{E}^{(1\to6)}$ and $\mathcal{L}^{(1\to6)}$ are symmetric second order strain and stress tensors, with spatially constant (nonzero) components. This loading is applied to a sample of microheterogeneous material. Each independent loading yields six different averaged stress components and hence provides six equations to determine the constitutive constants in $I\!\!E^*$. In order for such an analysis to be valid, i.e. to make the material data reliable, the sample of material must be small enough that it can be considered as a material point with respect to the size of the domain under analysis, but large enough to be a statistically representative sample of the microstructure.

If the effective response is assumed to be isotropic, then only one test loading (instead of usually six), containing non-zero dilatational ($\frac{tr\boldsymbol{\sigma}}{3}$ and $\frac{tr\boldsymbol{\epsilon}}{3}$) and deviatoric components ($\boldsymbol{\sigma}' \stackrel{\text{def}}{=} \boldsymbol{\sigma} - \frac{tr\boldsymbol{\sigma}}{3}\boldsymbol{1}$ and $\boldsymbol{\epsilon}' \stackrel{\text{def}}{=} \boldsymbol{\epsilon} - \frac{tr\boldsymbol{\epsilon}}{3}\boldsymbol{1}$), is necessary to determine the effective bulk (κ) and shear (μ) moduli:

$$3\kappa^* \stackrel{\text{def}}{=} \frac{\langle \frac{tr\boldsymbol{\sigma}}{3}\rangle_\Omega}{\langle \frac{tr\boldsymbol{\epsilon}}{3}\rangle_\Omega} \quad \text{and} \quad 2\mu^* \stackrel{\text{def}}{=} \sqrt{\frac{\langle \boldsymbol{\sigma}'\rangle_\Omega : \langle \boldsymbol{\sigma}'\rangle_\Omega}{\langle \boldsymbol{\epsilon}'\rangle_\Omega : \langle \boldsymbol{\epsilon}'\rangle_\Omega}}. \tag{8.3}$$

In general, in order to determine the material properties of a microheterogeneous material, one computes 36 constitutive constants[d] E^*_{ijkl} in the following relation between averages,

$$\left\{\begin{array}{c} \langle\sigma_{11}\rangle_\Omega \\ \langle\sigma_{22}\rangle_\Omega \\ \langle\sigma_{33}\rangle_\Omega \\ \langle\sigma_{12}\rangle_\Omega \\ \langle\sigma_{23}\rangle_\Omega \\ \langle\sigma_{13}\rangle_\Omega \end{array}\right\} = \left[\begin{array}{cccccc} E^*_{1111} & E^*_{1122} & E^*_{1133} & E^*_{1112} & E^*_{1123} & E^*_{1113} \\ E^*_{2211} & E^*_{2222} & E^*_{2233} & E^*_{2212} & E^*_{2223} & E^*_{2213} \\ E^*_{3311} & E^*_{3322} & E^*_{3333} & E^*_{3312} & E^*_{3323} & E^*_{3313} \\ E^*_{1211} & E^*_{1222} & E^*_{1233} & E^*_{1212} & E^*_{1223} & E^*_{1213} \\ E^*_{2311} & E^*_{2322} & E^*_{2333} & E^*_{2312} & E^*_{2323} & E^*_{2313} \\ E^*_{1311} & E^*_{1322} & E^*_{1333} & E^*_{1312} & E^*_{1323} & E^*_{1313} \end{array}\right] \left\{\begin{array}{c} \langle\epsilon_{11}\rangle_\Omega \\ \langle\epsilon_{22}\rangle_\Omega \\ \langle\epsilon_{33}\rangle_\Omega \\ 2\langle\epsilon_{12}\rangle_\Omega \\ 2\langle\epsilon_{23}\rangle_\Omega \\ 2\langle\epsilon_{13}\rangle_\Omega \end{array}\right\}. \tag{8.4}$$

As mentioned before, each independent loading leads to six equations and hence in total 36 equations are generated by the independent loadings, which are used to determine the tensor relation between average stress and strain, $I\!\!E^*$. $I\!\!E^*$ *is exactly what appears in engineering literature as the "property" of a material.* The usual choices for the six independent load cases are

$$\mathcal{E} \text{ or } \mathcal{L} = \begin{bmatrix} \beta & 0 & 0 \\ 0 & 0 & 0 \\ 0 & 0 & 0 \end{bmatrix}, \begin{bmatrix} 0 & 0 & 0 \\ 0 & \beta & 0 \\ 0 & 0 & 0 \end{bmatrix}, \begin{bmatrix} 0 & 0 & 0 \\ 0 & 0 & 0 \\ 0 & 0 & \beta \end{bmatrix}, \begin{bmatrix} 0 & \beta & 0 \\ \beta & 0 & 0 \\ 0 & 0 & 0 \end{bmatrix}, \begin{bmatrix} 0 & 0 & 0 \\ 0 & 0 & \beta \\ 0 & \beta & 0 \end{bmatrix}, \begin{bmatrix} 0 & 0 & \beta \\ 0 & 0 & 0 \\ \beta & 0 & 0 \end{bmatrix}, \tag{8.5}$$

where β is a load parameter. For completeness, we record a few related fundamental results, which are useful in micro-macro mechanical analysis.

---

[d]There are, of course, only 21 constants, since $I\!\!E^*$ is symmetric.

8.4.2 *The Average Strain Theorem*

If a heterogeneous body, see Fig. 8.4, has the following uniform loading on its surface: $\boldsymbol{u}|_{\partial\Omega} = \mathcal{E}\cdot\boldsymbol{x}$, then

$$
\begin{aligned}
\langle\epsilon\rangle_\Omega &= \frac{1}{2|\Omega|}\int_\Omega (\nabla\boldsymbol{u} + (\nabla\boldsymbol{u})^T)\,d\Omega \\
&= \frac{1}{2|\Omega|}\left(\int_{\Omega_1}(\nabla\boldsymbol{u} + (\nabla\boldsymbol{u})^T)\,d\Omega + \int_{\Omega_2}(\nabla\boldsymbol{u} + (\nabla\boldsymbol{u})^T)\,d\Omega\right) \\
&= \frac{1}{2|\Omega|}\left(\int_{\partial\Omega_1}(\boldsymbol{u}\otimes\boldsymbol{n} + \boldsymbol{n}\otimes\boldsymbol{u})\,dA + \int_{\partial\Omega_2}(\boldsymbol{u}\otimes\boldsymbol{n} + \boldsymbol{n}\otimes\boldsymbol{u})\,dA\right) \\
&= \frac{1}{2|\Omega|}\left(\int_{\partial\Omega}((\mathcal{E}\cdot\boldsymbol{x})\otimes\boldsymbol{n} + \boldsymbol{n}\otimes(\mathcal{E}\cdot\boldsymbol{x}))\,dA + \int_{\partial\Omega_1\cap\partial\Omega_2}([\![\boldsymbol{u}]\!]\otimes\boldsymbol{n} + \boldsymbol{n}\otimes[\![\boldsymbol{u}]\!])\,dA\right) \\
&= \frac{1}{2|\Omega|}\left(\int_\Omega(\nabla(\mathcal{E}\cdot\boldsymbol{x}) + \nabla(\mathcal{E}\cdot\boldsymbol{x})^T)\,d\Omega + \int_{\partial\Omega_1\cap\partial\Omega_2}([\![\boldsymbol{u}]\!]\otimes\boldsymbol{n} + \boldsymbol{n}\otimes[\![\boldsymbol{u}]\!])\,dA\right) \\
&= \mathcal{E} + \frac{1}{2|\Omega|}\int_{\partial\Omega_1\cap\partial\Omega_2}([\![\boldsymbol{u}]\!]\otimes\boldsymbol{n} + \boldsymbol{n}\otimes[\![\boldsymbol{u}]\!])\,dA, \tag{8.6}
\end{aligned}
$$

where $(\boldsymbol{u}\otimes\boldsymbol{n}\stackrel{\text{def}}{=}u_i n_j)$ is a tensor product of the vector $\boldsymbol{u}$ and vector $\boldsymbol{n}$. $[\![\boldsymbol{u}]\!]$ describes the displacement jumps at the interfaces between Ω_1 and Ω_2. *Therefore, only if the material is perfectly bonded, then* $\langle\epsilon\rangle_\Omega = \mathcal{E}$. Note that the presence of finite body forces does not affect this result. Also note that the third line in Equation (8.6) is not an outcome of the divergence theorem, but of a generalization that can be found in a variety of books, for example Chandrasekharaiah and Debnath.[13]

8.4.3 *The Average Stress Theorem*

Again we consider a body (in static equilibrium) with $\boldsymbol{t}|_{\partial\Omega} = \mathcal{L}\cdot\boldsymbol{n}$, where $\mathcal{L}$ is a constant tensor. We make use of the identity $\nabla\cdot(\boldsymbol{\sigma}\otimes\boldsymbol{x}) = (\nabla\cdot\boldsymbol{\sigma})\otimes\boldsymbol{x} + \boldsymbol{\sigma}\cdot\nabla\boldsymbol{x} = -\boldsymbol{f}\otimes\boldsymbol{x} + \boldsymbol{\sigma}$, where $\boldsymbol{f}$ represents the body forces. Substituting this into the definition of the average stress yields

$$
\begin{aligned}
\langle\boldsymbol{\sigma}\rangle_\Omega &= \frac{1}{|\Omega|}\int_\Omega \nabla\cdot(\boldsymbol{\sigma}\otimes\boldsymbol{x})\,d\Omega + \frac{1}{|\Omega|}\int_\Omega(\boldsymbol{f}\otimes\boldsymbol{x})\,d\Omega \\
&= \frac{1}{|\Omega|}\int_{\partial\Omega}(\boldsymbol{\sigma}\otimes\boldsymbol{x})\cdot\boldsymbol{n}\,dA + \frac{1}{|\Omega|}\int_\Omega(\boldsymbol{f}\otimes\boldsymbol{x})\,d\Omega \\
&= \frac{1}{|\Omega|}\int_{\partial\Omega}(\mathcal{L}\otimes\boldsymbol{x})\cdot\boldsymbol{n}\,dA + \frac{1}{|\Omega|}\int_\Omega(\boldsymbol{f}\otimes\boldsymbol{x})\,d\Omega \\
&= \mathcal{L} + \frac{1}{|\Omega|}\int_\Omega(\boldsymbol{f}\otimes\boldsymbol{x})\,d\Omega. \tag{8.7}
\end{aligned}
$$

If there are no body forces, $\boldsymbol{f} = \boldsymbol{0}$, then $\langle\boldsymbol{\sigma}\rangle_\Omega = \mathcal{L}$. *Note that debonding (interface separation) does not change this result.*

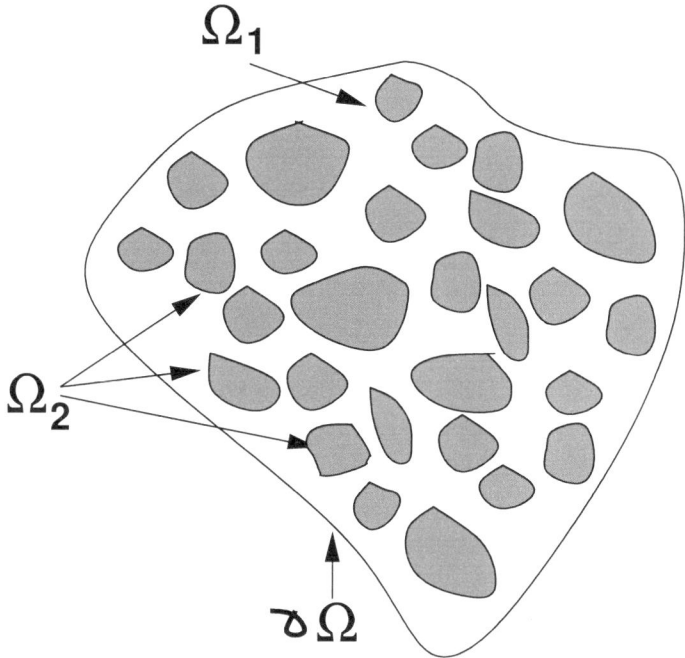

Fig. 8.4. Nomenclature for the averaging theorems (for a general body, Zohdi[143]).

8.4.4 *Satisfaction of Hill's Energy Condition*

Consider a body (in static equilibrium) with a perfectly bonded microstructure and $f = 0$. This condition yields

$$\int_{\partial\Omega} u \cdot t \, dA = \int_{\partial\Omega} u \cdot \sigma \cdot n \, dA = \int_{\Omega} \nabla \cdot (u \cdot \sigma) \, d\Omega. \qquad (8.8)$$

With $\nabla \cdot \sigma = 0$, it follows that $\int_{\Omega} \nabla \cdot (u \cdot \sigma) \, d\Omega = \int_{\Omega} \nabla u : \sigma \, d\Omega = \int_{\Omega} \epsilon : \sigma \, d\Omega.$
If $u|_{\partial\Omega} = \mathcal{E} \cdot x$ and $f = 0$, then

$$\int_{\partial\Omega} u \cdot t \, dA = \int_{\partial\Omega} \mathcal{E} \cdot x \cdot \sigma \cdot n \, dA = \int_{\Omega} \nabla \cdot (\mathcal{E} \cdot x \cdot \sigma) \, d\Omega \qquad (8.9)$$

$$= \int_{\Omega} \nabla(\mathcal{E} \cdot x) : \sigma \, d\Omega = \mathcal{E} : \langle\sigma\rangle_{\Omega}|\Omega|.$$

Noting that $\langle\epsilon\rangle_{\Omega} = \mathcal{E}$, we have $\langle\epsilon\rangle_{\Omega} : \langle\sigma\rangle_{\Omega} = \langle\epsilon : \sigma\rangle_{\Omega}$. If $t|_{\partial\Omega} = \mathcal{L} \cdot n$ and $f = 0$, then $\int_{\partial\Omega} u \cdot t \, dA = \int_{\partial\Omega} u \cdot \mathcal{L} \cdot n \, dA = \int_{\Omega} \nabla \cdot (u \cdot \mathcal{L}) \, d\Omega = \int_{\Omega} \nabla u : \mathcal{L} \, d\Omega = \mathcal{L} :$

$\int_{\Omega} \epsilon \, d\Omega$. Therefore, since $\langle \sigma \rangle_{\Omega} = \mathcal{L}$, as before we have $\langle \epsilon \rangle_{\Omega} : \langle \sigma \rangle_{\Omega} = \langle \epsilon : \sigma \rangle_{\Omega}$. Satisfaction of Hill's condition guarantees that the microscopic and macroscopic energies will be the same, and it implies the use of the two mentioned test boundary conditions on sufficiently large samples of material.

8.4.5 *The Hill-Reuss-Voigt Bounds*

Until recently, the direct computation of micromaterial responses was very difficult. Classical approaches have sought to approximate or bound the effective material responses. Many classical approaches start by splitting the stress field within a sample into a volume average and a purely fluctuating part, $\epsilon = \langle \epsilon \rangle_{\Omega} + \tilde{\epsilon}$, and we directly obtain

$$0 \leq \int_{\Omega} \tilde{\epsilon} : \mathbf{IE} : \tilde{\epsilon} \, d\Omega = \int_{\Omega} (\epsilon : \mathbf{IE} : \epsilon - 2\langle \epsilon \rangle_{\Omega} : \mathbf{IE} : \epsilon + \langle \epsilon \rangle_{\Omega} : \mathbf{IE} : \langle \epsilon \rangle_{\Omega}) \, d\Omega$$
$$= (\langle \epsilon \rangle_{\Omega} : \mathbf{IE}^* : \langle \epsilon \rangle_{\Omega} - 2\langle \epsilon \rangle_{\Omega} : \langle \sigma \rangle_{\Omega} + \langle \epsilon \rangle_{\Omega} : \langle \mathbf{IE} \rangle_{\Omega} : \langle \epsilon \rangle_{\Omega})|\Omega|$$
$$= \langle \epsilon \rangle_{\Omega} : (\langle \mathbf{IE} \rangle_{\Omega} - \mathbf{IE}^*) : \langle \epsilon \rangle_{\Omega}|\Omega|. \tag{8.10}$$

Similarly, for the complementary case, with $\sigma = \langle \sigma \rangle_{\Omega} + \tilde{\sigma}$, and the following assumption (microscopic energy equals the macroscopic energy)

$$\underbrace{\langle \sigma : \mathbf{IE}^{-1} : \sigma \rangle_{\Omega}}_{\text{micro energy}} = \underbrace{\langle \sigma \rangle_{\Omega} : \mathbf{IE}^{*-1} : \langle \sigma \rangle_{\Omega}}_{\text{macro energy}}, \quad \text{where} \quad \langle \epsilon \rangle_{\Omega} = \mathbf{IE}^{*-1} : \langle \sigma \rangle_{\Omega}, \tag{8.11}$$

we have

$$0 \leq \int_{\Omega} \tilde{\sigma} : \mathbf{IE}^{-1} : \tilde{\sigma} \, d\Omega$$
$$= \int_{\Omega} (\sigma : \mathbf{IE}^{-1} : \sigma - 2\langle \sigma \rangle_{\Omega} : \mathbf{IE}^{-1} : \sigma + \langle \sigma \rangle_{\Omega} : \mathbf{IE}^{-1} : \langle \sigma \rangle_{\Omega}) \, d\Omega$$
$$= (\langle \sigma \rangle_{\Omega} : \mathbf{IE}^{*-1} : \langle \sigma \rangle_{\Omega} - 2\langle \epsilon \rangle_{\Omega} : \langle \sigma \rangle_{\Omega} + \langle \sigma \rangle_{\Omega} : \langle \mathbf{IE}^{-1} \rangle_{\Omega} : \langle \sigma \rangle_{\Omega})|\Omega|$$
$$= \langle \sigma \rangle_{\Omega} : (\langle \mathbf{IE}^{-1} \rangle_{\Omega} - \mathbf{IE}^{*-1}) : \langle \sigma \rangle_{\Omega}|\Omega|. \tag{8.12}$$

Invoking Hill's condition, which is loading-independent in this form, we have

$$\underbrace{\langle \mathbf{IE}^{-1} \rangle_{\Omega}^{-1}}_{\text{Reuss}} \leq \mathbf{IE}^* \leq \underbrace{\langle \mathbf{IE} \rangle_{\Omega}}_{\text{Voigt}}. \tag{8.13}$$

This inequality means that the eigenvalues of the tensors $\mathbf{IE}^* - \langle \mathbf{IE}^{-1} \rangle_{\Omega}^{-1}$ and $\langle \mathbf{IE} \rangle_{\Omega} - \mathbf{IE}^*$ are non-negative. The practical outcome of the analysis is that bounds on effective properties are obtained. These bounds are commonly known as the Hill-Reuss-Voigt bounds, for historical reasons. Voigt,[100] in 1889, assumed that the strain field within a sample of aggregate of polycrystalline material was uniform (constant), under uniform strain exterior loading. If the constant strain Voigt field is assumed within the RVE, $\epsilon = \epsilon^0$, then $\langle \sigma \rangle_{\Omega} = \langle \mathbf{IE} : \epsilon \rangle_{\Omega} = \langle \mathbf{IE} \rangle_{\Omega} : \epsilon^0$, which implies $\boldsymbol{I\!E}^* = \langle \boldsymbol{I\!E} \rangle_\Omega$. The dual assumption was made by Reuss,[87] in 1929, who approximated the stress fields within the aggregate of polycrystalline material as uniform (constant), $\boldsymbol{\sigma} = \boldsymbol{\sigma}^0$, leading to $\langle \boldsymbol{\epsilon} \rangle_\Omega = \langle \boldsymbol{I\!E}^{-1} : \boldsymbol{\sigma} \rangle_\Omega = \langle \boldsymbol{I\!E}^{-1} \rangle_\Omega : \boldsymbol{\sigma}^0$, and thus to $\boldsymbol{I\!E}^* = \langle \boldsymbol{I\!E}^{-1} \rangle_\Omega^{-1}$.

Remark: Different boundary conditions (compared to the standard ones specified earlier) are often used in computational homogenization analysis. For example, periodic boundary conditions are sometimes employed. Although periodic conditions are really only appropriate for perfectly periodic media for many cases, it has been shown that, in some cases, their use can provide better effective responses than either linear displacement or uniform traction boundary conditions (for example, see Terada et al.[98] or Segurado and Llorca). Periodic boundary conditions also satisfy Hill's condition a priori. Another related type of boundary condition is the so-called "uniform-mixed" type, whereby tractions are applied on some parts of the boundary and displacements on other parts, generating, in some cases, effective properties that match those produced with uniform boundary conditions, but with smaller sample sizes (for example, see Hazanov and Huet[39]). Another approach is "framing", whereby the traction or displacement boundary conditions are applied to a large sample of material, with the averaging computed on an interior subsample to avoid possible boundary-layer effects. This method is similar to exploiting a St. Venant-type of effect, commonly used in solid mechanics, to avoid boundary layers. The approach provides a way of determining what the microstructure really experiences, without "bias" from the boundary loading. However, generally, the advantages of one boundary condition over another diminish as the sample size increases.

8.4.6 *Improved Estimates*

Over the last half-century, improved estimates have been pursued, with a notable contribution being the Hashin-Shtrikman bounds.[36−38] The Hashin-Shtrikman bounds are the tightest possible bounds on isotropic effective responses, with isotropic microstructures, when the volume fractions and phase contrasts of the constituents are the only data known. For isotropic materials with isotropic effective (mechanical) responses, the Hashin-Shtrikman bounds (for a two-phase material) are as follows for the bulk modulus

$$
\kappa^{*,-} \stackrel{\text{def}}{=} \kappa_1 + \frac{v_2}{\frac{1}{\kappa_2 - \kappa_1} + \frac{3(1-v_2)}{3\kappa_1 + 4\mu_1}} \leq \kappa^* \leq \kappa_2 + \frac{1 - v_2}{\frac{1}{\kappa_1 - \kappa_2} + \frac{3v_2}{3\kappa_2 + 4\mu_2}} \stackrel{\text{def}}{=} \kappa^{*,+}
$$

$$(8.14)$$

and for the shear modulus

$$\mu^{*,-} \stackrel{\text{def}}{=} \mu_1 + \frac{v_2}{\frac{1}{\mu_2-\mu_1} + \frac{6(1-v_2)(\kappa_1+2\mu_1)}{5\mu_1(3\kappa_1+4\mu_1)}} \leq \mu^* \leq \mu_2 + \frac{(1-v_2)}{\frac{1}{\mu_1-\mu_2} + \frac{6v_2(\kappa_2+2\mu_2)}{5\mu_2(3\kappa_2+4\mu_2)}} \stackrel{\text{def}}{=} \mu^{*,+},$$

$$(8.15)$$

where κ_2 and κ_1 are the bulk moduli and μ_2 and μ_1 are the shear moduli of the respective phases ($\kappa_2 \geq \kappa_1$ and $\mu_2 \geq \mu_1$), and where v_2 is the second phase volume fraction. Note that no geometric or other microstructural information is required for the bounds.

Remark 1: There exist a multitude of other approaches which seek to estimate or bound the aggregate responses of microheterogeneous materials. A complete survey is outside the scope of the present work. We refer the reader to the works of Hashin,[38] Mura,[65] Aboudi,[1] Nemat-Nasser and Hori[69] and, recently, Torquato[99] for such reviews.

Remark 2: Numerical methods have become a valuable tool in determining micro-macro relations, with the caveat being that local fields in the microstructure are resolved, which is important in being able to quantify the intensity of the loads experienced by the microstructure. This is important for ascertaining failure of the material. In particular, Finite Element-based methods are extremely popular for micro-macro calculations. Applying such methods entails generating a sample of material microstructure, meshing it to sufficient resolution for tolerable numerical accuracy and solving a series of boundary value problems with different test loadings. The effective properties can be determined by post processing (averaging over the RVE). For an extensive review of this topic, see Zohdi and Wriggers.[105–134] We also refer the reader to that work for more extensive mathematical details and background information.

Remark 3: If needed, one can post-process the effective bulk and shear modulus to obtain the effective Poisson ratio $\nu^* = \frac{3\kappa^*-2\mu^*}{2(3\kappa^*+\mu^*)}$ and the effective Young's modulus $E^* = 2\mu^*(1+\nu^*) = 3\kappa^*(1-2\nu^*)$.

8.5 Combining Bounds

The typical use of the bounds from the previous chapter is to make an estimate of the effective properties by forming a convex combination of them in the following manner:

$$\kappa^* \approx \phi\kappa^{*,+} + (1-\phi)\kappa^{*,-} \qquad (8.16)$$

and

$$\mu^* \approx \phi\mu^{*,+} + (1-\phi)\mu^{*,-}, \qquad (8.17)$$

where $0 \leq \phi \leq 1$ is a parameter such that:

    If $\phi = 0$ we have the lower bound,

    If $\phi = 1$ we have the upper bound,

    If $\phi = 1/2$ we have the average of the bounds.

ϕ is a function of the microstructure, and must be calibrated.

A critical observation is that the lower bound is more accurate when the material is composed of stiff particles that are surrounded by a soft matrix (denoted case 1) and the upper bound is more accurate for a stiff matrix surrounding soft particles (denoted case 2). This can be explained by considering two cases of material combinations, one with 50 % soft material and 50 % stiff material. A material with a continuous soft binder (50 %) will isolate the stiff particles (50 %), and the overall system will not be stiff (this is case 1 and the lower bound is more accurate), while a material formed by a continuous stiff binder (50 %) surrounding soft particles (50 %, case 2) will, in an overall sense, are stiffer than case 1. Thus, case 2 is more closely approximated by the upper bound and case 1 is closer to the lower bound.

As mentioned, for stiff spherical particles, at low volume fractions, for example under 15 %, where the particles are not making contact, the lower bound is more accurate. Thus, one would pick $\phi = \phi^s \leq 0.5$ to bias the estimate to the lower bound. However, if we take the same volume fraction of particles, but make the flat flakes, they will certainly touch, and produce stiff pathways. Their overall stiffness will be higher than those of sphere at the same volume fraction. Thus, one would pick $\phi = \phi^f > \phi^s$. One can calibrate ϕ by comparing it to different experiments. This was done before, for example for mechanical properties in the Zohdi et al.[104] Essentially, the more the particles interact, for example physically touch, the more the upper bound becomes relevant. The general trends are (a) for cases where the upper bound is more accurate, $\phi > \frac{1}{2}$ and (b) for cases when the lower bound is more accurate, $\phi < \frac{1}{2}$. ϕ indicates the degree of interaction of the particulate constituents.

Remark: This same trend holds for electrical and thermal properties (discussed later).

8.6   Local Fields: Stresses and Strains

The determination of the average load sharing between phases at the microstructural scale can be obtained from the overall effective mechanical properties of the microheterogeneous material, for example, comprised of particles suspended in a binding matrix.

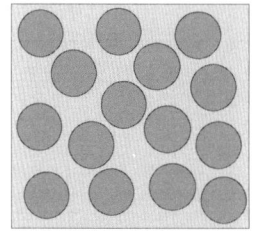 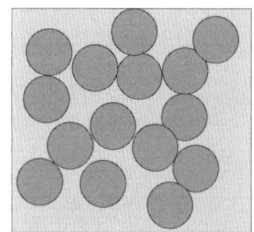

PARTICLES WELL SEPARATED   **PARTICLES TOUCHING**

Fig. 8.5. Comparing microstructures with the same volume fractions. Flakes touch more, and thus need a higher value of ϕ (Zohdi[143]).

The load carried by each phase in the microstructure is characterized via stress and strain concentration tensors, which we now discuss. These provide a measure of the deviation away from the mean fields throughout the material. One can decompose averages of an arbitrary quantity over Ω into averages over the each of the phases in the following manner: $\langle A \rangle_\Omega = (1/|\Omega|)\left(\int_{\Omega_1} A \, d\Omega + \int_{\Omega_2} A \, d\Omega\right) = v_1\langle A \rangle_{\Omega_1} + v_2\langle A \rangle_{\Omega_2}$. If we make use of this decomposition, we have

$$
\begin{aligned}
\langle \sigma \rangle_\Omega &= v_1\langle \sigma \rangle_{\Omega_1} + v_2\langle \sigma \rangle_{\Omega_2} \\
&= v_1 \mathbf{IE}_1 : \langle \epsilon \rangle_{\Omega_1} + v_2 \mathbf{IE}_2 : \langle \epsilon \rangle_{\Omega_2} \\
&= \mathbf{IE}_1 : (\langle \epsilon \rangle_\Omega - v_2\langle \epsilon \rangle_{\Omega_2}) + v_2 \mathbf{IE}_2 : \langle \epsilon \rangle_{\Omega_2} \\
&= \left(\mathbf{IE}_1 + v_2(\mathbf{IE}_2 - \mathbf{IE}_1) : \boldsymbol{C}^{\epsilon,2}\right) : \langle \epsilon \rangle_\Omega,
\end{aligned}
\tag{8.18}
$$

where $\boldsymbol{C}^{\epsilon,2} \stackrel{\text{def}}{=} \left(\frac{1}{v_2}(\mathbf{IE}_2 - \mathbf{IE}_1)^{-1} : (\mathbf{IE}^* - \mathbf{IE}_1)\right)$ with $\boldsymbol{C}^{\epsilon,2} : \langle \epsilon \rangle_\Omega = \langle \epsilon \rangle_{\Omega_2}$. The strain concentration tensor $\boldsymbol{C}^{\epsilon,2}$ relates the average strain over the particle phase (2) to the average strain over all phases. Similarly, for the variation in the stress we have $\boldsymbol{C}^{\epsilon,2} : \mathbf{IE}^{*-1} : \langle \sigma \rangle_\Omega = \mathbf{IE}_2^{-1} : \langle \sigma \rangle_{\Omega_2}$, which reduces to $\mathbf{IE}_2 : \boldsymbol{C}^{\epsilon,2} : \mathbf{IE}^{*-1} : \langle \sigma \rangle_\Omega \stackrel{\text{def}}{=} \boldsymbol{C}^{\sigma,2} : \langle \sigma \rangle_\Omega = \langle \sigma \rangle_{\Omega_2}$. $\boldsymbol{C}^{\sigma,2}$ is known as the stress concentration tensor; it relates the average stress in the particle phase to that in the whole RVE. Note that once either $\boldsymbol{C}^{\epsilon,2}$ or $\mathbf{IE}^*$ are known, the other can be determined. In the case of isotropy we may write

$$
C_\kappa^{\sigma,2} \stackrel{\text{def}}{=} \frac{1}{v_2} \frac{\kappa_2}{\kappa^*} \frac{\kappa^* - \kappa_1}{\kappa_2 - \kappa_1} \qquad \text{and} \qquad C_\mu^{\sigma,2} \stackrel{\text{def}}{=} \frac{1}{v_2} \frac{\mu_2}{\mu^*} \frac{\mu^* - \mu_1}{\mu_2 - \mu_1}
\tag{8.19}
$$

where $C_\kappa^{\sigma,2}\langle \frac{tr\sigma}{3} \rangle_\Omega = \langle \frac{tr\sigma}{3} \rangle_{\Omega_2}$ and where $C_\mu^{\sigma,2}\langle \sigma' \rangle_\Omega = \langle \sigma' \rangle_{\Omega_2}$. Clearly, the microstress fields are minimally distorted when $C_\kappa^{\sigma,2} = C_\mu^{\sigma,2} = 1$; there are no stress concentrations in a homogeneous material. For the matrix,

$$\langle\boldsymbol{\sigma}\rangle_{\Omega_1} = \frac{\langle\boldsymbol{\sigma}\rangle_{\Omega} - v_2\langle\boldsymbol{\sigma}\rangle_{\Omega_2}}{v_1} = \frac{\langle\boldsymbol{\sigma}\rangle_{\Omega} - v_2\boldsymbol{C}^{\sigma,2} : \langle\boldsymbol{\sigma}\rangle_{\Omega}}{v_1} = \frac{(1 - v_2\boldsymbol{C}^{\sigma,2}) : \langle\boldsymbol{\sigma}\rangle_{\Omega}}{v_1} \stackrel{\text{def}}{=} \boldsymbol{C}^{\sigma,1} : \langle\boldsymbol{\sigma}\rangle_{\Omega}.$$

(8.20)

Therefore, in the case of isotropy,

$$C_\kappa^{\sigma,1} \stackrel{\text{def}}{=} \frac{1}{v_1}(1 - v_2 C_\kappa^{\sigma,2}) \quad \text{and} \quad C_\mu^{\sigma,1} \stackrel{\text{def}}{=} \frac{1}{v_1}(1 - v_2 C_\mu^{\sigma,2}).$$

(8.21)

The fraction of the total stress carried by each phase can be determined by multiplying the concentration factors by the corresponding volume fractions

$$\langle\boldsymbol{\sigma}\rangle_{\Omega} = v_1\langle\boldsymbol{\sigma}\rangle_{\Omega_1} + v_2\langle\boldsymbol{\sigma}\rangle_{\Omega_2}$$
$$= v_1\boldsymbol{C}^{\sigma,1} : \langle\boldsymbol{\sigma}\rangle_{\Omega} + v_2\boldsymbol{C}^{\sigma,2} : \langle\boldsymbol{\sigma}\rangle_{\Omega}.$$

(8.22)

Remark: Similar to the stress, for the strain, we have for the matrix,

$$\langle\boldsymbol{\epsilon}\rangle_{\Omega_1} = \frac{\langle\boldsymbol{\epsilon}\rangle_{\Omega} - v_2\langle\boldsymbol{\epsilon}\rangle_{\Omega_2}}{v_1} = \frac{\langle\boldsymbol{\epsilon}\rangle_{\Omega} - v_2\boldsymbol{C}^{\epsilon,2} : \langle\boldsymbol{\epsilon}\rangle_{\Omega}}{v_1} = \frac{(1 - v_2\boldsymbol{C}^{\epsilon,2}) : \langle\boldsymbol{\epsilon}\rangle_{\Omega}}{v_1} \stackrel{\text{def}}{=} \boldsymbol{C}^{\epsilon,1} : \langle\boldsymbol{\epsilon}\rangle_{\Omega}.$$

(8.23)

Therefore, in the case of isotropy,

$$C_\kappa^{\epsilon,1} \stackrel{\text{def}}{=} \frac{1}{v_1}(1 - v_2 C_\kappa^{\epsilon,2}) \quad \text{and} \quad C_\mu^{\epsilon,1} \stackrel{\text{def}}{=} \frac{1}{v_1}(1 - v_2 C_\mu^{\epsilon,2}).$$

(8.24)

The fraction of the total strain carried by each phase can be determined by multiplying the concentration factors by the corresponding volume fractions

$$\langle\boldsymbol{\epsilon}\rangle_{\Omega} = v_1\langle\boldsymbol{\epsilon}\rangle_{\Omega_1} + v_2\langle\boldsymbol{\epsilon}\rangle_{\Omega_2}$$
$$= v_1\boldsymbol{C}^{\epsilon,1} : \langle\boldsymbol{\epsilon}\rangle_{\Omega} + v_2\boldsymbol{C}^{\epsilon,2} : \langle\boldsymbol{\epsilon}\rangle_{\Omega}.$$

(8.25)

8.7 Optimization: Formulation of a Cost-function

The deviation in the particulate stress fields from the mean value is

$$\left|\frac{\langle tr\boldsymbol{\sigma}\rangle_{\Omega_2} - \langle tr\boldsymbol{\sigma}\rangle_{\Omega}}{\langle tr\boldsymbol{\sigma}\rangle_{\Omega}}\right| = |C_\kappa^{\sigma,2} - 1|$$

(8.26)

and

$$\sqrt{\frac{(\langle\boldsymbol{\sigma}'\rangle_{\Omega_2} - \langle\boldsymbol{\sigma}'\rangle_{\Omega}) : (\langle\boldsymbol{\sigma}'\rangle_{\Omega_2} - \langle\boldsymbol{\sigma}'\rangle_{\Omega})}{\langle\boldsymbol{\sigma}'\rangle_{\Omega} : \langle\boldsymbol{\sigma}'\rangle_{\Omega}}} = |C_\mu^{\sigma,2} - 1|,$$

(8.27)

and for the matrix material

$$\left| \frac{\langle tr\boldsymbol{\sigma}\rangle_{\Omega_1} - \langle tr\boldsymbol{\sigma}\rangle_{\Omega}}{\langle tr\boldsymbol{\sigma}\rangle_{\Omega}} \right| = |C_\kappa^{\sigma,1} - 1| \tag{8.28}$$

and

$$\sqrt{\frac{(\langle\boldsymbol{\sigma}'\rangle_{\Omega_1} - \langle\boldsymbol{\sigma}'\rangle_{\Omega}) : (\langle\boldsymbol{\sigma}'\rangle_{\Omega_1} - \langle\boldsymbol{\sigma}'\rangle_{\Omega})}{\langle\boldsymbol{\sigma}'\rangle_{\Omega} : \langle\boldsymbol{\sigma}'\rangle_{\Omega}}} = |C_\mu^{\sigma,1} - 1|. \tag{8.29}$$

In order to incorporate the deviation into a cost function, we introduce a tolerance where, ideally,

$$|C_\kappa^{\sigma,2} - 1| \le TOL_\kappa \quad \text{and} \quad |C_\mu^{\sigma,2} - 1| \le TOL_\mu \tag{8.30}$$

and

$$|C_\kappa^{\sigma,1} - 1| \le TOL_\kappa \quad \text{and} \quad |C_\mu^{\sigma,1} - 1| \le TOL_\mu. \tag{8.31}$$

If the normalized deviation exceeds the corresponding TOL, then the level of violation is incorporated as a multilateral constraint to the macroscopic objectives. As an example, our immediate goal is to computationally design the macroscale effective bulk and shear moduli κ^* and μ^*, using convex combinations of the Hashin-Shtrikman bounds as approximations for the effective moduli $\kappa^* \approx \phi\kappa^{*,+} + (1-\phi)\kappa^{*,-}$ and $\mu^* \approx \phi\mu^{*,+} + (1-\phi)\mu^{*,-}$, where $0 \le \phi \le 1$. The micro-macro objective function is

$$\Pi = w_1 \left| \frac{\kappa^*}{\kappa^{*,D}} - 1 \right|^2 + w_2 \left| \frac{\mu^*}{\mu^{*,D}} - 1 \right|^2$$

$$+ \hat{w}_3 \left(|C_\kappa^{\sigma,2} - 1| - TOL_\kappa \right)^2 + \hat{w}_4 \left(|C_\mu^{\sigma,2} - 1| - TOL_\mu \right)^2$$

$$+ \hat{w}_5 \left(|C_\kappa^{\sigma,1} - 1| - TOL_\kappa \right)^2 + \hat{w}_6 \left(|C_\mu^{\sigma,1} - 1| - TOL_\mu \right)^2,$$

where (I) if $|C_\kappa^{\sigma,2} - 1| \le TOL_\kappa$, then $\hat{w}_3 = 0$, (II) if $|C_\kappa^{\sigma,2} - 1| > TOL_\kappa$, then $\hat{w}_3 = w_3$, (III) if $|C_\mu^{\sigma,2} - 1| \le TOL_\mu$, then $\hat{w}_4 = 0$, (IV) if $|C_\mu^{\sigma,2} - 1| > TOL_\mu$, then $\hat{w}_4 = w_4$, (V) if $|C_\kappa^{\sigma,1} - 1| \le TOL_\kappa$, then $\hat{w}_5 = 0$, (VI) if $|C_\kappa^{\sigma,1} - 1| > TOL_\kappa$, then $\hat{w}_5 = w_5$, (VII) if $|C_\mu^{\sigma,1} - 1| \le TOL_\mu$, then $\hat{w}_6 = 0$, (VIII) if $|C_\mu^{\sigma,1} - 1| > TOL_\mu$, then $\hat{w}_6 = w_6$. Here the design variables are $\boldsymbol{\Lambda} = \{\kappa_2, \mu_2\, v_2\}$, and their constrained ranges are $\kappa_2^{(-)} \le \kappa_2 \le \kappa_2^{(+)}$, $\mu_2^{(-)} \le \mu_2 \le \mu_2^{(+)}$ and $v_2^{(-)} \le v_2 \le v_2^{(+)}$. There are two characteristics of such a formulation which make the application of standard gradient type minimization schemes, such as Newton's method, difficult:

(I) the incorporation of limits on the microfield behavior, as well as design search space restrictions, renders the objective function not continuously differentiable in design space and

(II) the objective function is nonconvex, i.e. the system Hessian is not positive definite (invertible) throughout design space.

One way to minimize such objective functions is by following a two stage approach whereby (1) one determines promising optimal regions in parameter space using (non-derivative) algorithms (such are evolutionary "genetic" algorithms, simulated annealing, etc.) and then (2) applies classical gradient-based schemes in locally convex regions, if the objective functions are smooth, since they are generally extremely efficient for the minimization of smooth convex functions. As indicated, the search for convex "pockets" of Π can be achieved by using "genetic" algorithms (GA), before applying classical gradient-based schemes.[e]

Genetic algorithms are search methods based on the principles of natural selection, employing concepts of species evolution, such as reproduction, mutation and crossover. Implementation typically involves a randomly generated population of fixed-length elemental strings, "genetic" information, each of which represents a specific choice of system parameters. The population of individuals undergoes "mating sequences" and other biologically-inspired events in order to find promising regions of the search space. Such methods can be traced back, at least, to the work of John Holland (Holland[42]). For reviews of such methods, see, for example, Goldberg,[32] Davis,[18] Onwubiko,[75] Kennedy and Eberhart[50] Lagaros et al.,[54] Papadrakakis et al.[76−79] and Goldberg and Deb.[33] In Zohdi[105−134] a genetic algorithm has been developed to treat a wide variety of nonconvex inverse problems involving various aspects of multi-particle mechanics, and we refer the interested reader to that work. Specifically, the central idea is that the system parameters form a genetic string and a survival of the fittest algorithm is applied to a population of such strings.

The overall process is: (a) a population (S) of different parameter sets are generated at random within the parameter space, each represented by a ("genetic") string of the system (N) parameters, (b) the performance of each parameter set is tested, (c) the parameter sets are ranked from top to bottom according to their performance, (d) the best parameter sets (parents) are mated pairwise producing two offspring (children), i.e. each best pair exchanges information by taking random convex combinations of the parameter set components of the parents' genetic strings and (e) the worst performing genetic strings are eliminated, then new replacement parameter sets (genetic strings) are introduced into the remaining population of best

---

[e]An exhaustive review of these methods can be found in the texts of Luenberger[59] and Gill, Murray and Wright,[31] while a state of the art can be found in Papadrakakis et al.[80]

performing genetic strings and the process (a-e) is then repeated.

The term "fitness" of a genetic string is used to indicate the value of the objective function. The most fit genetic string is the one with the smallest objective function. The retention of the top fit genetic strings from a previous generation (parents) is critical, since if the objective functions are highly nonconvex (the present case), there exists a clear possibility that the inferior offspring will replace superior parents. When the top parents are retained, the minimization of the cost function is guaranteed to be monotone (guaranteed improvement) with increasing generations. There is no guarantee of successive improvement if the top parents are not retained, even though nonretention of parents allows more new genetic strings to be evaluated in the next generation. In the scientific literature, numerical studies imply that, for sufficiently large populations, the benefits of parent retention outweigh this advantage and any disadvantages of "inbreeding", i.e. a stagnant population. For more details on this so-called "inheritance property" see Davis[18] or Kennedy and Eberhart.[50] In the upcoming algorithm, inbreeding is mitigated since, with each new generation, new parameter sets, selected at random within the parameter space, are added to the population. Previous numerical studies of the author (Zohdi[105−134]) have indicated that not retaining the parents is suboptimal due to the possibility that inferior offspring will replace superior parents. Additionally, parent retention is computationally less expensive, since these parameter sets do not have to be reevaluated (or ranked) in the next generation. An implementation of such ideas is as follows (Zohdi[105−134]):

- **STEP 1:** Randomly generate a population of S starting genetic strings, $\mathbf{\Lambda}^i, (i = 1, ..., S)$:

$$\mathbf{\Lambda}^i \stackrel{\text{def}}{=} \{\Lambda^i_1, \Lambda^i_2, \Lambda^i_3, \Lambda^i_4, ...\Lambda^i_N\} \stackrel{\text{def}}{=} \{\kappa^i_2, \mu^i_2, v^i_2, ...\}$$

- **STEP 2:** Compute fitness of each string $\Pi(\mathbf{\Lambda}^i)$, (i=1, ..., S)
- **STEP 3:** Rank genetic strings: $\mathbf{\Lambda}^i$, (i=1, ..., S)
- **STEP 4:** Mate nearest pairs and produce two offspring, (i=1, ..., S)
  $$\mathbf{\lambda}^i \stackrel{\text{def}}{=} \Phi^{(I)}\mathbf{\Lambda}^i + (1 - \Phi^{(I)})\mathbf{\Lambda}^{i+1}, \quad \mathbf{\lambda}^{i+1} \stackrel{\text{def}}{=} \Phi^{(II)}\mathbf{\Lambda}^i + (1 - \Phi^{(II)})\mathbf{\Lambda}^{i+1}$$
- **NOTE:** $\Phi^{(I)}$ and $\Phi^{(II)}$ are random numbers, such that $0 \leq \Phi^{(I)}, \Phi^{(II)} \leq 1$, which are different for each component of each genetic string
- **STEP 5:** Kill off bottom $M < S$ strings and keep top $K < N$ parents and top K offspring (K offspring+K parents+$M=S$)
- **STEP 6:** Repeat STEPS 1-6 with top gene pool (K offspring and K parents), plus M new, randomly generated, strings
- **Option:** Rescale and restart search around best performing parameter set every few generations

Remark 1: STEPS 1-6, which are associated with the genetic part of the overall algorithm, attempt to collect multiple local minima.[f] At first glance, it seems somewhat superfluous to retain even the top parents in such an algorithm. However, many studies have shown that the retention of the top old fit genetic strings is critical for proper convergence. As alluded to earlier, by observing Fig. 8.6 one sees that if the objective functions are highly nonconvex, there exists a strong possibility that the inferior offspring will replace superior parents. Therefore, retaining the top parents is not only less computationally expensive, since these designs do not have to be reevaluated, it is theoretically superior. With parent retention, the minimization of the cost function is guaranteed to be monotone with increasing generations, i.e. $\Pi(\Lambda^{opt,I}) \geq \Pi(\Lambda^{opt,I+1})$, where $\Lambda^{opt,I+1}$ and $\Lambda^{opt,I}$ are the best genetic strings from generations $I+1$ and I respectively. There is no such guarantee if the top parents are not retained. While the nonretention of parents allows more newer genetic strings to be evaluated in the next generation, numerical studies conducted thus far imply, for sufficiently large populations, that the benefits of parent retention outweigh this advantage, as well as any disadvantages of "inbreeding", i.e. a stagnant population. The case of inbreeding is circumvented in the current algorithm due to the fact that, with each new generation, new material designs, selected at random within the design space, are introduced into the population. Not retaining the parents is suboptimal due to the possibility that inferior offspring will replace superior parents.

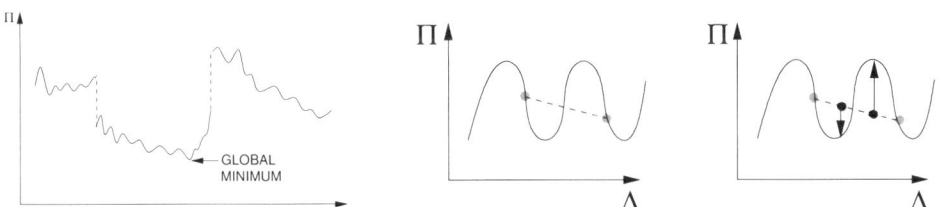

Fig. 8.6. Left: A characterization of the class of objective functions of interest. Right: A loss of superior older genetic strings if the top parents are not retained (Zohdi[143]).

Remark 2: After application of such a global search algorithm, one can apply a gradient-based method, if the objective function is sufficiently smooth in that region of the parameter space. In other words, if one has located a convex portion of the parameter space with a global genetic search, one can employ gradient-based procedures locally to minimize the objective function further, since they are generally much more efficient for convex optimization of smooth functions. In such procedures, in order to obtain a new directional step for Λ, one must solve the

---

[f]It is remarked that if the function Φ is allowed to be greater than unity, one can consider the resulting convex combination (offspring) as a "mutation".

following system

$$[\boldsymbol{I\!H}]\{\Delta\boldsymbol{\Lambda}\} = -\{\boldsymbol{g}\}, \tag{8.32}$$

where $[\boldsymbol{I\!H}]$ is the Hessian matrix ($N \times N$), where $\{\Delta\boldsymbol{\Lambda}\}$ is the parameter increment ($N \times 1$), and $\{\boldsymbol{g}\}$ is the gradient ($N \times 1$). We shall not employ this second (post-genetic) stage in this work. Specifically, this is determined by forcing the gradient of $\nabla_{\boldsymbol{\Lambda}}\Pi(\boldsymbol{\Lambda}) = \boldsymbol{0}$. Expanding (linearizing) around a first guess $\boldsymbol{\Lambda}^i$ yields

$$\nabla_{\boldsymbol{\Lambda}}\Pi(\boldsymbol{\Lambda}^{i+1}) \approx \nabla_{\boldsymbol{\Lambda}}\Pi(\boldsymbol{\Lambda}^i) + \nabla\left(\nabla_{\boldsymbol{\Lambda}}\Pi(\boldsymbol{\Lambda}^i)\right) \cdot (\boldsymbol{\Lambda}^{i+1} - \boldsymbol{\Lambda}^i)$$
$$+ higher - order - terms \approx \boldsymbol{0} \tag{8.33}$$

or in more streamlined matrix notation, defining the Hessian, $[\boldsymbol{I\!H}] = \nabla\left(\nabla_{\boldsymbol{\Lambda}}\Pi(\boldsymbol{\Lambda})\right)$ and $\{\boldsymbol{g}\} = \nabla_{\boldsymbol{\Lambda}}\Pi(\boldsymbol{\Lambda})$, thus

$$[\boldsymbol{I\!H}]\{\Delta\boldsymbol{\Lambda}\} + \{\boldsymbol{g}\} = \boldsymbol{0}. \tag{8.34}$$

Following a standard Newton-type multivariate search, a new design increment is computed,

$$\Delta = (\Delta\Lambda_1, \Delta\Lambda_2, ...\Delta\Lambda_N), \tag{8.35}$$

for a design vector, $\boldsymbol{\Lambda}$, by solving the following system, $[\boldsymbol{I\!H}]\{\Delta\boldsymbol{\Lambda}\} = -\{\boldsymbol{g}\}$, where $[\boldsymbol{I\!H}]$ is the Hessian matrix ($N \times N$), with components

$$H_{ij} = \frac{\partial^2\Pi(\boldsymbol{\Lambda})}{\partial\Lambda_i\partial\Lambda_j}, \tag{8.36}$$

$\{\boldsymbol{g}\}$ is the gradient ($N \times 1$), with components

$$g_i = \frac{\partial\Pi(\boldsymbol{\Lambda})}{\partial\Lambda_i} \tag{8.37}$$

and where $\{\Delta\boldsymbol{\Lambda}\}$ is the design increment ($N \times 1$), with components $\Delta\Lambda_i$. After the design increment has been solved for, one then forms an updated design vector, $\boldsymbol{\Lambda}^{new} = \boldsymbol{\Lambda}^{old} + \Delta\boldsymbol{\Lambda}$, and the process is repeated until $||\Pi|| \leq TOL$. Explicitly, the incremental system is

$$\begin{bmatrix} \frac{\partial^2\Pi(\Lambda)}{\partial\Lambda_1\partial\Lambda_1} & \frac{\partial^2\Pi(\Lambda)}{\partial\Lambda_1\partial\Lambda_2} & \frac{\partial^2\Pi(\Lambda)}{\partial\Lambda_1\partial\Lambda_3} & \frac{\partial^2\Pi(\Lambda)}{\partial\Lambda_1\partial\Lambda_4} & \cdots \\ \frac{\partial^2\Pi(\Lambda)}{\partial\Lambda_2\partial\Lambda_1} & \frac{\partial^2\Pi(\Lambda)}{\partial\Lambda_2\partial\Lambda_2} & \frac{\partial^2\Pi(\Lambda)}{\partial\Lambda_2\partial\Lambda_3} & \frac{\partial^2\Pi(\Lambda)}{\partial\Lambda_2\partial\Lambda_4} & \cdots \\ \frac{\partial^2\Pi(\Lambda)}{\partial\Lambda_3\partial\Lambda_1} & \frac{\partial^2\Pi(\Lambda)}{\partial\Lambda_3\partial\Lambda_2} & \frac{\partial^2\Pi(\Lambda)}{\partial\Lambda_3\partial\Lambda_3} & \frac{\partial^2\Pi(\Lambda)}{\partial\Lambda_3\partial\Lambda_4} & \cdots \\ \frac{\partial^2\Pi(\Lambda)}{\partial\Lambda_4\partial\Lambda_1} & \frac{\partial^2\Pi(\Lambda)}{\partial\Lambda_4\partial\Lambda_2} & \frac{\partial^2\Pi(\Lambda)}{\partial\Lambda_4\partial\Lambda_3} & \frac{\partial^2\Pi(\Lambda)}{\partial\Lambda_4\partial\Lambda_4} & \cdots \\ \cdots & \cdots & \cdots & \cdots & \cdots \\ \cdots & \cdots & \cdots & \cdots & \cdots \\ \frac{\partial^2\Pi(\Lambda)}{\partial\Lambda_N\partial\Lambda_1} & \frac{\partial^2\Pi(\Lambda)}{\partial\Lambda_N\partial\Lambda_2} & \frac{\partial^2\Pi(\Lambda)}{\partial\Lambda_N\partial\Lambda_3} & \frac{\partial^2\Pi(\Lambda)}{\partial\Lambda_N\partial\Lambda_4} & \cdots \end{bmatrix} \begin{Bmatrix} \Delta\Lambda_1 \\ \Delta\Lambda_2 \\ \Delta\Lambda_3 \\ \Delta\Lambda_4 \\ \cdots \\ \cdots \\ \Delta\Lambda_N \end{Bmatrix} = - \begin{Bmatrix} \frac{\partial\Pi(\Lambda)}{\partial\Lambda_1} \\ \frac{\partial\Pi(\Lambda)}{\partial\Lambda_2} \\ \frac{\partial\Pi(\Lambda)}{\partial\Lambda_3} \\ \frac{\partial\Pi(\Lambda)}{\partial\Lambda_4} \\ \cdots \\ \cdots \\ \frac{\partial\Pi(\Lambda)}{\partial\Lambda_N} \end{Bmatrix}. \tag{8.38}$$

The derivatives must often be computed numerically:

- For the first derivative of Π at $(\Lambda_1, \Lambda_2, \Lambda_3)$:

$$\frac{\partial \Pi}{\partial \Lambda_1} \approx \frac{\Pi(\Lambda_1 + \Delta\Lambda_1, \Lambda_2, \Lambda_3) - \Pi(\Lambda_1 - \Delta\Lambda_1, \Lambda_2, \Lambda_3)}{2\Delta\Lambda_1} \qquad (8.39)$$

- For the second derivative at $(\Lambda_1, \Lambda_2, \Lambda_3)$:

$$\frac{\partial}{\partial \Lambda_1}\left(\frac{\partial \Pi}{\partial \Lambda_1}\right) \approx \frac{\left(\frac{\partial \Pi}{\partial \Lambda_1}\right)|_{\Lambda_1 + \frac{\Delta\Lambda_1}{2}, \Lambda_2, \Lambda_3} - \left(\frac{\partial \Pi}{\partial \Lambda_1}\right)|_{\Lambda_1 - \frac{\Delta\Lambda_1}{2}, \Lambda_2, \Lambda_3}}{\Delta\Lambda_1} \qquad (8.40)$$

$$= \frac{1}{\Delta\Lambda_1}\left(\left(\frac{\Pi(\Lambda_1 + \Delta\Lambda_1), \Lambda_2, \Lambda_3 - \Pi(\Lambda_1, \Lambda_2, \Lambda_3)}{\Delta\Lambda_1}\right) - \left(\frac{\Pi(\Lambda_1, \Lambda_2, \Lambda_3) - \Pi(\Lambda_1 - \Delta\Lambda_1, \Lambda_2, \Lambda_3)}{\Delta\Lambda_1}\right)\right).$$

- For the cross-derivative at (Λ_1, Λ_2):

$$\frac{\partial}{\partial \Lambda_2}\left(\frac{\partial \Pi}{\partial \Lambda_1}\right) \approx \frac{\partial}{\partial \Lambda_2}\left(\frac{\Pi(\Lambda_1 + \Delta\Lambda_1, \Lambda_2, \Lambda_3) - \Pi(\Lambda_1 - \Delta\Lambda_1, \Lambda_2, \Lambda_3)}{2\Delta\Lambda_1}\right)$$

$$\approx \frac{1}{4\Delta\Lambda_1\Delta\Lambda_2}\left(\Pi(\Lambda_1 + \Delta\Lambda_1, \Lambda_2 + \Delta\Lambda_2, \Lambda_3) - \Pi(\Lambda_1 - \Delta\Lambda_1, \Lambda_2 + \Delta\Lambda_2, \Lambda_3)\right)$$

$$- \left(\Pi(\Lambda_1 + \Delta\Lambda_1, \Lambda_2 - \Delta\Lambda_2, \Lambda_3) - \Pi(\Lambda_1 - \Delta\Lambda_1, \Lambda_2 - \Delta\Lambda_2, \Lambda_3)\right). \qquad (8.41)$$

An exhaustive review of these methods can be found in the texts of Luenberger[59] and Gill, et al.,[31] while a state of the art can be found in Papadrakakis et al.[80]

8.8   Effective Electro-magnetic Properties of Mixtures

A critical aspect of additive manufacturing, in particular printed electronics, is the estimation of the effective properties of particle-functionalized dielectric materials. One of the primary properties of interest is the overall "effective" electrical conductivity, defined via Ohm's Law:

$$\langle \boldsymbol{J}\rangle_\Omega = \boldsymbol{\sigma}^* \cdot \langle \boldsymbol{E}\rangle_\Omega, \qquad (8.42)$$

where $\boldsymbol{\sigma}^*$ is the effective conductivity for the mixture, $\langle \boldsymbol{E}\rangle_\Omega$ is the volume averaged electric field, $\langle \boldsymbol{J}\rangle_\Omega$ is the volume averaged current, the averaging operator is defined as $\langle \cdot \rangle_\Omega \overset{\text{def}}{=} \frac{1}{|\Omega|}\int_\Omega (\cdot)\, d\Omega$ over a statistically representative volume element with domain Ω. Other properties, although not needed immediately in the analysis are:

- Overall electrical permittivity:

$$\langle \boldsymbol{D} \rangle_\Omega = \epsilon^* \cdot \langle \boldsymbol{E} \rangle_\Omega, \tag{8.43}$$

where ϵ^* is the effective electrical permittivity for the mixture, $\langle \boldsymbol{E} \rangle_\Omega$ is the volume averaged electric field, $\langle \boldsymbol{D} \rangle_\Omega$ is the volume averaged electric field flux,
- Overall magnetic permeability:

$$\langle \boldsymbol{B} \rangle_\Omega = \mu^* \cdot \langle \boldsymbol{H} \rangle_\Omega, \tag{8.44}$$

where μ^* is the effective magnetic permeability for the mixture, $\langle \boldsymbol{H} \rangle_\Omega$ is the volume averaged magnetic field, $\langle \boldsymbol{B} \rangle_\Omega$ is the volume averaged magnetic field flux,

As with mechanical properties, in order to make estimates of the overall electrical properties of a mixture, we consider the widely used Hashin and Shtrikman bounds (see previous chapter) for isotropic materials with isotropic effective responses. These estimates provide one with upper and lower bounds on the overall response of the material. For example, for the electrical conductivity, for two isotropic materials with an overall isotropic response one can utilize

$$\underbrace{\sigma_1 + \frac{v_2}{\frac{1}{\sigma_2 - \sigma_1} + \frac{1 - v_2}{3\sigma_1}}}_{\sigma^{*,-}} \leq \sigma^* \leq \underbrace{\sigma_2 + \frac{1 - v_2}{\frac{1}{\sigma_1 - \sigma_2} + \frac{v_2}{3\sigma_2}}}_{\sigma^{*,+}}, \tag{8.45}$$

where the conductivity of phase 2 (with volume fraction v_2) is larger than phase 1 ($\sigma_2 \geq \sigma_1$). Usually, v_2 corresponds to the particle material, although there can be applications where the matrix is more conductive than the particles. In that case, v_2 would correspond to the matrix material. Provided that the volume fractions and constituent conductivities are the only known information about the microstructure, the expressions are the tightest bounds for the overall isotropic effective responses for two phase media, where the constituents are both isotropic. A critical observation is that the lower bound is more accurate when the material is composed of high conductivity particles that are surrounded by a low conductivity matrix (denoted case 1) and the upper bound is more accurate for a high conductivity matrix surrounding low conductivity particles (denoted case 2).

Remark: As mentioned before, this can be explained by considering two cases of material combinations, one with 50 % low conductivity material and 50 % high conductivity material. A material with a continuous low conductivity (fine-scale powder) binder (50 %) will isolate the high conductivity particles ((50 %), and the overall system will not conduct electricity well (this is case 1 and the lower bound is more accurate), while a material formed by a continuous high conductivity (fine-scale powder) binder (50 %) surrounding low conductivity particles (50 %, case 2) will, in an overall sense, conduct electricity better than case 1. Thus, case 2 is more closely approximated by the upper bound and case 1 is closer to the lower bound. Since the true effective property lies between the upper and lower bounds, one can construct the following approximation

$$\sigma^* \approx \phi\sigma^{*,+} + (1 - \phi)\sigma^{*,-}, \tag{8.46}$$

where $0 \leq \phi \leq 1$. ϕ *is a function of the microstructure, and must be calibrated.* As mentioned, for high conductivity spherical particles, at low volume fractions, for example under 15 %, where the particles are not making contact, the lower bound is more accurate. Thus, one would pick $\phi = \phi^s \leq 0.5$ to bias the estimate to the lower bound. However, if we take the same volume fraction of particles, but make the flat flakes, they will certainly touch, and produce high-conductivity pathways. Their overall conductivity will be higher than those of sphere at the same volume fraction. Thus, one would pick $\phi = \phi^f > \phi^s$. One can calibrate ϕ by comparing it to different experiments.

8.9   Summary

The previous expressions provide a good way to estimate and optimize material combinations for desired effective properties. However, in order to probe the response of a given material combination more deeply, in particular the time-dependent behavior when it is deposited and/or thermoformed, one must resort to numerical methods. Generally, the most practical strategy is to:

- Use estimates (for example based on bounds) to determine proposed optimal combinations of materials and
- Use numerical discretizations of the continuum to determine the performance of the proposed optimal designs.

Accordingly, advanced methods are concerned with the computational characterization of the evolution of the material response and residual stresses in materials with microstructures that arise from heated (or curing) deposited mixtures of particles. Residual stresses arise because the hot bonded materials cannot freely contract to their stress-free state, when cooled, due to their interaction with other components in the system and the surrounding environment to which they are joined. The objective of many methods found in Zohdi[105−134,135−141,143] is to develop straight-forward computational frameworks that researchers in the field can easily implement and use as computationally-efficient design tools. Generally speaking, there is thermo-mechanical multifield coupling present, along with material changes associated with material hardening, elasto-plasticity and mechanical damage–thus numerical methods are essential. Furthermore, although we focussed on the final deposited material properties, we mention in passing that to model the dynamics of the deposition of particle systems, reduced-order particle-based or discrete element-based models, which treat such systems as multibody dynamical groups, are often used. They are advantageous in dealing with domains that break apart or coalesce, as compared to traditional continuum based finite difference and finite element methods, which have limitations when dealing with dynamic discontinua. For reviews see, for example, Duran, Pöschel and Schwager,[83] Onate et al.,[72,73] Rojek et al.,[70] Carbonell et al.,[11] Labra and Onate,[53] Leonardi et al.,[55] Cante et al.,[12] Rojek,[89] Onate et al.,[74] Bolintineanu et al.,[7] Campello and Zohdi,[9,10] Avci and Wriggers[6] and Zohdi.[105−134] In many cases, the deposition of these materials is the first stage of a multistep process which may involve, among other processes, compaction. Compaction is also somewhat outside the scope of the present work, and we refer the reader to Akisanya et al.,[3] Anand and Gu,[5] Brown and Abou-Chedid,[8] Domas,[22] Fleck,[24] Gethin et al.,[28] Gu et al.,[34] Lewis et al.,[56] Ransing et al.,[85] Tatzel,[97] Ganeriwala and Zohdi[27] and Zohdi.[105−134]

In closing, many of the challenges facing additive manufacturing processes utilize complex materials. For example, currently, rapid printing does not allow for precise control over the structure of the printed lines. This often results in lines with scalloped edges or non-uniform width, and offer only limited control over the height of the printed features. See Sirringhaus et al.,[96] Wang et al.,[101] Huang et al.,[44] Choi et al.[14−17] and Demko et al.[20,21] for details. Recently, nanoimprint lithography has been proposed as a means of decreasing the feature size of patterned nanoparticles while allowing more precise control over the structure of the printed lines (Park et al.,[81] Ko et al.[51,52] and Park et al.[82]). In this fabrication method, the nanoparticle inks are patterned by pressing with an elastomer mold and the particles are dried into their final configuration. While the resolution of nanoimprint lithography is improved over inkjet printing, there exists a residual layer on the substrate that must be etched away after patterning. Control over the height of features can be corrupted by capillary action between the mold and the drying ink, in particular along the length of longer features. Thus, as a possible alternative to nanoimprint lithography, nanoparticle self-assembly methods, based on capillary filling of photoresist templates have been proposed Demko et al.,[20,21] and appear to be promising. This leads to an obvious fact, namely, additive manufacturing alone is inadequate, and needs to be combined with classical manufacturing processes. *Explicitly stated, despite the attractive features of additive manufac-*

turing, it alone rarely produces the surface quality needed for structural integrity. Often classical subtractive, intermediate, high-precision surface milling is needed to create components with acceptable toughness and fatigue life. It is imperative that additive and subtractive processes be combined, guided by simulation software for deposition and removal of material with sensitivity to part quality, dimension, tolerances and surface finish. This must draw on a combination of existing and new numerical methods for additive manufacturing, multi-axis machines and computational modeling of process performance.

Organizations such as the American Society of Precision Engineers (ASPE) are exploring a number of issues that will need to be addressed if additive manufacturing is to realize its full potential for real components used in critical systems. These include issues related to: (a) Dimensional control needed for AM to be used in precision applications, (b) Design for manufacturing including design rules for additive manufacturing and the impact of dimensional errors on structures designed using optimization methodologies, (c) Standards including certifying additive manufacturing equipment capabilities and artifacts for assessing machine performance, (d) Using AM-fabricated components in precision assemblies and component-to-component relationships, stack-up tolerances, friction, robotic grip-ability and (e) Metrology and quality of external surfaces and internal features including materials validation. Some of these issues can be addressed by hybridizing AM with appropriate (subtractive, imprint, etc) advanced manufacturing technologies. However, due to the multistage complexity of combining many disparate processes, this must involve simulation-based strategies to guide these processes, loosely referred to as industrial "digitalization". Accordingly, manufacturers have been working continuously to quickly blend new simulation software with emerging technologies. In this vein, Machine Learning is critical for multistage advanced hybrid additive manufacturing processes to function properly, and is central to be able to achieve a level of precision and systemization that makes manufacturers comfortable (see Zohdi[138,141]).

On a broadly economic level, such technologies can benefit the world economy through three principal paths. First, manufacturing enterprises that adopt such technologies will realize energy savings and a smaller carbon footprint, which translates to reduced operating costs, increased market share, reinvestment of capital, and job creation. Second, commercial use of digitized technologies will increase demand for new digitized products and services, which will expand markets, exports, and jobs. Finally, manufacturing enterprises that adopt digitized technologies will realize improved productivity and margins on their products, resulting in more competitive pricing and greater market share. These production input cost savings can be reinvested or captured as profits by the enterprise, and spending of the savings generates new economic output jobs. In summary, by themselves, additive manufacturing processes rarely can deliver the degree of precision needed in industry–*they must be combined with subtractive processes.* In this vein, we refer the reader to the US National Academies Report of Zohdi and Dornfeld[142] for overviews. The combination of additive and subtractive processes is under current investigation by the author.

References

1. J. Aboudi, *Mechanics of Composite Materials: A Unified Micromechanical Approach.* Elsevier (1992).
2. Z. Ahmad, M. Rasekh and M. Edirisinghe, Electrohydrodynamic direct writing of biomedical polymers and composites, *Macromolecular Materials and Engineering* **295**, 315–319 (2010).
3. A.R. Akisanya A.C.F. Cocks and N.A. Fleck, The yield behavior of metal powders, *International Journal of Mechanical Sciences* **39**, 1315–1324 (1997)
4. P. Alivisatos, The use of nanocrystals in biological detection, *Nature Biotechnology* **22**(1), 47–52 (2004).
5. L. Anand and C. Gu, Granular materials: Constitutive equations and shear localization, *Journal of the Mechanics and Physics of Solids* **48**, 1701–1733 (2000).
6. B. Avci and P. Wriggers, A DEM-FEM coupling approach for the direct numerical simulation of 3D particulate flows, Journal of Applied Mechanics **79**(1–7), 010901 (2012).
7. D.S. Bolintineanu, G.S. Grest, J.B. Lechman, F. Pierce, S.J. Plimpton and P.R. Schunk, Particle dynamics modeling methods for colloid suspensions, Computational Particle Mechanics 1(3), 321–356 (2014).
8. S. Brown and G. Abou-Chedid, Yield behavior of metal powder assemblages, *Journal of Mechanics and Physics of Solids* **42**, 383–398 (1994).
9. E.M.B. Campello and T.I. Zohdi, Design evaluation of a particle bombardment system to deliver substances into cells, *International Journal for Numerical Methods in Biomedical Engineering* **30**(11), 1132–1152 (2014).
10. E.M.B. Campello and T.I. Zohdi, Design evaluation of a particle bombardment system to deliver substances into cells, Computational Mechanics.
11. J.M. Carbonell, E. Onate and B. Suarez, Modeling of ground excavation with the particle finite element method, *Journal of Engineering Mechanics* **136**, 455–463 (2010).
12. J. Cante, C. Davalos, J.A. Hernandez, J. Oliver, P. Jonsen, G. Gustafsson, H.A. Haggblad, PFEM-based modeling of industrial granular flows, *Computational Particle Mechanics* **1**(1), 47–70 (2014).
13. D.S. Chandrasekharaiah and L. Debnath, *Continuum Mechanics.* Academic Press (1994).
14. S. Choi, I. Park, Z. Hao, H.Y. Holman, A.P. Pisano and T.I. Zohdi, Ultra-fast self-assembly of micro-scale particles by open channel flow, *Langmuir* **26**(7), 4661–4667 (2010).
15. S. Choi, S. Stassi, A.P. Pisano and T.I. Zohdi, Coffee-ring effect-based three dimensional patterning of micro, nanoparticle assembly with a single droplet, *Langmuir* **26**(14), 11690–11698 (2010).
16. S. Choi, A. Jamshidi, T.J. Seok, T.I. Zohdi, M.C. Wu and A.P. Pisano, Fast, high-throughput creation of size-tunable micro, nanoparticle clusters via evaporative self-assembly in picoliter-scale droplets of particle suspension, *Langmuir* **28**(6) 3102–3111 (2012).

17. S. Choi, A.P. Pisano and T.I. Zohdi, An analysis of evaporative self-assembly of micro particles in printed picoliter suspension droplets, *Journal of Thin Solid Films*, **537**(30), 180–189 (2013).

18. L. Davis, *Handbook of Genetic Algorithms*. Thompson Computer Press (1991).

19. C. Deckard, Method and apparatus for producing parts by selective sintering. US Patent 4,863,538 (1986).

20. M. Demko, S. Choi, T.I. Zohdi and A.P. Pisano, High resolution patterning of nanoparticles by evaporative self-assembly enabled by in-situ creation and mechanical lift-off of a polymer template, *Applied Physics Letters* **99**, 253102-1–253102-3 (2012).

21. M.T. Demko, J.C. Cheng and A.P. Pisano, High-resolution direct patterning of gold nanoparticles by the microfluidic molding process, *Langmuir* **26**(22), 16710–16714 (2010).

22. F. Domas, *Eigenschaft profile und Anwendungsubersicht von EPE und EPP*. Technical Report of the BASF Company (1997).

23. G. Dwivedi, T. Wentz, S. Sampath and T. Nakamura, Assessing process and coating reliability through monitoring of process and design relevant coating properties, *J. Thermal Spray Technology* **19**, 695–712 (2010).

24. N.A. Fleck, On the cold compaction of powders, *Journal of the Mechanics and Physics of Solids* **43**, 1409–1431 (1995).

25. S.B. Fuller, E.J. Wilhelm and J.M. Jacobson, Ink-jet printed nanoparticle microelectromechanical systems, *Journal of Microelectromechanical Systems* **11**, 54–60 (2002).

26. D. Gamota, P. Brazis, K. Kalyanasundaram and J. Zhang, *Printed Organic and Molecular Electronics*. Kluwer Academic Publishers (2004).

27. R. Ganeriwala and T.I. Zohdi, A coupled discrete element-finite difference model of selective laser sintering, *Granular Matter* **18** (2016). http://dx.doi.org/10.1007/s10035–016-0626-0

28. D.T. Gethin, R.W. Lewis and R.S. Ransing, A discrete deformable element approach for the compaction of powder systems, *Modelling and Simulation in Materials Science and Engineering* **11**(1), 101–114 (2003).

29. S. Ghosh, *Micromechanical Analysis and Multi-Scale Modeling Using the Voronoi Cell Finite Element Method*. CRC Press (2011).

30. S. Ghosh and D. Dimiduk, *Computational Methods for Microstructure-Property Relations*. Springer (2011).

31. P. Gill, W. Murray and M. Wright, *Practical Optimization*. Academic Press (1995).

32. D.E. Goldberg, *Genetic Algorithms in Search, Optimization and Machine Learning*. Addison-Wesley (1989).

33. D.E. Goldberg and K. Deb, Preface: special issue on genetic algorithms, *Computer Methods in Applied Mechanics and Engineering* **186**(2–4), 121–124 (2000).

34. C. Gu, M. Kim and L. Anand, Constitutive equations for metal powders: Application to powder forming processes, *The International Journal of Plasticity* **17**, 147–209 (2001).

35. M. Haruta, Catalysis of gold nanoparticles deposited on metal oxides, *Cattech* **6**(3), 102–115 (2002).

36. Z. Hashin and S. Shtrikman, On some variational principles in anisotropic and nonhomogeneous elasticity, *Journal of Mechanics and Physics of Solids* **10**, 335–342 (1962).

37. Z. Hashin and S. Shtrikman, A variational approach to the theory of the elastic behaviour of multiphase materials, *Journal of the Mechanics and Physics of Solids* **11**, 127–140 (1963).

38. Z. Hashin, Analysis of composite materials: A survey, *ASME Journal of Applied Mechanics* **50**, 481–505 (1983).

39. S. Hazanov and C. Huet, Order relationships for boundary conditions effect in heterogeneous bodies smaller than the representative volume, *Journal of the Mechanics and Physics of Solids* **42**, 1995–2011 (1994).

40. R. Hill, The elastic behaviour of a crystalline aggregate, *Proc. Phys. Soc. (Lond.)* **A65**, 349–354 (1952).

41. C. Ho, D. Steingart, J. Salminent, W. Sin, T. Rantala, J. Evans and P. Wright, Dispenser printed electrochemical capacitors for power management of millimeter scale lithium ion polymer microbatteries for wireless sensors. In *6th International Workshop on Micro and Nanotechnology for Power Generation and Energy Conversion Applications (PowerMEMS)*, Berkeley, CA (2006).

42. J.H. Holland, *Adaptation in Natural and Artificial Systems*. University of Michigan Press (1975).

43. Y. Huang, M.C. Leu, J. Mazumdar and A. Donmez, Additive manufacturing: Current state, future potential, gaps and needs, and recommendation, *Journal of Manufacturing Science and Engineering* **137**, 014001-1 (2015).

44. D. Huang, F. Liao, S. Molesa, D. Redinger and V. Subramanian, Plastic-compatible low-resistance printable gold nanoparticle conductors for flexible electronics, *Journal of the Electrochemical Society* **150**(7), G412–417 (2003).

45. C. Huet, Universal conditions for assimilation of a heterogeneous material to an effective medium, *Mechanics Research Communications* **9**(3), 165–170 (1982).

46. C. Huet, On the definition and experimental determination of effective constitutive equations for heterogeneous materials, *Mechanics Research Communications* **11**(3), 195–200 (1984).

47. R. Householder, Molding Process. US Patent 4,247,508 (1979).

48. C. Hull, Apparatus for Production of Three-Dimensional Objects by Stereolithography. US Patent 4,575,330 (1984).

49. V.V. Jikov, S.M. Kozlov and O.A. Olenik, *Homogenization of Differential Operators and Integral Functionals*. Springer-Verlag (1994).

50. J. Kennedy and R. Eberhart, *Swarm Intelligence*. Morgan Kaufmann (2001).

51. S.H. Ko, I. Park, H. Pan, C.P. Grigoropoulos, A.P. Pisano, C.K. Luscombe, J.M.J. Frechet, Direct nanoimprinting of metal nanoparticles for nanoscale electronics fabrication, *Nan Letters* **7**, 1869–1877 (2007).

52. S.H. Ko, I. Park, H. Pan, N. Misra, M.S. Rogers, C.P. Grogoropoulos, A.P. Pisano, ZnO nanowire network transistor fabrication by low temperature, all-inorganic nanoparticle solution process, *Applied Physics Letters* **92**, 154102 (2008).

53. C. Labra and E. Onate, High-density sphere packing for discrete element method simulations, *Communications in Numerical Methods in Engineering* **25**(7), 837–849 (2009).

54. N. Lagaros, M. Papadrakakis and G. Kokossalakis, Structural optimization using evolutionary algorithms, *Computers & Structures* **80**, 571–589 (2002).

55. A. Leonardi, F.K. Witel, M. Mendoza and H.J. Herrmann, Coupled DEM-LBM method for the free-surface simulation of heterogeneous suspensions, *Computational Particle Mechanics* **1**(1), 3–13 (2014).

56. R.W. Lewis, D.T. Gethin, X.S.S. Yang and R.C. Rowe, A combined finite-discrete element method for simulating pharmaceutical powder tableting, *Int. J. Numer. Methods Eng.* **62**, 853–869 (2005).

57. Y. Liu, T. Nakamura, G. Dwivedi, A. Valarezo and S. Sampath, Anelastic behavior of plasma sprayed zirconia coatings, *J. American Ceramic Society* **91**, 4036–4043 (2008).

58. Y. Liu, T. Nakamura, V. Srinivasan, A. Vaidya, A. Gouldstone and S. Sampath, Non-linear elastic properties of plasma sprayed zirconia coatings and associated relationships to processing conditions, *Acta Materialia* **55**, 4667–4678 (2007).

59. D. Luenberger, *Introduction to Linear & Nonlinear Programming*. Addison-Wesley (1974).

60. S.A. Maier and H.A. Atwater, Plasmonics: Localization and guiding of electromagnetic energy in metal/dielectric structures, *Journal of Applied Physics* **98**, 011101 (2005).

61. P. Martin, *Handbook of Deposition Technologies for Films and Coatings, 3rd Edition*. Elsevier (2009).

62. P. Martin, *Introduction to Surface Engineering and Functionally Engineered Materials*, Scrivener and Elsevier (2011).

63. J.C. Maxwell, On the dynamical theory of gases, *Philos. Trans. Soc. London.* **157**, 49 (1867).

64. J.C. Maxwell, *A Treatise on Electricity and Magnetism, 3rd Edition*. Clarendon Press (1873).

65. T. Mura, *Micromechanics of Defects in Solids, 2nd Edition*, Kluwer Academic Publishers (1993).

66. T. Nakamura and Y. Liu, Determination of nonlinear properties of thermal sprayed ceramic coatings via inverse analysis, *International Journal of Solids and Structures* **44**, 1990–2009 (2007).

67. T. Nakamura, G. Qian and C.C. Berndt, Effects of pores on mechanical properties of plasma sprayed ceramic coatings, *Journal of American Ceramic Society* **83**, 578–584 (2000).

68. H. Nakanishi, K.J.M. Bishop, B. Kowalczyk, A. Nitzan, E.A. Weiss, K.V. Tretiakov, M. Apodava, R. Klajn, J.F. Stoddart and B.A. Grzybowski, Photoconductance and inverse photoconductance in thin films of functionalized metal nanoparticles, *Nature* **460**(7253), 371–375 (2009).

69. S. Nemat-Nasser and M. Hori, Micromechanics: *Overall Properties of Heterogeneous Solids, 2nd Edition*. Elsevier (1999).

70. J. Rojek, C. Labra, O. Su and E. Onate, Comparative study of different discrete element models and evaluation of equivalent micromechanical parameters, *International Journal of Solids and Structures* **49**, 1497–1517 (2012). doi:10.1016/j.ijsolstr.2012.02.032.

71. S.R. Samarasinghe, I. Pastoriza-Santos, M.J. Edirisinghe, M.J. Reece, L.M. Liz-Marzan, Printing gold nanoparticles with an electrohydrodynamic direct write device, *Gold Bulletin* **39**, 48–53 (2006).

72. E. Onate, S.R. Idelsohn, M.A. Celigueta and R. Rossi, Advances in the particle finite element method for the analysis of fluid multibody interaction and bed erosion in free surface flows, *Computer Methods in Applied Mechanics and Engineering* **197**(19–20), 1777–1800 (2008).

73. E. Onate, M.A. Celigueta, S.R. Idelsohn, F. Salazar and B. Suarez, Possibilities of the particle finite element method for fluid-soil-structure interaction problems, *Computational Mechanics* **48**, 307–318 (2011).

74. E. Onate, M.A. Celigueta, S. Latorre, G. Casas, R. Rossi and J. Rojek, Lagrangian analysis of multiscale particulate flows with the particle finite element method, *Computational Particle Mechanics* **1**(1), 85–102 (2014).

75. C. Onwubiko, *Introduction to Engineering Design Optimization*. Prentice Hall (2000).

76. M. Papadrakakis, N. Lagaros, G. Thierauf and J. Cai, Advanced solution methods in structural optimization using evolution strategies, *Engineering Computational Journal* **15**(1), 12–34 (1998a).

77. M. Papadrakakis, N. Lagaros and Y. Tsompanakis, Structural optimization using evolution strategies and neutral networks, *Computer Methods in Applied Mechanics and Engineering* **156**(1), 309–335 (1998b).

78. M. Papadrakakis, N. Lagaros and Y. Tsompanakis, Optimization of large-scale 3D trusses using evolution strategies and neural networks, *Int. J. Space Structures* **14**(3), 211–223 (1999a).

79. M. Papadrakakis, J. Tsompanakis and N. Lagaros, Structural shapa optimization using evolution strategies, *Eng. Optimization.* **31**, 515–540 (1999b).

80. M. Papadrakakis, N. Lagaros, Y. Tsompanakis and V. Plevris, Large scale structural optimization: Computational methods and optimization algorithms, *Archives of Computational Methods in Engineering, State of the Art Review* **8**(3), 239–301 (2001).

81. J.-U. Park, M. Hardy, S.J. Kang, K. Barton, K. Adair, D.K. Mukhopadhyay, C.Y. Lee, M.S. Strano, A.G. Alleyne, J.G. Georgiadis, P.M. Ferreira and J.A. Rogers, High-resolution electrohydrodynamic jet printing, *Nature Materials* **6**, 782–789 (2007).

82. I. Park, S.H. Ko, H. Pan, C.P. Grigoropoulos, A.P. Pisano, J.M.J. Frechet, E.S. Lee, J.H. Jeong, Nanoscale patterning and electronics on flexible substrate by direct nanoimprinting of metallic nanoparticles, *Advanced Materials* **20**, 489 (2008).

83. T. Poschel and T. Schwager, *Computational Granular Dynamics*. Springer-Verlag (2004).

84. G. Qian, T. Nakamura and C.C. Berndt, Effects of thermal gradient and residual stresses on thermal barrier coating fracture, *Mechanics of Materials* **27**, 91–110 (1998).

85. R.S. Ransing, R.W. Lewis and D.T. Gethin, Using a deformable discrete-element technique to model the compaction behavior of mixed ductile and brittle particular systems, *Philosophical Transactions of the Royal Society - Series A: Mathematical, Physical and Engineering Sciences* **362**(1822), 1867–1884 (2004).

86. J.W. Rayleigh, On the influence of obstacles arranged in rectangular order upon properties of a medium, *Phil. Mag.* **32**, 481–491 (1892).

87. A. Reuss, Berechnung der Fliessgrenze con Mischkristallen auf Grund der Plastizitatsbedingung fur Einkristalle. Z. angew. *Math. Mech.* **9**, 49–58 (1929).

88. B.A. Ridley, B. Nivi, J.M. Jacobson, All-inorganic field effect transistors fabricated by printing, *Science* **286**, 746–749 (1999).

89. J. Rojek, Discrete element thermomechanical modelling of rock cutting with valuation of tool wear, *Computational Particle Mechanics* **1**(1), 71–84 (2014).

90. S.R. Samarasinghe, I. Pastoriza-Santos, M.J. Edirisinghe, M.J. Reece, L.M. Liz-Marzan, Printing gold nanoparticles with an electrohydrodynamic direct write device, *Gold Bulletin* **39**, 48–53 (2006).

91. J. Segurado and J. Llorea, A numerical approximation to the elastic properties of sphererenforced composites, Journal of the Mechanics and Physics of Solids **50**(10), 2107–2121 (2002).

92. I. Sevostianov and M. Kachanov, Modeling of the anisotropic elastic properties of plasmasprayedcoatings in relation to their microstructure, *Acta Materialia* **48**(6), 1361–1370 (2000).

93. I. Sevostianov and M. Kachanov, Thermal conductivity of plasma sprayed coatings in relation to their microstructure, *Journal of Thermal Spray Technology* **9**(4), 478–482 (2001).

94. I. Sevostianov and M. Kachanov, Plasma-sprayed ceramic coatings: Anisotropic elastic and conductive properties in relation to the microstructure; cross-property correlations, *Materials Science and Engineering: A* **297**(1–2), 235–243 (2001).

95. I. Sevostianov and M. Kachanov, Effective properties of heterogeneous materials: Proper application of the non-interaction and the "dilute limit" approximations, *The International Journal of Engineering Science* **58**, 124–128 (2012).

96. H. Sirringhaus, T. Kawase, R.H. Friend, T. Shimoda, M. Inbasekaran, W. Wu, E.P. Woo, High-resolution inkjet printing of all-polymer transistor circuits, *Science* **290**, 2123–2126 (2000).

97. H. Tatzel, Grundlagen der Verarbeitungstechnik von EPP-Bewahrte und neue Verfahren. Technical Report of the BASF Company (1996).

98. K. Terada, M. Hori, T. Kyoya and N. Kikuchi, Simulation of the multi-scale convergence in computational homogenization approaches, *The International Journal of Solids and Structures* **37**, 2229–2361 (2000).

99. S. Torquato, *Random Heterogeneous Materials: Microstructure and Macroscopic Properties.* Springer-Verlag (2002).

100. W. Voigt, Uber die beziehung zwischen den beiden elastizitatskonstanten isotroper korper, *Wied. Ann.* **38,** 573–587 (2010).

101. J.Z. Wang, Z.H. Zheng, H.W. Li, W.T.S. Huck and H. Sirringhaus, Dewetting of conducting polymer inkjet droplets on patterned surfaces, *Nature Materials* **3**, 171–176 (2004).

102. W.M. Winslow, Method and Means for Translating Electrical Impulses into Mechanical Force. US Patent 2,417,850 (1947).

103. W.M. Winslow, Induced fibration of suspensions, *J. Appl. Phys.* **20**(12), 1137–1140 (1949).

104. T.I. Zohdi, P.J.M. Monteiro and V. Lamour, Extraction of elastic moduli from granular compacts, *The International Journal of Fracture/Letters in Micromechanics* **115**, L49–L54 (2002).

105. T.I. Zohdi, Genetic design of solids possessing a random particulate microstructure, *Philosophical Transactions of the Royal Society: Mathematical, Physical and Engineering Sciences* **361**(1806), 1021–1043 (2003).

106. T.I. Zohdi, On the compaction of cohesive hyperelastic granules at finite strains. In *Proceedings of the Royal Society* **454**(2034), 1395–1401 (2003).

107. T.I. Zohdi, Computational design of swarms, *The International Journal of Numerical Methods in Engineering* **57**, 2205–2219 (2003).

108. T.I. Zohdi, Constrained inverse formulations in random material design, *Computer Methods in Applied Mechanics and Engineering* **192**(28–30), 3179–3194 (2003).

109. T.I. Zohdi, Staggering error control for a class of inelastic processes in random microheterogeneous solids, *The International Journal of Nonlinear Mechanics* **39**, 281–297 (2004).

110. T.I. Zohdi, Modeling and simulation of a class of coupled thermo-chemo-mechanical processes in multiphase solids, *Computer Methods in Applied Mechanics and Engineering* **193**(6–8), 679–699 (2004).

111. T.I. Zohdi, G.A. Holzapfel and S.A. Berger, A phenomenological model for atherosclerotic plaque growth and rupture, *The Journal of Theoretical Biology* **227**(3), 437–443.

112. T.I. Zohdi, Modeling and direct simulation of near-field granular flows, *The International Journal of Solids and Structures* **42**(2), 539–564 (2004).

113. T.I. Zohdi, A computational framework for agglomeration in thermo-chemically reacting granular flows. In *Proceedings of the Royal Society* **460**(2052), 3421–3445 (2004).

114. T.I. Zohdi, Statistical ensemble error bounds for homogenized microheterogeneous solids, *Journal of Applied Mathematics and Physics (Zeitschrift fur Angewandte Mathematik und Physik)* **56**(3), 497–515 (2005).

115. T.I. Zohdi, Charge-induced clustering in multifield particulate flow, *The International Journal of Numerical Methods in Engineering* **62**(7), 870–898 (2005).

116. T.I. Zohdi, Simulation of coupled microscale multiphysical fields in particulate-doped dielectrics with staggered adaptive FDTD, *Computer Methods in Applied Mechanics and Engineering* **199**, 79–101 (2010).

117. T.I. Zohdi, Numerical simulation of charged particulate cluster droplet impact on electrified surfaces, *Journal of Computational Physics* **233**, 509–526 (2013).

118. T.I. Zohdi, On inducing compressive residual stress in microscale print-lines for flexible electronics, *The International Journal of Engineering Science* **62**, 157–164 (2013).

119. T.I. Zohdi, Rapid simulation of laser processing of discrete particulate materials, *Archives of Computational Methods in Engineering* **20**, 309–325 (2013).

120. T.I. Zohdi, A direct particle-based computational framework for electrically-enhanced thermo-mechanical sintering of powdered materials, *Mathematics and Mechanics of Solids* **19**(1), 93–113 (2014).

121. T.I. Zohdi, On cross-correlation between thermal gradients and electric field, *The International Journal of Engineering Science* **74**, 143–150 (2014).

122. T.I. Zohdi, Mechanically-driven accumulation of microscale material at coupled solid-fluid interfaces in biological channels, *Journal of the Royal Society Interface* **11**(91), 20130922 (2014).

123. T.I. Zohdi, A computational modeling framework for heat transfer processes in laser-induced dermal tissue removal, *Computational Mechanics in Engineering and Sciences* **98**(3), 261–277 (2014).

124. T.I. Zohdi, Additive particle deposition and selective laser processing - A computational manufacturing framework, *Computational Mechanics* **54**, 171–191 (2014).

125. T.I. Zohdi, Embedded electromagnetically sensitive particle motion in functionalized fluids, *Computational Particle Mechanics* **1**, 27–45 (2014).

126. T.I. Zohdi, Impact and penetration resistance of network models of coated lightweight fabric shielding, *GAMM Mitteilungen* **37**(1), 124–150 (2014).

127. T.I. Zohdi, Rapid computation of statistically stable particle/feature ratios for consistent substrate stresses in printed flexible electronics, *Journal of Manufacturing Science and Engineering* **137**(2), 021019 (2015).

128. T.I. Zohdi, A computational modelling framework for high-frequency particulate obscurant cloud performance, *The International Journal of Engineering Science* **87**, 75–85 (2015).

129. T.I. Zohdi, On necessary pumping pressures for industrial process-driven particle-laden fluid flows, *Journal of Manufacturing Science and Technology* **183**(3), (2015).

130. T.I. Zohdi, On the thermal response of a laser-irradiated powder particle, *CIRP Journal of Manufacturing Science and Technology* **10**, 77–83 (2015).

131. T.I. Zohdi, Modeling and simulation of the post-impact trajectories of particles in oblique precision shot-peening, *Computational Particle Mechanics* **3**, 533–540 (2016).

132. T.I. Zohdi, Modeling and simulation of cooling-induced residual stresses in heated particulate mixture depositions, *Computational Mechanics* **56**, 613–630 (2015).

133. T.I. Zohdi, Modeling and efficient simulation of the deposition of particulate flows onto compliant substrates, *The International Journal of Engineering Science* **99**, 74–99 (2015). doi:10.1016/j.ijengsci.2015.10.012

134. T.I. Zohdi, Modeling and simulation of laser processing of particulate-functionalized materials, *Archives of Computational Methods in Engineering* **24**, 89–113 (2017).

135. T.I. Zohdi, Computational modeling of electrically-driven deposition of ionized poly-disperse particulate powder mixtures in advanced manufacturing processes, *Journal of Computational Physics* **340**, 309–329 (2017).

136. T.I. Zohdi, Construction of a rapid simulation design tool for thermal responses to laser-induced feature patterns, *Computational Mechanics* **62**, 393–409 (2018).

137. T.I. Zohdi, Laser-induced heating of dynamic particulate depositions in additive manufacturing, *Comput.Methods Appl. Mech. Engrg.* **331**, 232–258 (2018).

138. T.I. Zohdi, Dynamic thermomechanical modeling and simulation of the design of rapid free-form 3D printing processes with evolutionary machine learning, *Computer Methods in Applied Mechanics and Engineering* **331**, 343–362 (2018).

139. T.I. Zohdi, Ultra-fast laser-patterning computation for advanced manufacturing of powdered materials exploiting knowledge-based heat-kernels, *Computer Methods in Applied Mechanics and Engineering* **343**, 234–248 (2019).

140. T.I. Zohdi, Rapid voxel-based digital-computation for complex microstructured media, *Archives of Computational Methods in Engineering* **26**, 1379–1394 (2019).

141. T.I. Zohdi, Electrodynamic machine-learning-enhanced fault-tolerance of robotic free-form printing of complex mixtures, *Computational Mechanics* **63**, 913–929 (2019).

142. T.I. Zohdi and D.A. Dornfeld, Recent trends in mechanics: Future synergy between computational mechanics and advanced additive manufacturing, US National Academies Report. Available at: http://sites.nationalacademies.org/cs/groups/pgasite/documents/webpage/pga_166813.pdf (Accessed 2018).

143. T.I. Zohdi, *Modeling and Simulation of Functionalized Materials for Additive Manufacturing and 3D Printing: Continuous and Discrete Media.* Springer-Verlag (2018).

Chapter 9

Machine Learning for Quality Control in Additive Manufacturing

Qiang Huang[a]

9.1 Introduction – Shift in Manufacturing Paradigm[b]

Additive manufacturing (AM), or known as three-dimensional (3D) printing, refers to a new class of technologies associated with the direct fabrication of physical products from Computer-Aided Design (CAD) models by a layered manufacturing process. In contrast to traditional material removal processes, AM constructs products by adding material layer by layer without part-specific tooling and fixturing. Objects with complex shapes or geometries can therefore be formed by slicing 3D models into successive cross sections in Additive Manufacturing File (AMF) format such as STL (STereoLithography) format. It holds the promise of direct digital manufacturing of products with highly complex geometry, material compositions, and functionalities.[1–8] It is widely recognized as a disruptive technology, having the potential to fundamentally change the nature of future manufacturing. Indeed, the changes can amount to a *third industrial revolution*, according to a special report on manufacturing and innovation by *The Economist* magazine.[9]

With the advancement of Industrial Internet of Things and its Cyber-physical Systems as a backbone, future product creation and manufacturing environments will be hyper-connected and globalized. One important trend of this manufacturing revolution is cyber-enabled additive manufacturing, which has inspired the formation of entirely new Product-Service-Systems and cyber communities of additive manufacturers centered around creative design and fabrication of innovative products. Ultimately, these developments in AM will fundamentally transform the manner in which people interact with manufacturing, ushering in an era of cyber-

---

[a]Materials of this chapter are mainly based on the author's published work.
[b]Partially published in Colosimo et al.[1]

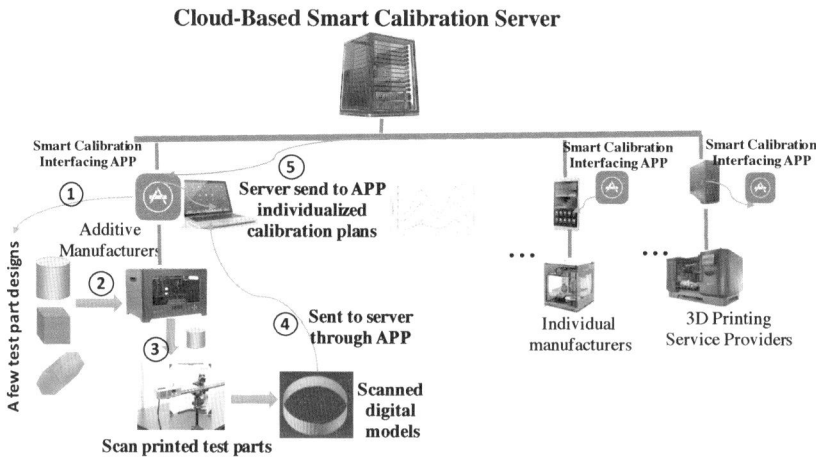

Fig. 9.1. Cyber-Physical Additive Manufacturing Systems.

manufacturing.

Under the new cyber infrastructures, computational design models are intentionally separated from specific AM machines to maximize design freedom for innovation. The connection between models and AM processes can be seamless through generation of AMF files and machine-oriented service layers. For example, given the same CAD model, products manufactured by a stereolithography (SLA) process and a fused deposition modeling (FDM) process can achieve the desired quality due to cloud-based quality services. As illustrated in Fig. 9.1, a calibration server residing in the cloud will communicate with individual AM machines through APPs by sending instructions and receiving data for machine learning. The learning engines are expected to customize the service layers for the needs of individual AM machines.

Currently, there exist huge barriers in the establishment of high-confidence Cyber-Physical Additive Manufacturing Systems (CPAMS) for seamless integration of design models with AM processes. Issues include, but not limited to,

- *High-fidelity and adaptable physical components:* In a CPAMS with a potentially large number of AM machines, a (shared) design model successfully constructed by a specific machine may not be manufactured with satisfactory precision or quality by another due to variations in materials, machines, processing techniques, and processing conditions. Trial-and-error or post-processing approaches to correct for nonconformities diminish the benefits of AM. This fact demands greater assurances of quality for

CPAMS. High-fidelity AM machines require adaptability to CAD models in the CPAMS to reliably manufacture high-quality products.

- *Extensible computational algorithms*: AM is targeted towards extremely low volume, personalized manufacturing of complex products with frequent design changes. Traditional computational algorithms are unable to facilitate the incorporation of complexity introduced by new CAD models, or adequately predict the quality of a new manufactured product. The existing library of design models and associated algorithms have to be extensible in a manner such that newly designed products can be fabricated with high fidelity in CPAMS.

- *System scalability*: Though individual AM machines are primarily limited to low volume production, CPAMS have the potential to mobilize all AM machines in the distributed system to achieve higher volume manufacturing. On the other hand, when CPAMS are scaled up by increasing the number of CAD models and AM machines to meet larger demand, the performance of the CPAMS measured by yield and quality should not deteriorate. Confounded with adaptability and extendability issues, current AM research focuses more on the component level by improving the accuracy of individual machines. Little research has been conducted to address the system scalability issue.

To catalyze the evolution of existing cyberinfrastructures into fully functioning CPAMS, new strategies and methodologies are needed to establish a new smart interfacing layer of infrastructure with computational algorithms intended to seamlessly integrate CAD design models with AM machines in CPAMS. The computational algorithms will dynamically and efficiently calibrate and recalibrate CPAMS through innovative modeling, learning, and control in order to achieve high-fidelity and adaptable physical AM machines, extensible computational design models, and scalable systems.

9.2 Paradigm Shift in Quality Control for AM[3]

Quality control (including calibration) is an indispensable service layer in the CPAMS. The scheme of quality control is dictated by the manufacturing paradigm. In the past, manufacturing paradigm experienced the revolutionary shift from craft production to mass production. Meanwhile, quality control scheme evolved from

---

[c]Ibid.

Table 9.1. Comparison of quality control in mass production and AM.[1]

Categories	Mass production	AM
Product volume N	Large N or $N \approx \infty$	Small N or $N \approx 1$
Product varieties V	Small V	$V = \infty$
Volume-to-variety ratio N/V	$N/V \approx \infty$	$N/V \approx 0$
Sample data for quality control	Sufficient for statistical analysis	$n = 1$ or $n = 0$
Quality definition	Inversely proportional to variation	Undefined yet
Quality improvement objective	Variation reduction	Undefined yet
Quality control methods	- Statical process control - Acceptance sampling - Design of experiments	To be developed

the practice of individual craftsmanship to the widely adopted statistical quality control (SQC). At present, AM holds the promise of fundamentally changing the nature of manufacturing and the changes can amount to a third industrial revolution.[9] A vital question arising is "what will be the corresponding quality control paradigm for AM, particularly for cyber-enabled AM?"

To understand the relevance of this question, we first conduct the comparison study of quality control for mass production and for AM. Unlike mass production, AM reduces tooling and intermediate steps for direct digital manufacturing. Complexity-free fabrication through layer-by-layer techniques enables individualized manufacturing of low-volume products with huge varieties and geometric complexity. As shown in Table 9.1, if we define a product "volume-to-variety" ratio or N/V ratio as

$$\frac{N}{V} = \begin{cases} \infty, & \text{mass product;} \\ 0, & \text{additive manufacturing,} \end{cases} \qquad (9.1)$$

then N/V approaches to ∞ in mass production, but to 0 in AM.

This huge disparity in N/V ratio has profound implications on methodology development for quality control. In mass production, the definition for quality naturally focuses on the variation of a large batch of products. The objective of quality control therefore aims at variation reduction.[10] Thanks to sufficient sample data, statistics-based quality control methods such as statistical process control, acceptance sampling, and design of experiments have been established for quality improvement. Mean and variance estimated from sample data are frequently utilized to characterize quality characteristics.

On the contrary, the near-zero N/V ratio in AM suggests that it is cost-prohibitive to collect sufficient sample data. Without enough sample data to build credible statistical distributions for quality characterization, the long-established concept of quality for mass production cannot be directly adopted for AM. For cybermanufacturing, big data can potentially be aggregated from a large community of AM users with interconnected AM machines and shared design models. However, the aggregated data can be highly heterogeneous due to variations in product designs, materials, AM processes, and process conditions. The independent and identically distributed assumption critical for SQC can hardly be satisfied. Furthermore, statistical approaches cannot be applied when no data can be aggregated for new products, materials, or processes not being tried before (i.e., $n = 0$).

With quality concept undefined for AM, the quality improvement objective is not clearly defined either. Consequently, cyber-enabled AM not only fundamentally changes the manufacturing paradigm, but also changes the premise of quality control. There is a paradigm shift in quality control for AM, which requires significant methodology development to establish an academic discipline complementary to SQC.

9.3   Challenges of Quality Control for Additive Manufacturing[4]

To establish a new quality control paradigm for AM, we first need to understand the complexities of quality control for AM, in particular, geometric accuracy control of AM built products:

- *Complexity of process physics*: A typical AM process generally involves material phase-changing, that is, liquid, paste or loose powder are selectively solidified into solid. The phase-changing processes and related shape deviation are therefore unavoidable. Due to the nature of layer-based fabrication, the actual deviation in an AM process is generally non-uniform, anisotropic, and non-linear. Accuracy prediction based on the first principle is a daunting task.
- *Complexity of product geometry*: AM theoretically makes the fabrication process free of geometric complexity. Nevertheless, the complexity of quality control is directly related to the geometric complexity of products. Using quality inspection as an example, one build of product in AM makes the inspection of intermediate steps or hidden features extremely difficult. The law of conservation of complexity in home-computer interaction states that every application has an inherent amount of complexity that can only be shifted, but cannot be removed or hidden. Clearly the complexity of

---

[d]Ibid.

fabrication has been shifted to quality control in AM.

- *Complexity of process calibration*: AM is most suitable for extremely low volume or even personalized manufacturing of complex products with frequent change of geometries. This presents a unique challenge to calibrate AM processes for high dimensional and geometric accuracy. In the mass production processes such as the injection molding process, weeks or even months can be afforded for fine-tuning process parameters and tooling setups because the same configuration remains unchanged for a large batch of identical products.

- *Complexity of data collection and analysis*: The predominant Statistical Quality Control (SQC) methods have mainly been developed for mass production. SQC relies on sufficient sample data to support statistical process control or acceptance sampling. Product quality is characterized by a statistical distribution and less variation among product units indicates better quality. In AM with frequent design changes and low volume production, it is often cost-prohibitive to collect sufficient sample data in order to establish credible statistical distributions for quality characterization. The existing SQC methodology therefore faces tremendous challenges to improve quality for AM.

Although AM has evolved from rapid prototyping to product manufacturing in the past 30 years, the bulk of current AM research is devoted to the novel development of CAD and process planning methods, finite element modeling (FEM), materials, processes, and machines suitable for AM.[2] Table 9.2 summarizes the related strategies and methods reported in the literature.

Table 9.2. Geometric quality control methods in AM.[11]

	Method and Strategy	Sample Literature
1.	Simulation study based on the first principles	12–14
2.	Offline optimization of process settings through experimentation	15–17
3.	Machine calibration through building test parts	15, 16, 18–21
4.	Part geometry calibration through extensive trial-build	4
5.	Adjustment of product design and process planning	11, 19–27
6.	Feedback control and online monitoring	28–34

Theoretical models for predicting shrinkage could potentially reduce experimental efforts. Simulation models based on the first principles have been developed, e.g., in the powder sintering process[12, 13] and metal injection molding.[14] Although numerical FEM simulation can then be developed to calculate the impact of shrinkage compensation, three-dimensional deviations and distortions in AM processes are still rather complicated. Improving part accuracy based purely on such simulation approaches is far from effective, and seldom used in practice.[2]

Similar to the calibration of CNC machines, the AM machine accuracy in x, y, z directions can be calibrated through building test cases,[18,20,21] or AM process settings such as layer thickness or hatch spacing are optimized through designed experiments.[16,19] The dimensional accuracy of AM products is anticipated to be ensured during full production. However, the position of AM light or laser exposure does not play the same dominant role as the tool tip position of CNC machines. Part geometry and shape, process planning, materials, and processing techniques jointly have complex effects on the profile accuracy. The calibration of the AM machine is therefore mostly limited to the scope of a family of products, specific types of material and machine, and process planning methods.

Besides machine calibration, another strategy is to apply either a shrinkage compensation factor uniformly to the entire product or different factors to the CAD design for each section of a product.[4] However, it is time-consuming to establish a library of compensation factors for all part shapes. The library is therefore often not inclusive. In addition, the interactions between different shapes or sections are not considered in this approach. Our study in Sabbaghi et al.[25] shows that the strategy of applying section-wise compensation may have detrimental effects on the overall shape due to "carryover effects", or interference between adjacent sections.

Three monitoring and feedback control strategies for AM are summarized in:[11] (i) control process variables based on the observed disturbance of process variables,[28,30,34] (ii) control process variables based on the observed product deviation,[31,32] and (iii) control input product geometry based on the observed product deviation.[11,20,21,24] However, a majority of the work relies on the collection of sufficient sample data in order to benchmark the normal process conditions, which is hardly applicable to untried products.

The review of the state of the art indicates that existing quality control methods are unable to serve the unique needs of AM. Due to multiple complex interacting physical and chemical phenomena, fabricating interchangeable parts using AM currently to certain degree relies on basic trial-and-error approaches. Post-processing with machine tools is then still required to meet design specifications, significantly negating the time and cost benefits of direct digital manufacturing. The challenges in quality control for AM call for new research to establish theoretical foundation enabling complexity-free quality control.

9.4   Machine Learning for Quality Control in Additive Manufacturing

Recent advances in computing and machine learning spur the heated interest of addressing quality control problems using data analytics. The remaining sections of this chapter present some initial development on this topic (not intended to be complete). Note that the unique challenges in AM demand new data analytics approaches, as opposed to applying existing ones.

9.4.1   *Quality Representation of AM Built Products*

Quality of AM built products includes geometric accuracy, surface finish, mechanical properties (e.g., stress, tensile strength), and structure properties (e.g., grain size, pores, cracks, etc). Quality presentation for data with complicated structures, for example, geometric shape accuracy of 3D objects, can be challenging. Assessment of geometric accuracy involves 3D measurement, which can be conducted by a Coordinate Measurement Machine (CMM) or a 3D scanner. The measurement data normally is in the form of point cloud with point coordinates defined in a Cartesian coordinate system (CCS). Representation of geometric accuracy based on point cloud data dictates control activities. A shape-dependent formulation would likely lead to shape-dependent accuracy control approaches, which would restrict learning and extrapolating from limited tested cases. Considering the nature of low-volume production, an unified formulation of deviation of any shape is desirable for quality control in AM, which will facilitate modeling and learning from limited sample data, and equally important, inferences on the prediction and compensation of untested shapes.

Shape representation has been extensively discussed in Computer Vision and Image Processing, where shape recognition, search, and discrimination are of major interest.[35–38] In AM, shapes of products are known while the boundary shape deviation after the build is of major concern. Quality representation of AM built products therefore involves more on the representation of shape deviations, as opposed to the shapes themselves.

9.4.1.1   *Two-dimensional shape deviation representation*[5]

The deviation of a 3D object can be classified as in-plane ($x - y$ plane) and out-of-plane (z direction) shape deviation, respectively. For illustrative purposes, we first

---

[e]Partially published in Huang et al.[24]

introduce the formulation of 2D shape deviation.

Under the assumption that the center is well-defined, the boundary of a 2D shape can be represented by a function $r_0(\theta)$ denoting the *nominal radius* at angle θ in the Polar Coordinates System (PCS).[24] The actual boundary after the build at angle θ is denoted as $r(\theta)$. The difference between the actual and nominal radius at an angle θ is essentially what defines the boundary deviation profile of an AM built product

$$\Delta r(\theta) = r(\theta) - r_0(\theta) \tag{9.2}$$

For product shapes like those in Fig. 9.2, we could locate the origin of the PCS in such a way that, for any given angle θ, there is only one unique point with radius $r(\theta)$ on the product boundary. The origin usually coincides with build center and measurement origin. The transformed shape deviation in the PCS will be a *single continuous profile*.

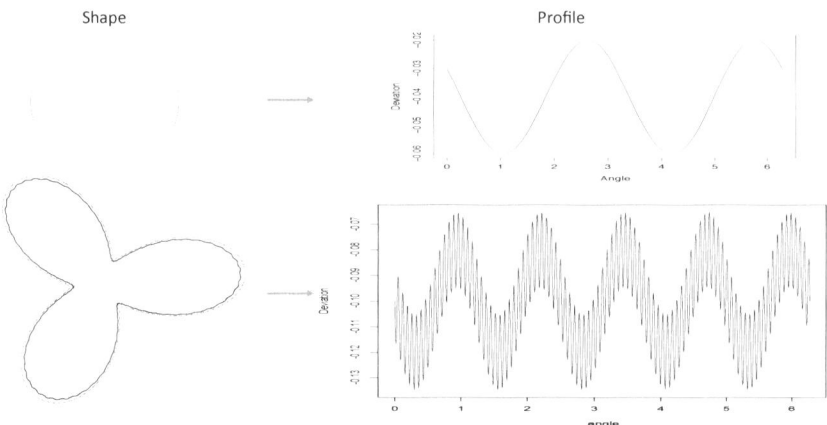

Fig. 9.2. Transform shape deviation to deviation profiles in the PCS.[39]

For more complex shapes, however, no matter how we translate and/or rotate the PCS, there exists more than one point on the product boundary for a given angle θ (Fig. 9.3(a)). We then have to establish multiple PCSs in which each profile is uniquely defined. A special case shown in Fig. 9.3(b) is a product with outer and inner boundaries defined in two PCSs and the transformation of shape deviation will yield multiple deviation profiles. For complicated concave shapes, segmentation may be needed to generate multiple deviation profiles as well. We should point out that the analysis herein will still apply except the complication arising from the inference between deviation profiles.[25]

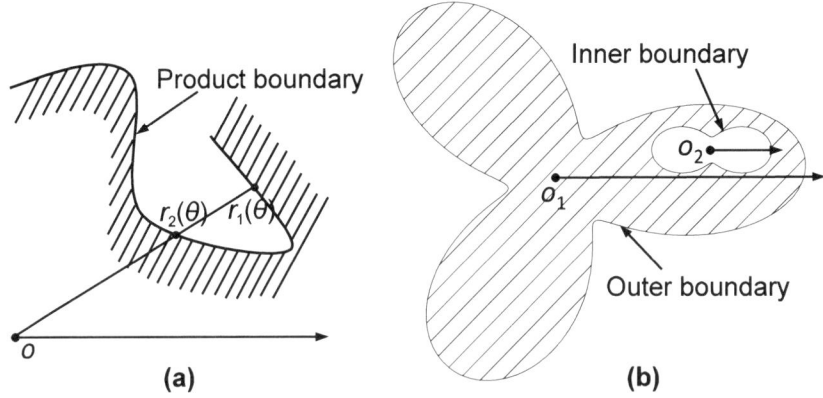

Fig. 9.3. Multiple profiles and PCSs.[39]

Like other manufacturing processes, the center of the product normally coincides with the origin of the part coordinate system defined by CAD software. For the convenience of building and measuring parts, we can also choose the center to be the origin of the machine or inspection coordinate system.

The Cartesian representation has been previously studied in the literature.[20,21] It faces a practical issue of correctly identifying shape deviation. As shown in Fig. 9.4, for a given nominal point $A(x, y)$, its final position A' is hard to be identified after deviation. A practical solution is to fix the x or y coordinate and study the deviation of the other coordinate (Δx or Δy in Fig. 9.4). Another method is to study deviation along three directions separately.[20,21] But the apparent correlation of deviation among the three directions cannot be captured, potentially leading to prediction error.

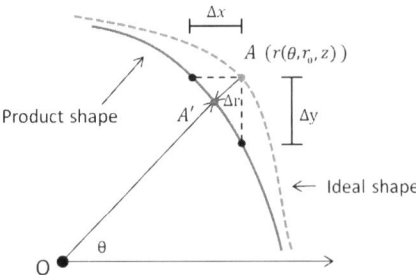

Fig. 9.4. Deviation representation under different coordinate systems.

In contrast, definition of radius deviation in Eq. (9.2) naturally captures deviation and is convenient for visualizing deviation patterns. As illustrated in Fig. 9.2, once the shape deviation $\Delta r(\theta, r_0(\theta))$ are presented in the PCS as deviation pro-files, modeling and analysis of geometric part errors are greatly alleviated from the original geometric complexity. The essence of this representation is to transform in-plane geometric errors into a functional profile defined on the interval $[0, 2\pi]$. This representation decouples the geometric shape complexity from the deviation modeling and a generic formulation of shape deviation can thus be achieved.

9.4.1.2 *Three-dimensional shape deviation representation*[f]

The geometric shape accuracy of 3D shapes can be represented in the Spherical Coordinate System (SCS)[39,40] as deviation surfaces. Consistent representation can be achieved for both 2D and 3D cases.

Under the SCS (r, θ, φ) both the in-plane and out-of-plane (z direction) de-viation are represented in a unified formulation as well. As can be seen later, it facilitates the representation of the out-of-plane error in the same way as the in-plane error. To illustrate this idea, we can first define the in-plane error and out-of-plane error in the SCS for the simple cylindrical shape. Denote $r(\theta, \varphi)$ the boundary shape of an AM built product with $r_0(\theta, \varphi)$ being the nominal shape. As shown in Fig. 9.5, for an arbitrary point $P_0(r_0, \theta_0, \varphi_0)$ at a given height $\varphi = \varphi_0$ or $z = r_0(\theta, \varphi)\cos(\varphi_0)$, the horizontal cross-section view of the product passing P_0 is given as $(r_0(\theta, \varphi)\sin(\varphi_0), \theta|\varphi_0)$, whose shape deviation $\Delta r(\theta, r_0(\theta)|\varphi_0)$ repre-sents the in-plane geometric error and its model formulation has been developed for 2D cases.[11,24]

Denote the in-plane deviation model $\Delta r(\theta|\varphi)$ as $h(r, \theta|\varphi)$. Let us define the expectation of the in-plane deviation $h(r, \theta|\varphi)$ over all φ:

$$\int_{\varphi} h(r, \theta|\varphi)d\varphi \tag{9.3}$$

Model (9.3) represents the average in-plane deviation over all layers. Our previ-ous models developed for cylindrical and polyhedron shapes can be viewed as the average in-plane deviation when out-of-plane deviation is not considered.

On the other hand, the out-of-plane error, which is the error in the vertical di-rection, can be represented in the vertical cross section containing P_0 (Fig. 9.6). Any point P_0 on the boundary of the vertical cross section is given as $(r, \varphi|\theta_0)$.

Denote the out-of-plane deviation model $\Delta r(\varphi|\theta)$ as $v(r, \varphi|\theta)$. Let us define the expectation of the out-of-plane deviation $v(r, \varphi|\theta)$ over all θ in the vertical cross section as:

---

[f]Partially published in Huang).[39]

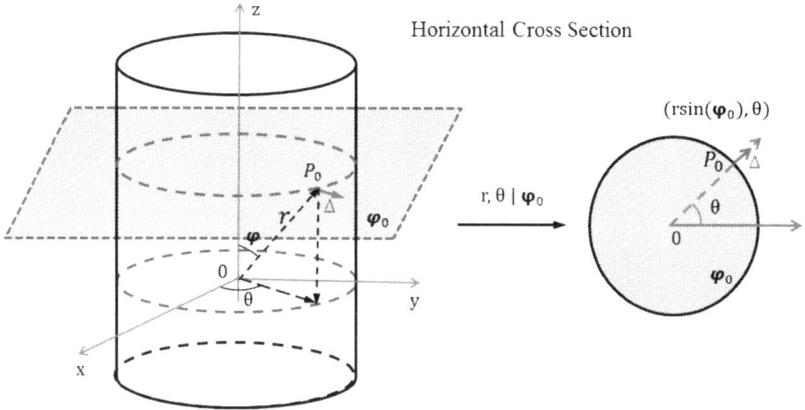

Fig. 9.5. In-plane deviation representation.

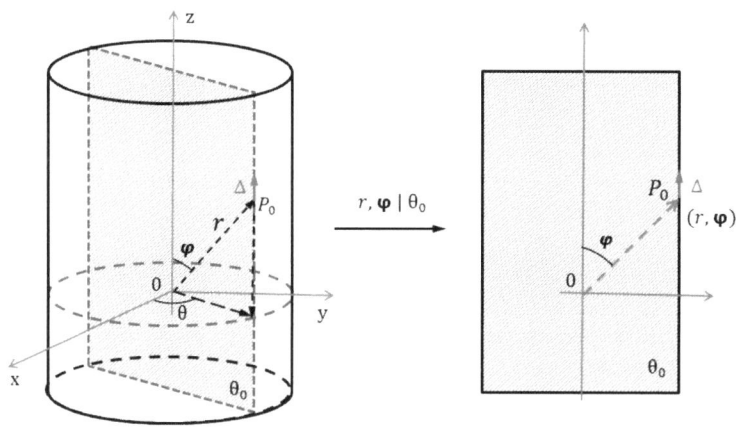

Fig. 9.6. Out-of-plane deviation representation.

$$\int_\theta v(r,\varphi|\theta)d\theta \qquad\qquad (9.4)$$

which can be interpreted as the average out-of-plane deviation in a similar fashion.

Mathematically, modeling of $v(r,\varphi|\theta)$ is essentially equivalent to modeling of $h(r,\varphi|\theta)$. This suggests that the mathematical formulation developed in our previous work for the in-plane errors can be borrowed under this new formulation in the SCS. In this way, 3D geometric errors can be described in a unified framework.

However, it should be noted that $v(r, \varphi|\theta)$ and $h(r, \theta|\varphi)$ may differ even if the horizontal and vertical cross section views share the same shape. Different from the in-plane deviation whose representation is along the radial direction, the out-of-plane deviation is defined along the vertical direction. The difference in representation leads to different error patterns. Furthermore, the vertical deviation is influenced by extra factors such as layer interactions and gravity, resulting in different deviation patterns.[41]

This representation, even for geometric accuracy, may not be the only option. For different applications and control objectives, quality representation has to be carefully designed to accommodate the shape complexity of 3D objects.

9.4.2 *Prediction of Quality of AM Built Product*[2]

Quality prediction for AM aims to predict the quality for both built and untried products based on limited number of test cases. For example, a limited number of test shapes and training data might be available in an AM process (Fig. 9.7). The geometric accuracy of a new product with completely different shape may have to be predicted for effective quality control geometric. We classify the modeling approaches as predicting modeling and prescriptive modeling. While traditional predictive modeling usually makes prediction within its experimental domains, e.g, a class or family of products, prescriptive modeling is able to make prediction of quality of new and untried categories of shapes beyond the experimental scope.

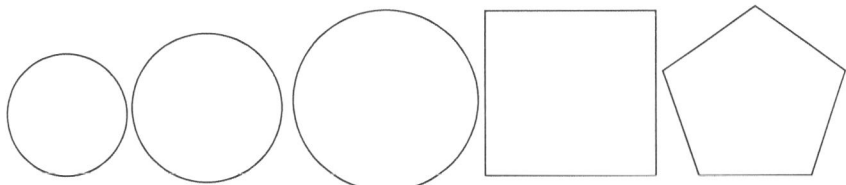

Fig. 9.7. Training data.

In a series of work,[11,24,25,39,40,42,43] Huang and co-authors intend to develop a new prescriptive modeling strategy with the ability of learning from a limited number of tested shapes and deriving compensation plans for new and untried products. Cylinders and polyhedrons are used as primitive shapes to model and predict in-plane ($x - y$ plane) shape deviation of freeform products. The model primitives identified include cylindrical base and the so-called cookie-cutter functions.[11,24] The cookie-cutter function connects the cylindrical shape model to polyhedron

---

[g]Materials of this sub section has been published in Luan and Huang.[42]

models by treating an in-plane polygon as being carved out from its circumcircle. With limited number of test shapes, dramatic in-plane accuracy improvement has been achieved through compensation for cylindrical shape (90%), polyhedrons and freeform shapes (> 50%).[11,24,40,42,43]

Below we will introduce the prediction of 2D freeform shape deviation by learning from a small training sample. The work is largely based on Luan and Huang[42] with the aim illustrated in Fig. 9.8.

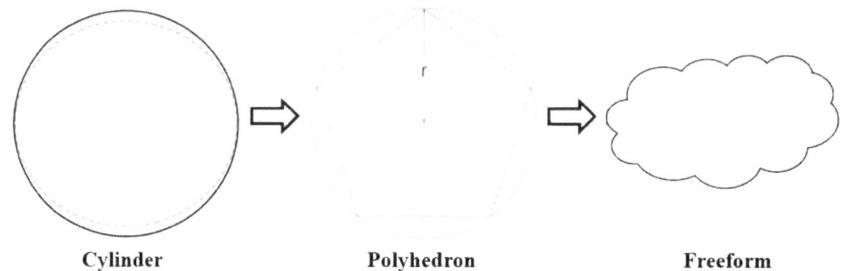

Cylinder   Polyhedron   Freeform

Fig. 9.8. Modeling extension from cylinder,[24] polyhedron,[11] to freeform.[42]

9.4.2.1 *Strategies for prescriptive methodology extension to freeform shapes*

The first challenging issue for methodology extension to freeform shapes is the functional representation of the in-plane geometric shape error for arbitrary shapes. A close-form representation is essential for implementing the optimal compensation policy derived in Huang et al.[24] Below we discuss the formulation of two possible strategies for representing the in-plane error of freeform products.

Strategy A: Polygon Approximation with Local Compensation (PALC): One observation is that any in-plane 2D freeform shape can be approximated by a polygon (Fig. 9.9). Since we have predictive in-plane shape deviation models established for cylindrical and polyhedron shapes,[11,24,25,43] one intuitive idea is therefore to approximate an arbitrary freeform shape by a polygon first and then improve the shape deviation model of that polygon by compensation. This line of thinking results in the first strategy: polygon approximation with local compensation (PALC).

Denote the in-plane error model for the fitted polygon as $f(\theta, r(\theta))$ defined in the Polar Coordinate System (PCS) (Different polygonal approximation approaches will be discussed shortly.). At angle θ, the approximation error is denoted as $x(\theta)$, which is the amount of compensation to be applied to improve prediction. The

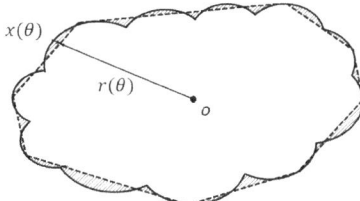

Fig. 9.9. Strategy of polygon approximation with local compensation (PALC) (solid line: freeform shape; dash line: approximated polygon shape; shadow area: amount of compensation to approach the freeform from the polygon shape).

predictive in-plane shape deviation model for the freeform can thus be derived as $f(\theta, r(\theta) + x(\theta))$. In this way, the previous in-plane deviation modeling for polyhedrons can be extended to arbitrary freeform shapes.

Strategy B: Circular Approximation with Selective Cornering (CASC): An alternative strategy, which is not that intuitive, is based on the observation that any in-plane 2D freeform shape can be approximated by the addition of a series of sectors with different radii (Fig. 9.10). To accommodate the potential sharp transitions or corners between adjacent sectors, we properly select corners and impose a cookie-cutter function[11] to the cylindrical base functions. This second strategy is called circular approximation with selective cornering (CASC).

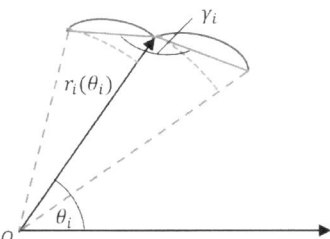

Fig. 9.10. Strategy of circular approximation with selective cornering (CASC) (solid arcs: actual radii of sectors; solid line segments: polygons sides).

The first step of CASC strategy is to obtain a series of sectors with different $r_i(\theta_i)$ at θ_i in the PCS (Fig. 9.10). A natural approach to derive $r_i(\theta_i)$ is to fit a polygon with a large number of sides to the freeform shape. The deviation of each sector can be predicted by $g(\theta_i, r_i(\theta_i))$, a modification of the cylindrical base model in[24] as a section of a circle.

The second step of CASC is to properly select corners to catch the transition points along the boundary of the freeform shape. The good approximation in step 1 of CASC results in a large number of sectors. Yet the sharp transition points along the freeform can be less than the number of sectors. We will present a method to design the cookie-cutter function to catch the transition points from the CAD design. Therefore, the two-step CASC strategy presents another path to predict the in-plane shape deviation of a freeform product.

It can be seen that polygon approximation to an in-plane freeform shape is an important step in both strategies. The related research has been widely reported in Image Processing, Computer Vision and Pattern Reorganization with methods classified into three groups: (1) split based approaches;[44,45] (2) merge based approaches;[46,47] and (3) dynamic programming based optimal approximation.[48,49] Both the split and merge based approaches demonstrate good approximation of convex and concave freeform. Not a focus of this chapter, the merge based approach is chosen for polygon approximation in our study.

9.4.2.2 *Strategy evaluation and selection through experimental investigation*

To evaluate the two plausible strategies proposed in Section 9.4.2.1, we conduct experimental investigation in order to make the final selection.

Our null hypothesis is that the intuitive PALC strategy works better than CASC. If this hypothesis is true, the main deviation pattern of a freeform should be captured by the deviation pattern of a polygon that approximates the freeform (cf. Fig. 9.9). Otherwise, CASC strategy has to be considered.

(a) Freeform with circumcircle radius = 2 ″   (b) Approximated similar polyhedron with circumcircle radius = 1 ″

Fig. 9.11. Printed freeform shape and corresponding similar polyhedron.

To test this hypothesis, we build one freeform product and one polyhedron with its in-plane polygon shape approximating the in-plane freeform shape. As shown in Fig. 9.11, the freeform product takes arbitrary shape with "circumcircle" radius being 2″. Here the radius of minimum circumcircle that contains a part shape is used to determine its size. Using the merge based approach, a polygon is generated to approximate the freeform shape. Using the same polygon shape at a smaller size (circumcircle radius = 1″), a polyhedron is built as well. All test parts have the same thickness of 0.25″.

It should be noted that size of the polygon is not a concern to us because our previous studies[11] show that the deviation pattern of a polygon does not vary with its size, though the magnitude differs.

Table 9.3. MP-SLA process settings.

Thickness of the product	0.25″
Thickness of each layer	0.00197″
Resolution of the mask	1920×1200
Dimension of each pixel	0.005″
Illuminating time of each layer	7s
Average waiting time between layers	15s
Type of the resin	SI500

To facilitate the identification of the orientation of test parts during or after the building process, a non-symmetric sunk cross with line thickness of 0.02″ is designed on the top surface. The 3D CAD models are exported to STL format files and then sent to the Mask Image Projection SLA (MIP-SLA) machine. The MIP-SLA platform in this experiment is the ULTRA® machine from EnvisionTec. Specification of the manufacturing process is shown in Table 9.3.

All test parts are measured using a Micro-Vu precision machine. Same measurement procedure is followed for each test part to reduce errors. For convenience and consistency, the center of the cross is chosen as the origin of measurement coordinate system in the Micro-Vu machine. Boundary profile is fitted using splines in the metrology software associated with the machine. For the two printed parts shown in Fig. 9.11, their deviation profiles defined in the PCS are shown as solid curve with high-order roughness in Fig. 9.12.

Comparing these two deviation profiles, their basic deviation patterns show significant discrepancies in two aspects: (1) The trends of deviation profiles do not match, and (2) The polygon deviation profile has more sharp deviation transitions and the positions of sharp changes are inconsistent with its freeform counterpart.

The immediate conclusion from the experimental investigation suggests that our null hypothesis is not true, i.e., the intuitive PALC strategy will not work well. We believe major reasons causing the discrepancies can be as follows.

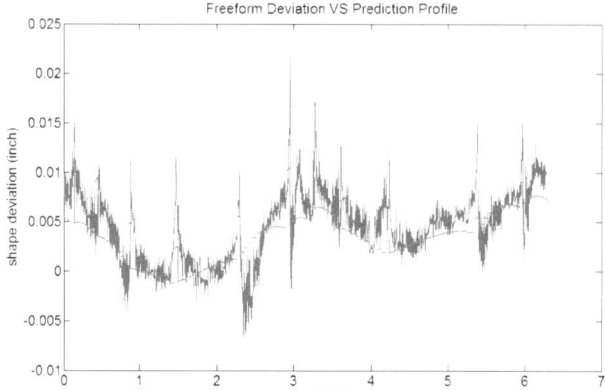

(a) Deviation profile (solid curve with high-order roughness) and model prediction (smooth dash curve) of convex freeform shape with circumcircle radius = 2″

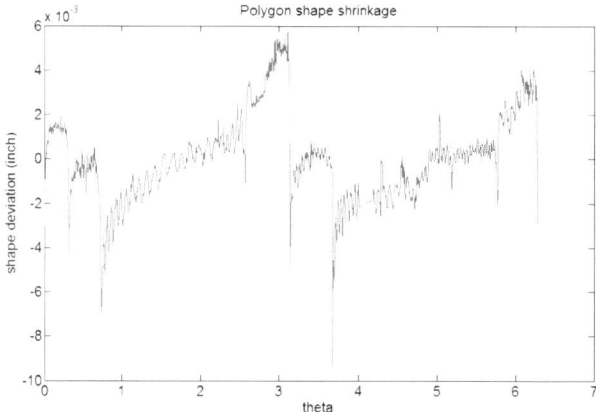

(b) Deviation profile of approximated convex polygon with circumcircle radius = 1 ″

Fig. 9.12. Deviation profiles of convex freeform and corresponding polygon approximated.

- Comparing the two parts in Fig. 9.11, the amount of approximation errors vary with location θ. This different level of approximation accuracy can alter the trend of deviation profile for the approximated polygon.
- Although increasing the number of sides can improve the approximation, a more serious issue is that vertices in approximated polygon and its corresponding freeform do not match. It is unavoidable that a vertex can be identified by the merged based method at a smooth segment of the freeform. Additionally chances exist that a line segment passing through a vertex of the freeform can be fitted for the polygon.

Both these two reasons lead to vertex mismatch. Therefore, if we directly use the shape deviation model for polygons to represent the major shape deviation pattern of the freeform, large errors can be introduced. Compensation for sharp changes could be introduced at wrong locations and real sharp transitions of the freeform shape may be missed.

Based on experimental result and analysis, CASC is deemed to be more suitable strategy for methodology extension.

9.4.2.3 *Prescriptive modeling of freeform deviation through Circular Approximation with Selective Cornering (CASC)*

The experimental disapproval of PALC strategy raises critical issues for CASC strategy:

(i) How to present cylindrical basic function appropriately to capture the correct trend pattern of freeform deviation?
(ii) How to detect actual vertices or sharp corners in freeform by adding cookie-cutter function?

Prescriptive Modeling of Freeform Deviation via CASC: It is worthwhile to first review the predictive model developed in Huang et al.[11] for in-plane polygon shapes. Denote $f^P(\theta, r(\theta))$ as the deviation profile of a polygon defined in the PCS. It is generally formulated as

$$f^P(\theta, r(\theta)) = g_1(\theta, r(\theta)) + g_2(\theta, r(\theta)) + g_3(\theta, r(\theta)) + \epsilon \qquad (9.5)$$

where g_1 depicts the deviation profile of cylindrical shapes, g_2 is the cookie-cutter function trimming the cylindrical base, g_3 is the remaining deviation along each side not captured by g_1 and g_2, and ϵ represents the un-modeled term.

Since the PALC strategy of directly applying $f^P(\theta, r(\theta) + x(\theta))$ to freeform prediction has been invalidated by experiments and analysis, we need to modify the model $f^P(\theta, r(\theta))$ in Eq. (9.5) so that it is consistent with strategy CASC.

According to CASC (Fig. 9.10), a large number of sectors with different radii $r_i(\theta_i)$ at θ_i will be adopted to capture the curvature and trend of freeform deviation profiles. This suggests that, rather than using a single and whole cylinder base, we need to modify $g_1(\theta, r(\theta))$ as a combination of cylindrical bases with each of them only representing a portion or a sector, e.g., $g_1(\theta_i, r_i(\theta))$ for the ith sector falling between (θ_{i-1}, θ_i).

An example of cylindrical basis model $g_1(\theta, r(\theta))$ has been established for MIP-SLA process in:[24]

$$g_1(\theta, r(\theta)) = x_0 + \alpha(r_0 + x_0)^a + \beta(r_0 + x_0)^b \cos(2\theta) \qquad (9.6)$$

where $\cos(2\theta)$ term represents the dominant deviation pattern, x_0 is a constant effect of over or under exposure for the MIP-SLA process, which is equivalent to a default compensation x_0 applied to original CAD model.

However, as has been mentioned in Huang et al.,[11] the MIP-SLA machine settings were changed after the experimentation in Huang et al.[24] Experiments on cylinders are therefore conducted again and new deviation profiles clearly show the difference between deviation of the upper and lower half of the cylinders. For the upper half ($\theta = 0 \sim \pi$) of the product, the cylindrical basis model $g_1(\theta, r(\theta))$ is modified as:

$$g_1(\theta, r(\theta)) = x_0^u + \alpha^u(r_0 + x_0^u)^{a^u} + \beta^u(r_0 + x_0^u)^{b^u} \cos(2(\theta + \pi/8)) \qquad (9.7)$$

For the lower half ($\theta = \pi \sim 2\pi$) of the product, the cylindrical basis model is modified as:

$$g_1(\theta, r(\theta)) = x_0^l + \alpha^l(r_0 + x_0^l)^{a^l} + \beta^l(r_0 + x_0^l)^{b^l}(-\sin(2\theta)) \qquad (9.8)$$

Note that superscripts u and l indicate parameters corresponding to upper and lower half of the product, respectively.

Following the CASC strategy, the upper cylindrical basis function in Eq. (9.7) is generalized for freeform shapes as:

$$g_1^F(\theta, r(\theta)) = x_0^u + \beta_0^u(r_i(\theta_i) + x_0^u)^{a^u}$$
$$+ \beta_1^u(r_i(\theta_i) + x_0^u)^{b^u} \cos(2(\theta + \pi/8))$$
$$for \ \theta_{i-1} < \theta < \theta_i, \ 1 \leq i \leq n^u, \ \theta_0 = \theta_n \quad (9.9)$$

And the lower cylindrical basis function in Eq. (9.8) is generalized for freeform shapes as:

$$g_1^F(\theta, r(\theta)) = x_0^l + \beta_0^l(r_i(\theta_i) + x_0^l)^{a^l}$$
$$+ \beta_1^l(r_i(\theta_i) + x_0^l)^{b^l}(-\sin(2\theta))$$
$$for \ \theta_{i-1} < \theta < \theta_i, \ n^u + 1 \leq i \leq n, \quad (9.10)$$

where n is the number of sides of the approximated polygon, n^u is the number of sides of the upper half polygon (θ from 0 to π).

In order to obtain n, n^u and θ_i in Eqs. (9.9) and (9.10), we adopt the merge based approach for polygon approximation. By linear scanning of a digital curve, merged based approach uses two thresholds regarding distance and area to determine whether each scanned point is a vertex of the approximated polygon. n is the number of all the selected vertices. When thresholds are small enough, the freeform boundary can be precisely approximated by a polygon with large number of sides (see in Fig. 9.13). θ_i can be obtained numerically as one outcome of the approximation procedure. In the approximation procedure, the boundary points are stored as digital coordinates. Therefore the angle θ_i in the PCS could be easily derived once the ith vertex is determined. For large n, the radius of each sector can be approximated by the distance from ith vertex to the origin of circumcircle, which is denoted as $r_i(\theta_i)$ (as shown in Fig. 9.10). Figure 9.13 shows that both convex and concave freeform shapes can be well approximated.

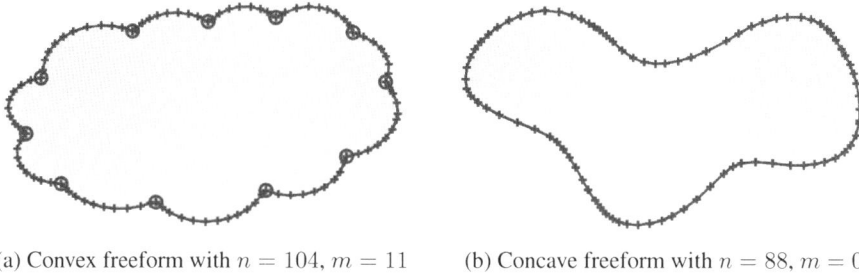

(a) Convex freeform with $n = 104, m = 11$ (b) Concave freeform with $n = 88, m = 0$

Fig. 9.13. Polygon approximation of freeform shapes.

With the generalized cylindrical basis model, the second issue to be addressed for CASC strategy is the proper determination of vertices or corners with sharp transitions along the boundary of a freeform shape. The cookie-cutter model or $g_2(\theta, r(\theta))$ in Eq. (9.6) is first developed in Huang et al.[11] to capture the sharp transitions of polygon shapes.

$$g_2(\theta, r(\theta)) = \beta_2(r_0 + x_0)^\alpha cookie.cutter(\theta - \phi_0) \qquad (9.11)$$

One example of cookie-cutter functions is the square wave model:

$$sign[cos(n(\theta - \phi_0)/2)]$$

where n is the number of sides of a polygon and ϕ_0 is a phase variable to shift the cutting position in the PCS. The sawtooth cookie-cutter model is another alternative:

$$saw.tooth(\theta - \phi_0) = (\theta - \phi_0) \text{ MOD } (2\pi/n)$$

where x MOD y = remainder of (x/y).

These two cookie-cutter functions apply to regular polygons well where the transitions of the square wave or sawtooth wave can be easily identified. For a freeform shape, however, we already point out in Section III that the vertices of the approximated polygons cannot correctly describe the transition points of the freeform boundary. In addition, the fitted polygons from merged based approach can be irregular in general. We therefore have to develop a method to design the cookie-cutter function to catch the transition points of the freeform from its CAD design.

Notice that when an interior angle γ_i of the fitted polygon (Fig. 9.10) is close to π or $|\gamma_i - \pi| \leq \delta_{critical}$, the corresponding ith vertex positioned at θ_i is not likely to produce a sharp transition in the deviation profile at location θ_i. Since each boundary point is stored as coordinates, the interior angle γ_i could be derived accordingly. Here $\delta_{critical}$ is a threshold value, e.g., $\delta_{critical}$ is less than $(1/6)\pi$ from the experimental studies in Huang et al.[11]

Based on this observation and proposed criterion, only vertices with sharp transitions in the fitted polygon will be selected for the cookie-cutter function to alternate the function amplitude. Let m be the number of selected vertices for the cookie-cutter function with $m \ll n$ (the small circles in Fig. 9.13 show m vertices finally selected), and the angle of m vertices be: ϑ_k, $k = 1, 2, ..., m$. Both m and ϑ_k are obtained by the polygon approximation procedure. Note that the cookie-cutter will only be applied to sectors whose vertices have sharp transitions. Then the square wave cookie-cutter function $sign[cos(n(\theta - \phi_0)/2)]$ is extended for a freeform shape as:

$$g_2^F(\theta, r(\theta)) = \beta_2(r_j(\theta_j) + x_0)^\alpha sign[cos(\frac{(2 + (-1)^j)\pi\theta}{2\theta_j})]$$

$$for \ \theta_{j-1} < \theta < \theta_j, \ \theta_0 = \theta_n, \ if \ \theta_j = \vartheta_k, \ 1 \leq k \leq m \quad (9.12)$$

where the term $(2 + (-1)^j)$ is adopted to guarantee that $sign(\cdot)$ changes between -1 and $+1$ alternatively to build the cookie-cutter function.

The sawtooth wave cookie-cutter model is also extended as:

$$g_2^F(\theta, r(\theta)) = \beta_2(r_j(\theta_j) + x_0)^\alpha saw.tooth(\theta)$$

$$= \beta_2(r_j(\theta_j) + x_0)^\alpha \frac{\pi(\theta - \theta_{j-1}) \ MOD \ (\theta_j - \theta_{j-1})}{2(\theta_j - \theta_{j-1})}$$

$$for \ \theta_{j-1} < \theta < \theta_j, \ \theta_0 = \theta_n, \ if \ \theta_j = \vartheta_k, \ 1 \leq k \leq m \quad (9.13)$$

We still adopt the sawtooth cookie-cutter function proposed in.[11] Note that not all freeform shapes have sharp transitions required for cookie-cutter functions. For example, the freeform in Fig. 9.13(b) does not contain sharp transition points, and the cookie-cutter function is not needed in prediction.

With the generalized cylindrical basis model and extended cookie-cutter model, the precriptive learning model $f^F(\theta, r(\theta))$ for freeform deviation is extended from the polyhedron model in Eq. (9.5) as

$$f^F(\theta, r(\theta)) = g_1^F(\theta, r(\theta)) + g_2^F(\theta, r(\theta)) + \epsilon \tag{9.14}$$

Prescriptive Model Estimation based on Limited Trial Shapes: Model (9.14) is deemed to be prescriptive if its model parameters can be estimated based on limited number of trial shapes and it can be applied to predict the deviation of a freeform shape given the product CAD design. This section illustrates model parameter estimation based on six test parts: three cylindrical and three polyhedron shapes in various sizes.

Three cylinders are fabricated with circumcircle radii 0.5 ″, 1.5 ″ and 3 ″, respectively. Fig. 9.14(a) shows their deviation profiles. The maximum likelihood estimation (MLE) of cylindrical basis model is given in Table 9.4, while the predicted cylindrical deviation profiles are shown as smooth lines in Fig. 9.14(a). Note that we assume $\epsilon \sim N(0, \sigma^2)$ and the initial parameter values for the numerical MLE estimation are chosen as $\alpha^u = \alpha^l = 0.01$, $a^u = a^l = 0.5$, $\beta^u = \beta^l = 0.001$, $b^u = b^l = 0.5$, $x_0^u = x_0^l = -0.01$, which are based on our previous analysis in Huang et al.[24]

Table 9.4. Estimated parameters for cylinder model.

Model of the upper half cylinder		
Parameter	Estimated value	Standard deviation
α^u	0.0076	9.593×10^{-5}
a^u	0.7878	0.0069
β^u	0.0015	1.670×10^{-5}
b^u	-0.1043	0.0125
x_0^u	-0.0056	9.398×10^{-5}
Model of the lower half cylinder		
Parameter	Estimated value	Standard deviation
α^l	0.0312	8.060×10^{-4}
a^l	0.2198	0.0054
β^l	0.0018	1.804×10^{-5}
b^l	0.7110	0.0096
x_0^l	-0.0264	7.931×10^{-4}

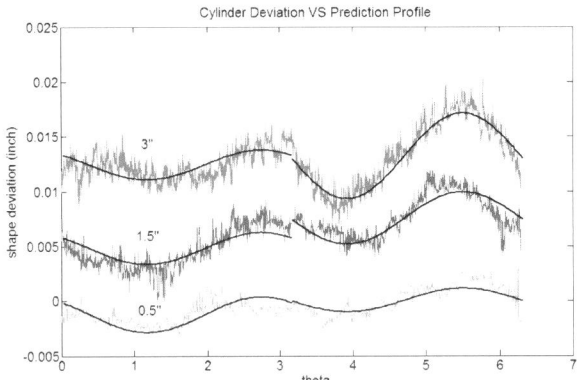

(a) Deviation profiles (curves with high-order roughness) and prediction (smooth dark lines) of three cylinders

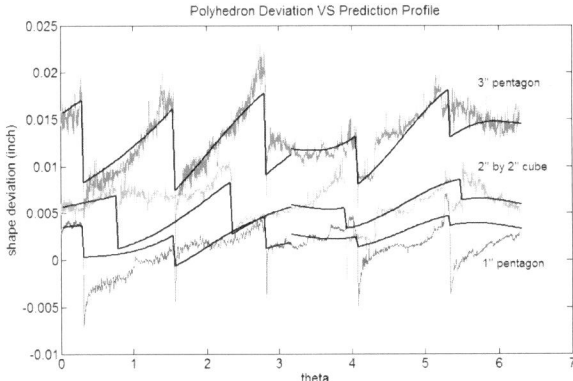

(b) Deviation profiles (curves with high-order roughness) and prediction (dark lines) of three polyhedrons

Fig. 9.14. Deviation and prediction profiles of simple trial shapes.

Three regular polyhedrons, that is, a 2″ by 2″ cube, a pentagon with circumcircle radius 1″, and a pentagon with circumcircle radius 3″, are fabricated to establish the in-plane polygon model. Figure 9.14(b) shows the in-plane shape deviation profiles of these polyhedrons.

Since the polyhedron is regarded as being trimmed from a cylinder, the cylindrical part in polyhedron model should be consistent with the cylindrical model. Therefore, we plug the six fitted parameters x_0^u, x_0^l, a^u, a^l, b^u, and b^l in the cylindrical model to polyhedron model directly. The MLE parameter estimation of remaining parameters is shown in Table 9.5, and the smooth black lines in Fig. 9.14(b) show the predicted shape deviation profiles of three polyhedrons. The initial parameter values for numerical iteration are set as: $\beta_0^u = \beta_0^l = 0.01$, $\beta_1^u = \beta_1^l = 0.01$, $\beta_2^u = \beta_2^l = 0.001$, $\alpha^u = \alpha^l = 1.$[11]

Table 9.5. Estimated parameters for polyhedron model.

Model of the upper half polyhedron		
Parameter	Estimated value	Standard deviation
β_0^u	0.0058	1.313×10^{-5}
β_1^u	0.0011	2.117×10^{-5}
β_2^u	0.0027	2.870×10^{-5}
α^u	0.8778	0.0094
Model of the upper half polyhedron		
Parameter	Estimated value	Standard deviation
β_0^l	0.0291	2.141×10^{-5}
β_1^l	0.0012	1.004×10^{-5}
β_2^l	0.0009	2.712×10^{-5}
α^l	1.4067	0.0250

Prescriptive Model Validation through Experimentation: The model (9.14) is obtained by learning from a limited number of tested shapes (cylinders, cubes, and pentagons) in Huang et al.[11,24] The model is deemed to be prescriptive if it can predict new and untried products. To validate the prescriptive power of model (9.14) for arbitrary shapes, we build one convex and one concave freeform shape with circumcircle radius 2 ", which are shown in Figs. 9.11(a) and 9.15(a), respectively. We will compare the prediction of model (9.14) to the measured shape deviation of two freeform shapes.

To obtain model prediction of freeform shapes, we first plug the estimated parameters for polyhedron model in Huang et al.[11,24] into the generalized cylindrical basis model Huang et al.(9.9) and (9.10), and the extended cookie-cutter model (9.13). Secondly, following CASC strategy, the polygon approximation is applied to the freeform CAD models to obtain n, m, θ_i, r_i γ_i and ϑ_k (Fig. 9.13), $i = 1, ..., n$, $k = 1, ..., m$. They are plugged into the extended freefrom model to predict the deviation for each sector then combined together to form the prediction. In merge based approach, the distance threshold is 3 pixel, the area threshold is 10 pixel. The threshold in catching transition points is $\delta_{critical} = \pi/6$, which is not a strict constraint. For the convex freeform (Fig. 9.11(a)), the approximated number of sides $n = 104$, and the number of vertices selected in cookie-cutter function $m = 11$ (Fig. 9.13(a)), which demonstrates that all the true vertices are founded without mistake. The concave freeform shape (Fig. 9.15(a)) has parameters $n = 88$ and $m = 0$ (Fig. 9.13(b)). The smooth shape has no transition point selected.

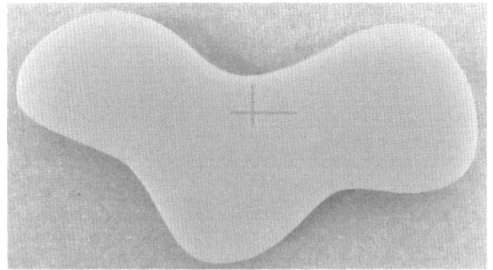

(a) Concave freeform shape with circumcircle radius
= 2 ″

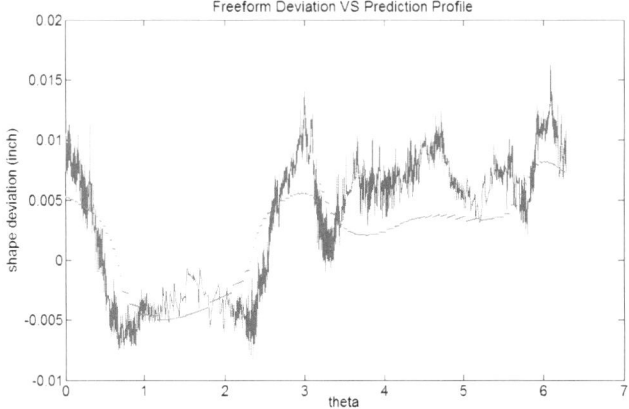

(b) Deviation profile (solid curve with high-order roughness) and model prediction (smooth dash curve) of concave freeform with circumcircle radius = 2″

Fig. 9.15. Deviation and prediction profiles for concave freeform shape.

The observed deviation profiles of the two freeform parts are shown as the solid blue lines in Figs. 9.12(a) and 9.15(b), respectively. The predicted deviation profiles for two freeform shapes using model (9.14) are represented as red dash lines in each figure. Considering the limited data used for model fitting and errors involved in freeform shape measurement (e.g., curve registration and boundary approximation in measurement), the prescriptive prediction of model (9.14) is remarkably close to the measurement. The encouraging result suggests:

- Our methodology of predicting in-plane error of AM built product is generic, which can be directly extended from cylinder and polyhedrons to freeform shapes;
- Our methodology has the capability of predicting new and untried products by learning from a limited number of tested shapes.

As a result, the in-plane shape deviation profile of any arbitrary freeform can be derived directly from CAD design. This provides a great opportunity to implement the optimal compensation policy established in Huang et al.[24] to improve the geometric accuracy of AM built products.

It should be mentioned that parameters in Table 9.5 are estimated from experimentation with regular polyhedrons, while most of the approximated polygons are irregular. Our model prediction accuracy can be greatly improved if irregular polyhedron data are available for model learning. This indirectly demonstrates the robustness of model (9.14).

9.4.2.4 *Prescriptive compensation and experimental validation*

One direct way to take advantage of the prescriptive model (9.14) is to improve printing accuracy by compensating the CAD design model of an arbitrary freeform shape even before fabricating the product. Work in Tong et al.[20, 21] first put forward the machine parametric error model to evaluate the part dimensional accuracy and model the parametric error functions. The negative values of predicted errors will be added to CAD design directly to compensate the product deviation.[50] further extend this error compensation method through employing conical sockets for probe measurement. The compensation strategy, however, is not optimal.

We first establish the optimal compensation policy for 2D shapes in Huang et al.[24] and 3D shapes in Huang.[39] Based on the result in Huang et al.,[24] the optimal amount of compensation $x^*(\theta)$ for 2D freeform shape is:

$$x^*(\theta) = -\frac{g_1^F(\theta, r(\theta)) + g_2^F(\theta, r(\theta))}{1 + g_1^{F'}(\theta, r(\theta)) + g_2^{F'}(\theta, r(\theta))} \tag{9.15}$$

Therefore, with the freeform model and its input information, the optimal compensation for each approximated sector is calculated then combined together to achieve the whole compensation plan.

Two new experiments are conducted to fabricate two freeform products by modifying original freeform CAD designs based on model (9.15). Figure 9.16 compares the deviation profiles of two freeform products before (blue lines) and after (red lines) compensation implemented. We also generate Fig. 9.17 by removing the sharp spikes in Fig. 9.16.

It is clear from these two figures that the whole deviation profiles after compensation are more centering around zero line. Table 9.6 summarizes the compensation performance. The average deviation is reduced nearly 60% and the absolute average deviation is reduced nearly 50%. Therefore, on average the deviation reduction

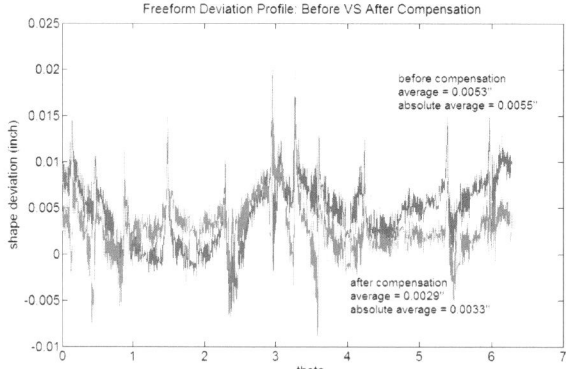

(a) Deviation profile of convex freeform shape: before and after compensation

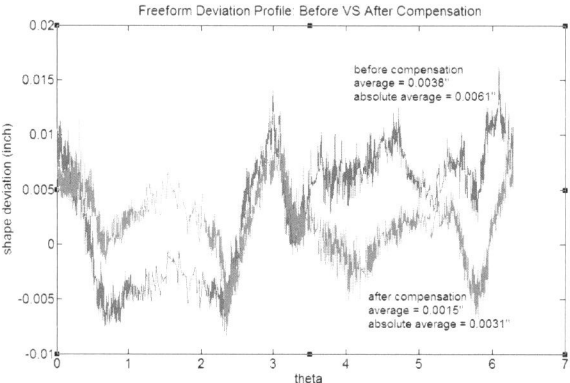

(b) Deviation profile of concave freeform shape: before and after compensation

Fig. 9.16. Freeform shape deviation profiles: before and after compensation.

is remarkable considering the limited data used to predict the complicated freeform shapes.

Aiming at improving geometric accuracy of AM built freeform products, this work makes a breakthrough by establishing a generic and prescriptive methodology to predict the in-plane ($x - y$ plane) shape deviation of arbitrary shapes. Building upon our previous predictive model for cylinder and polyhedron shapes, we propose circular approximation with selective cornering (CASC) strategy to extend the polyhedron model to arbitrary freeform shapes. This strategy essentially approximates a freeform with a series of circular sectors with different radii first and then further improve the approximation by imposing a generalized cookie-cutter function. The series of circular sectors are modeled as generalized cylindrical basis functions.

Table 9.6. Compensation performance.

Convex freeform shape			
Deviation	Before	After	Reduction
Average	0.0053″	0.0029″	45%
Absolute average	0.0055″	0.0033″	40%
Concave freeform shape			
Deviation	Before	After	Reduction
Average	0.0038″	0.0015″	61%
Absolute average	0.0061″	0.0031″	49%

The prescriptive model is estimated/calibrated based on a limited number of trials shapes including cylinders, cubes, and pentagons. Experimental observations of both convex and concave freeform shapes match the model prediction well. It demonstrates the prescriptive capability of predicting new and untried products by learning from a limited number of tested shapes. As a result, the in-plane shape deviation profile of any arbitrary freeform can be derived directly from CAD design.

To take advantage of prescriptive power of the developed model, we apply optimal compensation policy to freeform shapes. The geometric accuracy of compensated freeform products improves by 50% on average. This experimental validation further indicates that the developed prescriptive methodology has the capability of predicting new and untried products by learning from a limited number of tested shapes.

However, quality prediction for freeform shapes has large room to improve prediction accuracy. Full extension to prediction of 3D freeform shapes have not been accomplished. Fundamental research is needed to understand the model building blocks/elements to enable more informed model selection.

Furthermore, extension to 3D shapes involves more complicated mechanisms such as inter-layer interactions and deviation accumulation. Figure 9.18 shows the deviation patterns of four cylinders with different radii built horizontally. Their in-plane ($x - y$ plane) deviation, mainly caused by non-uniformly material shrinkage, exhibits similar deviation pattern. In contrast, Fig. 9.19 illustrates the out-of-plane deviation of four half-cylinders with different radii built vertically. Clearly, the layer-to-layer building process generates inter-layer bonding effect and the curl distortion, which leads to different deviation patterns even for the same type of shape. More advanced modeling approaches have to be developed to address these issues.

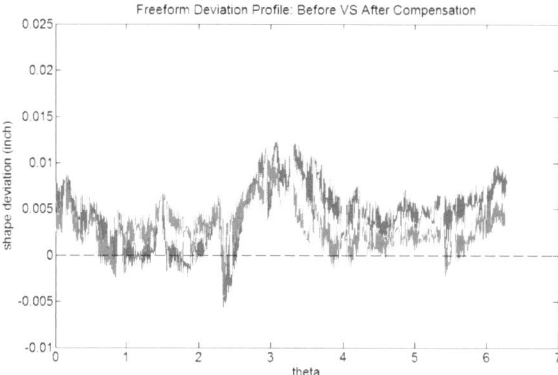

(a) Main trend of deviation profile of convex freeform shape

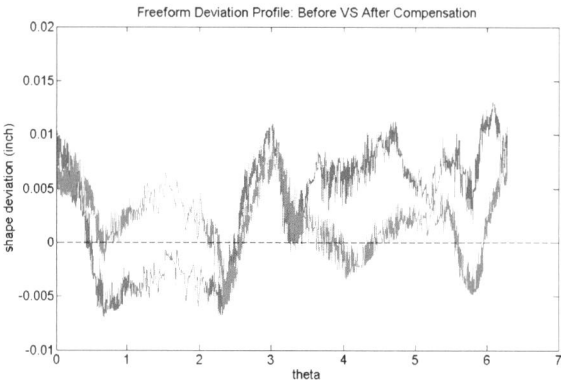

(b) Main trend of deviation profile of concave freeform shape

Fig. 9.17. Trend of freeform shape deviation profiles: before and after compensation.

9.4.3 *Offline Compensation of Shape Deviation Based on Learned Models*[9]

While predicting shape deviation belongs to the forward problem, minimizing shape deviation of AM built products is the challenging inverse problem due to geometric complexity, product varieties, material phase-changing and shrinkage, interlayer bonding, and limited sample data. Given measurement data regarding the deviation of an AM built product, one viable and efficient approach for accuracy control is through compensation of the product design to offset the geometric shape deviation. As shown in Fig. 9.20, once the deviation between the actual and nominal boundaries or profiles of a product is known through either measurement or model prediction, the adjustment of the CAD model can be applied in such a way

---

[i]Materials of this sub section has been published in Huang et al. and Huang.[24,39]

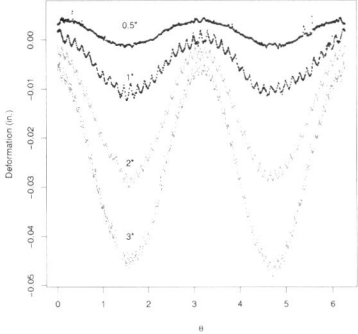

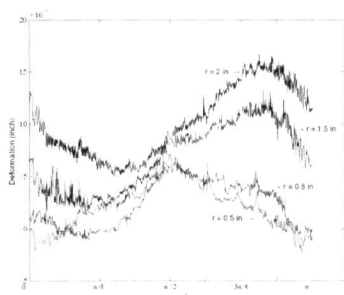

Fig. 9.18. In-plane deviation.[24]   Fig. 9.19. Out-of-plane deviation.[51]

that the new actual profile will match the original nominal profile as much as possible. The key issue is therefore to determine the optimal amount of compensation based on measured or predicted shape deviation for both 2D and 3D cases.

Compensation Strategy

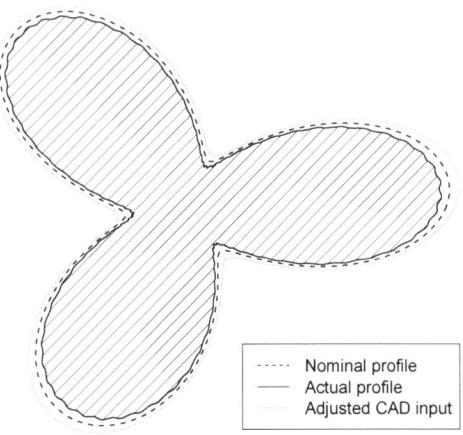

Fig. 9.20. Adjust CAD model.

The prevalent method of determining compensation in practice is shrinkage compensation factor approach, which takes its root in the material shrinkage study in casting and injection molding processes. This approach applies a shrinkage compensation factor uniformly to the entire product or different factors to the CAD model for each section of a product.[4] This method implicitly assumes the shape deviation is uniform in the section where the compensation factor is applied. Since products built via AM often have complex shapes, this assumption does not hold for general cases. The strategy of applying section-wise compensation may have detrimental effects on the overall shape due to "carryover effects", or interference between adjacent sections.[25] Compensation factor approach is thus far from being optimal for AM.

Huang and co-authors[24] establish an optimal compensation policy for in-plane ($x - y$ plane) shape deviation. The optimal amount of compensation for reducing deviations at each point on the product boundary is derived for high-precision AM. An analytical approach is further developed in Huang[39] where minimum area deviation (MAD) criterion and minimum volume deviation (MVD) criterion are proposed to derive close-form solutions for compensating 2D and 3D shape deviation, respectively. Furthermore, MAD and MVD criteria provide convenient quality measure or quality index for AM built products that facilitate online monitoring and feedback control of shape geometric accuracy. Experimental validation shows the compensation method is able to improve accuracy by an order of magnitude for cylindrical products,[24] by at least 75% for polyhedrons,[11] and by at least 50% for freeform shapes.[42,52]

This section provides an analytical foundation to achieve optimal compensation for high-precision AM based on the work in Huang.[39] We first present the optimal compensation policy or the optimal amount of compensation for 2D shape deformation. By analyzing its optimality property, we propose the minimum area deviation (MAD) criterion to offset 2D shape deformation. This result is then generalized by establishing the minimum volume deviation (MVD) criterion and by deriving the optimal amount of compensation for 3D shape deformation. Furthermore, MAD and MVD criteria provide convenient quality measure or quality index for AM built products that facilitate online monitoring and feedback control of shape geometric accuracy.

9.4.3.1 *Optimal compensation of 2D geometric shape deviation*

In real applications, a thin product or a section of a product with small thickness can be approximated with a 2D shape where only in-plane geometric error is of major concern. Therefore, it is meaningful for us to start with the discussion of optimal compensation for 2D shape deformation.

Let $\Delta r(\theta)) = f(\theta, r_0(\theta)) + \varepsilon$, with $f(\theta, r_0(\theta))$ being the model predicting shape deformation with error ε. Modeling of $f(\theta, r_0(\theta))$ can be achieved either through finite element modeling simulation or through data-driven approaches presented previously. In this section $f(\theta, r_0(\theta))$ is assumed to be known.

We aim to reduce deformation of manufactured products by direct compensation to the CAD model. Specifically, we revise the CAD model according to prediction of deformation, obtained through an understanding of the effect of compensation to the boundary of the CAD model. Under the polar coordinates system, a compensation of $x(\theta)$ units at location θ can be represented as an extension of the product's radius by $x(\theta)$ units in that specific direction θ. Clearly, we want an optimal compensation function that results in elimination of systematic shrinkage at all angles. To obtain such a function, we need to extend the previous model for prediction to accommodate the effect of compensation.

We first generalize the notation in Eq. (9.2). Let $r(\theta, x(\theta))$ denote the actual radius at angle θ when compensation $x(\theta)$ is applied at that location. Since the amount of compensation is relatively small in comparison to the nominal radius $r_0(\theta)$ (e.g., 1% to 2%), we can reasonably assume dynamics of the manufacturing and deformation process remains the same under compensation as compared to the entire process without compensation. Therefore, the predictive model $f(\cdot, \cdot)$ is still valid given the new design input $r_0(\theta) + x(\theta)$. Hence we have

$$r(\theta, x(\theta)) - (r_0(\theta) + x(\theta)) = f(\theta, r_0(\theta) + x(\theta))$$

Note that $f(\theta, r_0(\theta))$ without compensation is a special case of $f(\theta, r_0(\theta) + x(\theta))$ with $f(\theta, r_0(\theta) + 0) + 0$, and $\Delta r(\theta)$ is $\Delta r(\theta, 0)$.

Since the reference or nominal shape remains to be $r_0(\theta)$, the shape deformation at an angle θ is

$$\Delta r(\theta, x(\theta)) = r(\theta, x(\theta)) - r_0(\theta)$$
$$= f(\theta, r_0(\theta) + x(\theta)) + x(\theta). \tag{9.16}$$

The optimal amount of compensation $x^*(\theta)$ will minimize $\Delta r(\theta, x(\theta))$ or $\Delta r(\theta, x(\theta)) = 0$ for any θ. To find the solution of $x^*(\theta)$, Taylor series expansion of $f(\theta, r_0(\theta) + x(\theta))$ at $r_0(\theta)$ is applied in.[24] We have from Eq. (9.16) that

$$\Delta r(\theta, x(\theta)) \approx f(\theta, r_0(\theta)) + f'(\theta, r_0(\theta))x(\theta) + x(\theta) \tag{9.17}$$

where $f'(\theta, r_0(\theta))$ is the derivative with respect to $r_0(\theta)$.

By equating $\Delta r(\theta, r_0(\theta), x(\theta))$ to zero, the optimal compensation function $x^*(\theta)$ can be obtained as

$$x^*(\theta) = -\frac{f(\theta, r_0(\theta))}{1 + f'(\theta, r_0(\theta))} \tag{9.18}$$

Remark 1. *The result in Eq. (9.18) shows that the optimal compensation is not simply to apply the negative value of the observed or predicted shape deformation, i.e., $x^*(\theta) = -f(\theta, r_0(\theta)) = -\Delta r(\theta, r_0(\theta))$, which is essentially the shrinkage compensation factor approach widely used in practice. The compensation factor approach can only be optimal when $f'(\theta, r_0(\theta)) = 0$, i.e., deformation is uniform everywhere.*

In consideration of accuracy control for AM, we view AM as one-of-a-kind manufacturing with frequent change of designs. Under this problem setting, accuracy control for products never being built before will be frequently encountered. Therefore, we only take the first-order Taylor series expansion for robustness consideration.

Minimum Area Deviation (MAD) Criterion and Optimality: This subsection aims to understand the optimality of the optimal compensation policy given in Eq. (9.18). First, we introduce the following optimality criterion.

Definition 1 (Minimum Area Deviation (MAD) Criterion). *For a 2D shape deviating from its intended design model, the minimum area deviation (MAD) criterion is satisfied if the total absolute area change of the deformed shape is the smallest.*

The optimal amount of compensation given in Eq. (9.18) intends to minimize the deviation of every point along the boundary of a product. Intuitively we can expect the optimal compensation will result in the minimum area change of the nominal shape. Below we give a theorem and its proof.

Theorem 1 (Minimum-Area-Deviation Compensation). *The optimal compensation policy in Eq. (9.18) satisfies MAD criterion.*

Proof: As shown in Fig. 9.21(a), the area deviation at location θ due to $\Delta r(\theta, r_0(\theta))$ is

$$\Delta S(r(\theta, r_0(\theta)), \theta) \approx [r_0(\theta)\Delta\theta]\Delta r(\theta, r_0(\theta)) \approx r_0(\theta)f(\theta, r_0(\theta))d\theta$$

The total area deviation before compensation is therefore

$$\Delta S = \int_0^{2\pi} r_0(\theta)|f(\theta, r_0(\theta))|d\theta$$

Absolute value $|f(\theta, r_0(\theta))|$ *is applied because shape deformation can be positive or negative at different locations.*

After applying compensation $x(\theta)$ *as shown in Fig. 9.21(b), the total area deviation becomes*

$$\Delta S(x(\theta)) = \int_0^{2\pi} r_0(\theta)\Big|f(\theta, r_0(\theta) + x(\theta)) + x(\theta)\Big| d\theta \qquad (9.19)$$

Note that $x(\theta)$ *in Fig. 9.21(b) is an unknown quantity to be determined. Its optimal value can be positive or negative depending on the shape deformation at location* θ. ΔS *before compensation is just a special case with* $x(\theta) = 0$ *or* $\Delta S(x(\theta) = 0)$.

The optimal compensation $x^*(\theta)$ *will minimize the area deviation from the nominal shape, i.e.,*

$$\begin{array}{c} \min_{x(\theta)} \quad \Delta S(x(\theta)) \\ \\ \text{for all} \quad 0 \le \theta \le 2\pi. \end{array} \qquad (9.20)$$

And the solution clearly has to satisfy

$$f(\theta, r_0(\theta) + x(\theta)) + x(\theta) = 0$$

Taylor series expansion of $f(\theta, r_0(\theta) + x(\theta))$ *at* $r_0(\theta)$ *will yield the same result in Eq. (9.18) and complete the proof.*

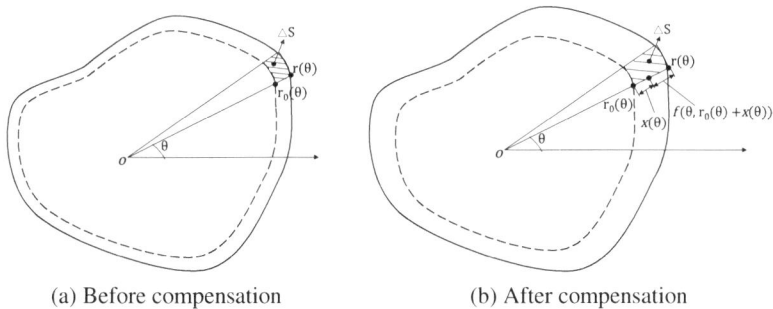

(a) Before compensation   (b) After compensation

Fig. 9.21. Shape area deviation before and after compensation.

Theorem 1 is not just a reinterpretation of the previously developed compensation policy Eq. (9.18). It has two important implications by providing (1) a convenient measure of in-plane error; and (2) a stepping stone to extend the 2D compensation policy to 3D case.

We first discuss the quality measure and defer the 3D extension to the next section.

Definition 2 (Equivalent Amount of Compensation). *Suppose an error source, either from design or manufacturing, causes a product deviating from its intended shape $r_0(\theta)$ to $r(\theta, r_0(\theta))$ by $\Delta r(\theta, r_0(\theta))$. The same amount of deviation $\Delta r(\theta, r_0(\theta))$ can be reproduced if we apply the correct amount of compensation $x(\theta)$ under normal condition (i.e., condition without assignable causes). Here, $x(\theta)$ is defined as the equivalent amount of compensation to the error source.*

This definition is based on the concept of error equivalence we developed for traditional manufacturing.[53–55] Two error sources are equivalent if they can generate the same error pattern. Once we establish the equivalence between two error sources, we can use one error source to compensate the other.

For AM processes, we represent the over-exposure error in the stereolithography process[24] and the extruder positioning in the Fused Deposition Modeling process[43] by the equivalent amount of compensation, respectively. This concept will greatly simplify the predictive modeling and take the best use of available model $f(\theta, r_0(\theta))$. Given an error source and its equivalent amount of compensation $x(\theta)$, the product shape deformation can be predicted by $f(\theta, r_0(\theta) + x(\theta)) + x(\theta)$. Additionally, this concept can be utilized for error source management through deformation compensation.

Lemma 1 (In-plane Quality Measure). $\Delta S(x(\theta))$ *is a quality measure or quality index of in-plane shape deformation.*

Proof: Based on the definition of equivalent amount of compensation, the product shape deformation or the quality of two units can be represented as $f(\theta, r_0(\theta) + x_1(\theta)) + x_1(\theta)$ and $f(\theta, r_0(\theta) + x_2(\theta)) + x_2(\theta)$. From Theorem 1 and its proof, the shape deformation of two units will end up with area deviations $\Delta S(x_1(\theta))$ and $\Delta S(x_2(\theta))$. $x_1(\theta)$ is more optimal than $x_2(\theta)$ if $\Delta S(x_1(\theta)) \leq \Delta S(x_2(\theta))$. Therefore, $\Delta S(x(\theta))$ is a quantitative measure of shape quality, the smaller the better.

Note that the quality measure based on Eq. (9.18) requires to compare the deviation of every point along the boundary of a product, which ends up with a deviation profile. $\Delta S(x(\theta))$ is a much more convenient quality measure with theoretical support from Theorem 1.

Remark 2. *(MinMax Quality Criterion) Another candidate quality measure is the maximum shape deformation of AM built products, which is equivalent to the Hausdorff distance between the profile of AM built product and the profile of nominal design shape. As an example of the deviation profiles of cylindrical shapes shown* *in Fig. 9.22, the maximum deformation for the small cylinder with $r_0 = 0.5''$ is the peak value (expansion), while the maximum deformation for the rest of cylinders are at the valleys (shrinkage).*

The optimal compensation policy aiming to minimize the maximum shape deformation has to follow the MinMax criterion. Note that MinMax compensation policy is not simply to compensate the maximum shape deformation at one location because neighboring locations could become new peaks or valleys after compensation.

The optimal amount of compensation given in Eq. (9.18) is clearly not a Min-Max policy because equal weights are assigned to minimize all deviations on the deviation profile. New optimal compensation policy in addition to Eq. (9.18) has to be developed if MiniMax quality criterion is adopted. MAD criterion will minimize the overall or total shape deformation, not necessarily the maximum shape deviation.

9.4.3.2 *Example one – optimal compensation of in-plane deformation for SLA process*

Work in Huang et al.[24] introduces the deformation modeling of cylindrical shape and experimentation with Stereolithography (SLA) process. Four cylindrical parts with different sizes were built and their in-plane shape deviation profiles are shown in Fig. 9.22. This section demonstrates the optimal compensation policy with and without consideration of over-exposure effect.

The derived deviation model without considering over-exposure effect, $f(\theta, r_0(\theta))$ for cylinders with different sizes has the form:[24]

$$f(\theta, r_0(\theta)) = \alpha r_0{}^a + \beta r_0{}^b \cos(2\theta)$$

where α, β, a and b denote the model parameters.

Substituting this $f(\theta, r_0)$ into the general compensation model Eq. (9.16), we have the predicted deformation as

$$f(\theta, r_0, x(\theta)) = x(\theta) + \alpha(r_0 + x(\theta))^a + \beta(r_0 + x(\theta))^b \cos(2\theta), \qquad (9.21)$$

Further approximation by the first and the second terms of the Taylor expansion at point r_0 yields

$$f(\theta, r_0, x(\theta)) \approx \alpha r_0^a + \beta r_0^b \cos(2\theta) + (1 + a\alpha r_0^{a-1} + b\beta r_0^{b-1} \cos(2\theta))x(\theta). \quad (9.22)$$

Equation (9.22) technically serves as a description of the predicted deformation of cylinders at θ when compensation $x(\theta)$ applied to the boundary of the CAD model.

Observed Deformation

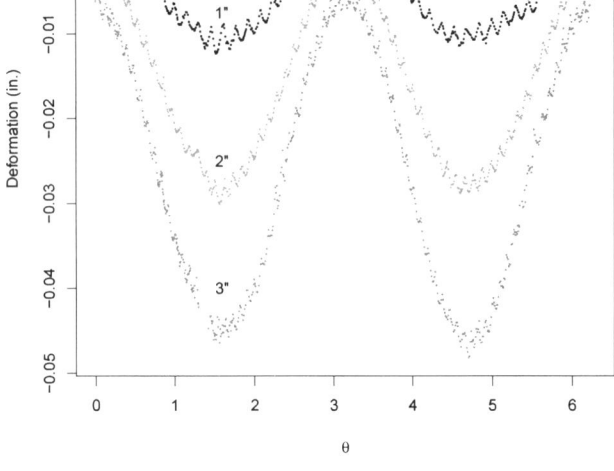

Fig. 9.22. Deviation profiles of $r_0 = 0.5''$, $1''$, $2''$, $3''$ cylinders.

By setting $f(\theta, r_0, x(\theta))$ to zero, we have the closed-form expression for the optimal compensation of cylinders with different sizes r_0:

$$x^*(\theta) = -\frac{\alpha r_0^a + \beta r_0^b \cos(2\theta)}{1 + a\alpha r_0^{a-1} + b\beta r_0^{b-1} \cos(2\theta)} \qquad (9.23)$$

To validate the effectiveness of the optimal compensation strategy in Eq. 9.18, a cylinder with nominal radius $1.0''$ is built with adjustment of CAD design based on Eq. (9.23), that is, $1.0'' + x^*(\theta)$. Command *law curve* in the CAD software *UG* is employed to construct the compensated CAD model according to Eq. (9.23). All manufacturing and measuring specifications remain the same as in the case of uncompensated cylinders. Parameters α, β, a, and b are set as the mean values.

The result for this cylinder is shown in Fig. 9.23. Deformation of the uncompensated cylinder is represented by the black line, and deformation of the compensated cylinder is the red line. The nominal value 0 is plotted as the dashed line. Obviously, absolute deformation has significantly decreased under compensation. The sinusoidal pattern of the original deformation has also been eliminated. We compute the average and standard deviation of deformation for both products. As can be seen in Table 9.7, the average and standard deviation of deformation have decreased to 10% of the original. This demonstrates that the compensation method

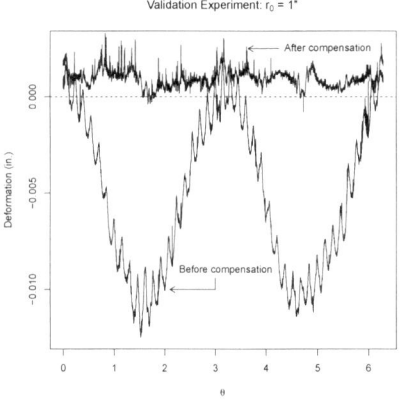

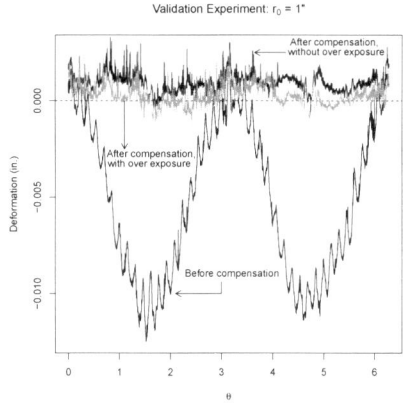

Fig. 9.23. Compensated $1.0''$ cylinder: without exposure term.

Fig. 9.24. Compensated $1.0''$ cylinder: with exposure term.

has effectively increased the accuracy of the product. However, we notice that the deformation under compensation is still above the desired value of 0, which indicates an overall bias of the compensation method. The source and solution to this bias is attributed to over-exposure.

Table 9.7. Deformation statistics for $1.0''$ cylinder, before and after compensation (in inches).

	Mean	SD
Before compensation	-5×10^{-3}	4×10^{-3}
After compensation	9×10^{-4}	4×10^{-4}

The error source of light over-exposure causes expansion of the illuminated shape due to the spread of light beams on the boundary of the product. Based on the definition of equivalent amount of compensation, the constant effect of over-exposure for all cylinders is equivalent to a default compensation x_0 applied to every angle in the original CAD model. According to Eq. (9.16), the predicted shrinkage model $f(\theta, r_0(\theta) + x(\theta)) + x(\theta)$ becomes

$$f(\theta, r_0(\theta) + x(\theta)) + x(\theta) = \alpha(r_0 + x_0)^a + \beta(r_0 + x_0)^b \cos(2\theta) + x_0. \quad (9.24)$$

The optimal compensation level under this new model is then derived as

$$x^*(\theta) = -\frac{\alpha r_0^a + \beta r_0^b \cos(2\theta)}{1 + a\alpha r_0^{a-1} + b\beta r_0^{b-1} \cos(2\theta)} - x_0. \quad (9.25)$$

This derivation acknowledges the fact that the amount of compensation x_0 will always be automatically added afterwards.

Alternatively, or more rigorously, we could view the nominal process input as $r_0 + x_0$, and perform the Taylor expansion at $r_0 + x_0$ instead of r_0. In this case, the compensation strategy will be

$$x^*(\theta) = -\frac{x_0 + \alpha(r_0 + x_0)^a + \beta(r_0 + x_0)^b \cos(2\theta)}{1 + a\alpha(r_0 + x_0)^{a-1} + b\beta(r_0 + x_0)^{b-1} \cos(2\theta)}. \qquad (9.26)$$

A comparison of the compensations in Eqs. (9.25) and (9.26) shows effectively no difference (details omitted). Consequently, we adopt the compensation strategy given by Eq. (9.26).

To validate the improved model with over-exposure effect, $1''$ and $2.5''$ cylinders are built with CAD design adjusted based on the optimal compensation plan in Eq. (9.26). The measured deformation results are shown in Figs. 9.24 and 9.25 respectively. In Fig. 9.24 a comparison of the uncompensated cylinder, compensated cylinder ignoring over-exposure, and compensated cylinder considering over-exposure, are demonstrated. Although both compensation methods decrease deformation substantially, the product compensated according to over-exposure apparently has uniformly smaller deformation: its deformation curve effectively shifted down closer to the nominal value 0, resolving the compensation bias problem discussed earlier.

Figure 9.25 shows the compensation effect for a cylinder with size $r_0 = 2.5''$ and its comparison with uncompensated cylinders with size $r_0 = 2''$, $3''$. Obviously deformation has been dramatically decreased, and the significant sinusoidal pattern has been eliminated. Note that the $2.5''$ cylinder has not been constructed before, and so this experiment demonstrates great predictability of our compensation model.

The three shaded areas in Fig. 9.25 also illustrate the quality measure $\Delta S(x(\theta))$ proposed in this work. Since $r_0(\theta)$ is a constant for a cylinder, $\Delta S(x(\theta))$ in Eq. (9.19) becomes

$$\Delta S(x(\theta)) = r_0 \int_0^{2\pi} \left| f(\theta, r_0(\theta) + x(\theta)) + x(\theta) \right| d\theta$$

with $\int_0^{2\pi} \left| f(\theta, r_0(\theta) + x(\theta)) + x(\theta) \right| d\theta$ corresponds to three shaded areas, respectively.

The quality measure for the cylinder of size $r_0 = 2''$, for instance, $\Delta S(r_0 = 2'', x(\theta) = 0)$ is simply the mesh area multiplied by size $r_0 = 2''$, a convenient index to assess product quality.

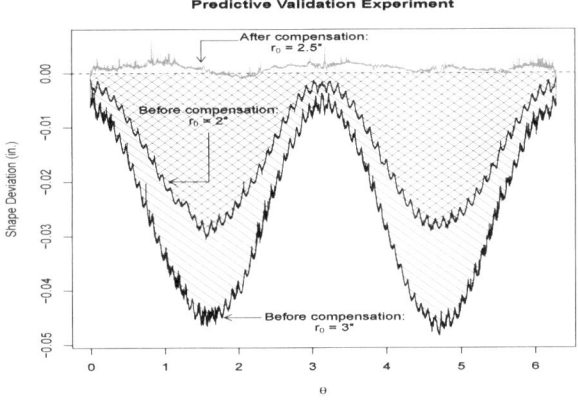

Fig. 9.25. Predictive validation for $2.5''$ cylinder.

9.4.3.3 *Optimal compensation of 3D geometric shape deformation*

To derive a optimal compensation policy for 3D shape deformation, we first briefly summarize the formulation of spatial shape deformation. We adopt the spherical coordinate system (SCS) (r, θ, φ) to depict both the in-plane and out-of-plane (z direction) deformation. Denote $r(\theta, \varphi)$ the boundary shape of an AM built product with $r_0(\theta, \varphi)$ being the nominal shape. At a given height $\varphi = \varphi_0$ or $z = r_0(\theta, \varphi) \cos(\varphi_0)$, $\Delta r(\theta, r_0(\theta)|\varphi_0) = h(r, \theta|\varphi)$ represents the in-plane geometric deformation of the horizontal cross-section view of the product. of the vertical cross-section of the product, denote the out-of-plane deformation $\Delta r(\varphi|\theta)$ as $v(r, \varphi|\theta)$.

Minimum Volume Deviation (MVD) Criterion and Optimality: Let us define the spatial deformation $\Delta r(\theta, \varphi)$ as

$$\Delta r(\theta, \varphi) = r(\theta, \varphi) - r_0(\theta, \varphi). \tag{9.27}$$

Let $f(\theta, \varphi, r_0(\theta, \varphi))$ be the given model predicting $\Delta r(\cdot)$, i.e., $\Delta r(\theta, \varphi) = f(\theta, \varphi, r_0(\theta, \varphi)) + \varepsilon$. Denote $r(\theta, \varphi, x(\theta, \varphi))$ the actual radius at (θ, φ) when compensation $x(\theta, \varphi)$ is applied at that location. By extending Theorem 1's MAD criterion to minimum volume deviation (MVD) criterion, we have the following result.

Theorem 2 (Minimum Volume Deviation). *The optimal compensation policy or the optimal amount of compensation $x^*(\theta, \varphi)$ for spatial shape deformation reduction is*

$$x^*(\theta, \varphi) = -\frac{f(\theta, \varphi, r_0(\theta, \varphi))}{1 + f'(\theta, \varphi, r_0(\theta, \varphi))} \tag{9.28}$$

which minimizes the volume deviation from its nominal shape, that is, it follows the minimum volume deviation (MVD) criterion.

Proof: As shown in Fig. 9.26, the volume deviation at location (θ, φ) due to $\Delta r(\theta, \varphi, r_0(\theta, \varphi))$ is

$$\Delta V(r, \theta, \varphi) \approx \Big(r_0(\theta, \varphi)\sin(\varphi)d\varphi\Big)\Big(r_0(\theta, \varphi)d\theta\Big)\Delta r(\theta, \varphi, r_0(\theta, \varphi))$$

$$\approx r_0^2(\theta, \varphi)\,\sin(\varphi)\,f(\theta, \varphi, r_0(\theta, \varphi))\,d\theta d\varphi$$

The total volume deviation before compensation is therefore

$$\Delta V = \iint r_0^2(\theta, \varphi)\Big|\sin(\varphi)\,f(\theta, \varphi, r_0(\theta, \varphi))\Big|\,d\theta d\varphi$$

Again, we consider absolute volume change by taking $|\sin(\varphi)f(\theta, \varphi, r_0(\theta, \varphi))|$ in the integral.

After applying compensation $x(\theta, \varphi)$, the shape deformation at (θ, φ) is

$$\Delta r(\theta, \varphi, r_0(\theta, \varphi), x(\theta, \varphi)) = r(\theta, \varphi, r_0(\theta, \varphi), x(\theta, \varphi)) - r_0(\theta, \varphi)$$

$$= f\Big(\theta, \varphi, r_0(\theta, \varphi) + x(\theta, \varphi)\Big) + x(\theta, \varphi)$$

Then the total volume deviation becomes

$$\Delta V(x(\theta, \varphi)) =$$

$$\iint r_0^2(\theta, \varphi)\Big|\sin(\varphi)\Big[f\Big(\theta, \varphi, r_0(\theta, \varphi) + x(\theta, \varphi)\Big) + x(\theta, \varphi)\Big]\Big|\,d\theta d\varphi \quad (9.29)$$

Note that ΔV before compensation is a special case with $x(\theta, \varphi) = 0$ or $\Delta V(x(\theta, \varphi)) = 0$.

The optimal compensation $x^*(\theta, \varphi)$ will minimize the volume deviation from the nominal shape, i.e.,

$$\min_{x(\theta, \varphi)} \quad \Delta V(x(\theta, \varphi))$$

$$\text{s.t.} \quad \theta_{min} \leq \theta \leq \theta_{max}; \text{ and } \varphi_{min} \leq \varphi \leq \varphi_{max} \quad (9.30)$$

A solution to minimize the volume deviation is

$$f\Big(\theta, \varphi, r_0(\theta, \varphi) + x(\theta, \varphi)\Big) + x(\theta, \varphi) = 0$$

Taylor series expansion of $f\Big(\theta, \varphi, r_0(\theta, \varphi) + x(\theta, \varphi)\Big)$ at $r_0(\theta, \varphi)$ will yield the result in Eq. (9.28) and complete the proof.

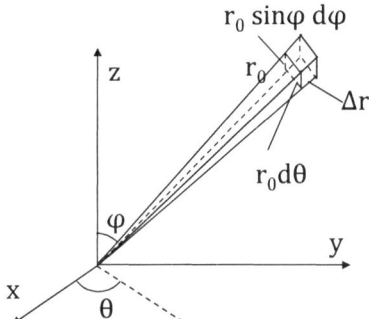

Fig. 9.26. Volume deviation at location (θ, φ).

Remark 3. *The result Eq.* (9.28) *based on MVD criterion essentially requires to minimize the deviation of every point on the boundary of the product, which is a strong condition. This is a direct extension from Eq.* (9.18) *for the compensation of in-plane error.*

Lemma 2 (3D Shape Quality Measure). $\Delta V(x(\theta, \varphi))$ *is a quality measure of 3D shape deformation.*

By extending the equivalent amount of compensation to 3D case, the proof can be obtained by following Theorem 2.

9.4.3.4 *Optimal policy for online compensation of 3D deformation*

Although Theorem 2 nicely extend the optimal compensation to 3D case, it relies on the model $f(\theta, \varphi, r_0(\theta, \varphi))$ to predict spatial deformation. Clearly, it is generally more challenging to establish $f(\theta, \varphi, r_0(\theta, \varphi))$ than the in-plane error model. Additionally, it is also demanding for AM machine to apply compensation three-dimensionally because a product is built layer by layer. To overcome this issue, we derive the following Lemma.

Lemma 3 (A Weaker Condition). *The optimal compensation policy* $x^*(\theta, \varphi)$

$$x^*(\theta, \varphi)sin(\varphi) = -\frac{f(\theta, \varphi, r_0(\theta, \varphi))sin(\varphi)}{1 + f'(\theta, \varphi, r_0(\theta, \varphi))} \tag{9.31}$$

is a weaker condition to minimize the volume deviation.

Proof: Other than the stronger condition of $f\left(\theta, \varphi, r_0(\theta, \varphi) + x(\theta, \varphi)\right) + x(\theta, \varphi) = 0$ *to minimize* $\Delta V(x(\theta, \varphi)) = 0$ *in Eq.* (9.29), *a weaker condition below will also minimize the volume deviation* $\Delta V(x(\theta, \varphi))$:

$$sin(\varphi)\left[f\left(\theta, \varphi, r_0(\theta, \varphi) + x(\theta, \varphi)\right) + x(\theta, \varphi)\right] = 0$$

Taylor series expansion of $f\left(\theta, \varphi, r_0(\theta, \varphi) + x(\theta, \varphi)\right)$ at $r_0(\theta, \varphi)$ and keeping the term $sin(\varphi)$ will give (9.31).

Remark 4. *The result (9.31) actually has powerful implications. Note that $x^*(\theta, \varphi)sin(\varphi)$ means the in-plane compensation by projecting $x^*(\theta, \varphi)$ to the $x - y$ plane at height φ. Following the notation in Fig. 9.5, let us denote $x^*(\theta, \varphi)sin(\varphi)$ as $x^*(\theta|\varphi)$, i.e., the optimal compensation for layer at height φ.*

Similarly, $f(\theta, \varphi, r_0(\theta, \varphi))sin(\varphi)$ means the in-plane deformation by projecting spatial deformation to the $x - y$ plane at height φ, which is $h(r, \theta|\varphi)$ or $\Delta r(\theta, r_0(\theta, \varphi)|\varphi)$ in Fig. 9.5. $f(\theta, \varphi, r_0(\theta, \varphi))$ can be expressed as $h(r, \theta|\varphi)/sin(\varphi)$. Then (9.31) can be rewritten as:

$$x^*(\theta|\varphi) = -\frac{h(r, \theta|\varphi)}{1 + h'(r, \theta|\varphi)/sin(\varphi)} \tag{9.32}$$

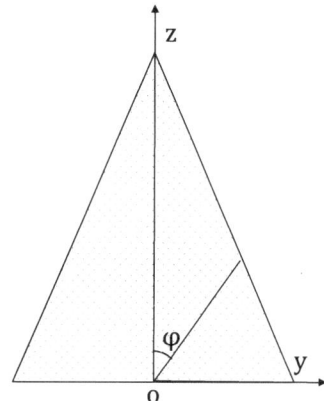

Fig. 9.27. Compensation at $\varphi = 0, \pi/2$.

This suggests that the optimal spatial compensation can be transformed to the in-plane compensation and can be implemented layer by layer. This is more consistent with the mechanism of AM itself and still satisfies the MVD criterion. It thus provides an theoretical base for online condition monitoring and online feedback control of 3D geometric shape deformation.

Remark 5. *It is interesting to compare the transformed optimal in-plane compensation (9.32) for spatial shape deformation with the optimal 2D shape deformation compensation in (9.18). Note that the result in Eq. (9.32) is expected to reduce both in-plane and out-of-plane errors, while Eq. (9.18) only considers in-plane* *errors. The two results will agree only when $\varphi = \pi/2$. This can be illustrated by a vertical cross-section of a pyramid example in Fig. 9.27. When $\varphi = \pi/2$, the AM process just builds the first layer of the product and this thin layer of product can be approximated as a 2D slab with out-of-plane errors being ignored. Therefore, the spatial optimal compensation literally degenerates to 2D case. This also proves that (9.18) is a special case of (9.32).*

It is also worthwhile to investigate the compensation at location $\varphi = 0$. Since $sin(\varphi = 0) = 0$, then $x^*(\theta|\varphi = 0) = 0$ at the tip, i.e., there is no need for compensation if the built pyramid has a tip. If there is a missing tip, then $\varphi > 0$ and the optimal compensation $x^*(\theta|\varphi) \neq 0$.

This section introduces the theoretical foundation of three-dimensional geometric error compensation for additive manufacturing processes. We theorize the optimal compensation result for 2D shape deformation by proposing the minimum area deviation (MAD) criterion. By extending the MAD criterion to the minimum volume deviation (MVD) criterion, we derive the optimal amount of compensation for 3D shape deformation in a consistent framework. Furthermore, MAD and MVD criteria provide convenient quality measures for AM built products that facilitate online monitoring and feedback control of shape geometric accuracy. The established analytical foundation fills the gap for the quality improvement in additive manufacturing.

Acknowledgment

The work has been partially supported by US National Science Foundation with grant # CMMI-1544917.

References

1. B. Colosimo, Q. Huang, T. Dasgupta and F. Tsung, Opportunities and challenges of quality engineering for additive manufacturing, *Journal of Quality Technology* **50**(3), 233–252 (2018).
2. D. L. Bourell, M. C. Leu and D. W. Rosen, Roadmap for additive manufacturing: Identifying the future of freeform processing, Tech. Rep., Sponsored by National Science Foundation and the Office of Naval Research (2009).
3. I. Gibson, D. Rosen and B. Stucker, *Additive Manufacturing Technologies: Rapid Prototyping to Direct Digital Manufacturing*. Springer Verlag (2009).
4. P. Hilton and P. Jacobs, *Rapid Tooling: Technologies and Industrial Applications*. CRC (2000).
5. F. Melchels, J. Feijen and D. Grijpma, A review on stereolithography and its applications in biomedical engineering, *Biomaterials* **31**(24), 6121–6130 (2010).

6. T. Campbell, C. Williams, O. Ivanova and B. Garrett, Could 3D printing change the world? Technologies, potential, and implications of additive manufacturing, Atlantic Council Strategic Foresight Report (2011).

7. C. Beyer, Strategic implications of current trends in additive manufacturing, *Journal of Manufacturing Science and Engineering* **136**(6), 064701 (2014).

8. W. Gao, Y. Zhang, K. R. Devarajan Ramanujan, Y. Chen, C. B. Williams, C. C. Wang, Y. C. Shin, S. Zhang and P. D. Zavattieri, The status, challenges, and future of additive manufacturing in engineering, *Computer-Aided Design* **69**, 65–89 (2015).

9. I. Whadcock, Manufacturing and innovation- A third industrial revolution (Special Report), *Economist* (2012).

10. D. C. Montgomery, *Statistical Quality Control, 7th edition*. John Wiley and Sons, Inc. (2012).

11. Q. Huang, H. Nouri, K. Xu, Y. Chen, S. Sosina and T. Dasgupta, Statistical predictive modeling and compensation of geometric deviations of three-dimensional printed products, *ASME Transactions, Journal of Manufacturing Science and Engineering* **136**(6), 061008 (2014).

12. B. Storåkers, N. Fleck and R. McMeeking, The viscoplastic compaction of composite powders, *Journal of the Mechanics and Physics of Solids* **47**, 785–815 (1999).

13. J. Secondi, Modeling powder compaction from a pressure-density law to continuum mechanics, *Powder Metallurgy* **45**(3), 213–217 (2002).

14. K. Mori, K. Osakada and S. Takaoka, Simplified three-dimensional simulation of non-isothermal filling in metal injection moulding by the finite element method, *Engineering Computations* **13**(2), 111–121 (1996).

15. W. Wang, C. Cheah, J. Fuh and L. Lu, Influence of process parameters on stereolithography part shrinkage, *Materials & Design* **17**(4), 205–213 (1996).

16. J. Zhou, D. Herscovici and C. Chen, Parametric process optimization to improve the accuracy of rapid prototyped stereolithography parts, *International Journal of Machine Tools and Manufacture* **40**(3), 363–379 (2000).

17. A. Sood, R. Ohdar and S. Mahapatra, Improving dimensional accuracy of fused deposition modelling processed part using grey taguchi method, *Materials & Design* **30**(10), 4243–4252 (2009).

18. X. Wang, Calibration of shrinkage and beam offset in sls process, *Rapid Prototyping Journal* **5**(3), 129–133 (1999).

19. C. Lynn-Charney and D. W. Rosen, Usage of accuracy models in stereolithography process planning, *Rapid Prototyping Journal* **6**(2), 77–87 (2000).

20. K. Tong, E. Lehtihet and S. Joshi, Parametric error modeling and software error compensation for rapid prototyping, *Rapid Prototyping Journal* **9**(5), 301–313 (2003).

21. K. Tong, S. Joshi and E. Lehtihet, Error compensation for fused deposition modeling (fdm) machine by correcting slice files, *Rapid Prototyping Journal* **14**(1), 4–14 (2008).

22. W. Cho, E. M. Sachs, N. M. Patrikalakis and D. E. Troxel, A dithering algorithm for local composition control with three-dimensional printing, *Computer-aided design* **35**(9), 851–867 (2003).

23. C. Zhou, Y. Chen and R. A. Waltz, Optimized mask image projection for solid freeform fabrication, *ASME Journal of Manufacturing Science and Engineering* **131**(6), 061004 (2009).

24. Q. Huang, J. Zhang, A. Sabbaghi and T. Dasgupta, Optimal offline compensation of shape shrinkage for 3d printing processes, *IIE Transactions on Quality and Reliability* **47**(5), 431–441 (2015).

25. A. Sabbaghi, T. Dasgupta, Q. Huang and J. Zhang, Inference for deformation and interference in 3D printing, *Annals of Applied Statistics* **8**(3), 1395–1415 (2014).

26. G. Moroni, W. P. Syam and S. Petrò, Towards early estimation of part accuracy in additive manufacturing, *Procedia CIRP* **21**, 300–305 (2014).

27. K. Xu and Y. Chen, Mask image planning for deformation control in projection-based stereolithography process, *Journal of Manufacturing Science and Engineering* **137**(3), 031014 (2015).

28. D. Hu, H. Mei and R. Kovacevic, Improving solid freeform fabrication by laser-based additive manufacturing, *Proceedings of the Institution of Mechanical Engineers, Part B: Journal of Engineering Manufacture* **216**(9), 1253–1264 (2002).

29. D. Hu and R. Kovacevic, Sensing, modeling and control for laser-based additive manufacturing, *International Journal of Machine Tools and Manufacture* **43**(1), 51–60 (2003).

30. L. Song and J. Mazumder, Feedback control of melt pool temperature during laser cladding process, *IEEE Transactions on Control Systems Technology* **19**(6), 1349–1356 (2011).

31. A. Heralic, A.-K. Christiansson and B. Lennartson, Height control of laser metal-wire deposition based on iterative learning control and 3D scanning, *Optics and Lasers in Engineering* **50**(9), 1230–1241 (2012).

32. D. L. Cohen and H. Lipson, Geometric feedback control of discrete-deposition SFF systems, *Rapid Prototyping Journal* **16**(5), 377–393 (2010).

33. L. Lu, J. Zheng and S. Mishra, A layer-to-layer model and feedback control of ink-jet 3D printing, *IEEE/ASME Transactions on Mechatronics* **20**(3), 1056–1068 (2015).

34. G. Tapia and A. Elwany, A review on process monitoring and control in metal-based additive manufacturing, *ASME Transactions, Journal of Manufacturing Science and Engineering* **136**(6), 060801–060810 (2014).

35. S. Loncaric, A survey of shape analysis techniques, *Pattern Recognition* **31**(8), 983–1001 (1998).

36. D. Zhang and G. Lu, Review of shape representation and description techniques, *Pattern Recognition* **37**(1), 1–19 (2004).

37. D. G. Kendall, D. Barden, T. K. Carne and H. Le., *Shape and Shape Theory*. John Wiley & Sons (2009).

38. I. L. Dryden and K. V. Mardia, *Statistical Shape Analysis: With Applications in R, 2nd edition*. John Wiley & Sons (2016).

39. Q. Huang, An analytical foundation for optimal compensation of three-dimensional shape deformation in additive manufacturing, *ASME Transactions, Journal of Manufacturing Science and Engineering* **138**(6), 061010 (8 pages) (2016).

40. Y. Jin, S. Qin and Q. Huang, Offline predictive control of out-of-plane geometric errors for additive manufacturing, *ASME Transactions on Manufacturing Science and Engineering* **138**(12), 121005(7 pages) (2016).

41. Y. Jin, S. Qin and Q. Huang, Out-of-plane geometric error prediction for additive manufacturing, 2015 IEEE International Conference on Automation Science and Engineering (CASE 2015), Gothenberg, Sweden. (2015).

42. H. Luan and Q. Huang, Prescriptive modeling and compensation of in-plane shape deformation for 3-d printed freeform products, *IEEE Transactions on Automation Science and Engineering* **14**(1), 73–82 (2017).

43. A. Wang, S. Song, Q. Huang and F. Tsung, In-plane shape-deviation modeling and compensation for fused deposition modeling processes, *IEEE Transactions on Automation Science and Engineering* **17**(2), 968–976 (2017).

44. R. Dinesh, S. S. Damle and D. Guru, A split-based method for polygonal approximation of shape curves. In *Pattern Recognition and Machine Intelligence Christopher Bishop (ed)*. Springer, (2005).

45. P.-Y. Yin, A discrete particle swarm algorithm for optimal polygonal approximation of digital curves, *Journal of visual communication and image representation* **15**(2), 241–260 (2004).

46. J.-S. Wu and J.-J. Leou, New polygonal approximation schemes for object shape representation, *Pattern Recognition* **26**(4), 471–484 (1993).

47. A. Pikaz *et al.*, An algorithm for polygonal approximation based on iterative point elimination, *Pattern Recognition Letters* **16**(6), 557–563 (1995).

48. J.-C. Perez and E. Vidal, Optimum polygonal approximation of digitized curves, *Pattern recognition letters* **15**(8), 743–750 (1994).

49. M. Salotti, An efficient algorithm for the optimal polygonal approximation of digitized curves, *Pattern Recognition Letters* **22**(2), 215–221 (2001).

50. C. Cajal, J. Santolaria, D. Samper and J. Velazquez, Efficient volumetric error compensation technique for additive manufacturing machines, *Rapid Prototyping Journal* **22**(1), 2–19 (2016).

51. Y. Jin, S. Qin and Q. Huang, Modeling inter-layer interactions for out-of-plane shape deviation reduction in additive manufacturing, *IISE Transactions on Design and Manufacturing* **52**(7), 721–731 (2020).

52. H. Luan and Q. Huang, Predictive modeling of in-plane geometric deviation for 3d printed freeform products, 2015 IEEE International Conference on Automation Science and Engineering (CASE 2015), Gothenberg, Sweden., 912–917 (2015).

53. H. Wang, Q. Huang and R. Katz, Multi-operational machining processes modeling for sequential root cause identification and measurement reduction, *ASME Transactions, Journal of Manufacturing Science and Engineering* **127**, 512–521 (2005).

54. H. Wang and Q. Huang, Error cancellation modeling and its application in machining process control, *IIE Transactions on Quality and Reliability* **38**, 379–388 (2006).

55. H. Wang and Q. Huang, Using error equivalence concept to automatically adjust discrete manufacturing processes for dimensional variation control, *ASME Transactions, Journal of Manufacturing Science and Engineering* **129**, 644–652 (2007).